REPARATIVE MEDICINE

GROWING TISSUES AND ORGANS

ANNALS OF THE NEW YORK ACADEMY OF SCIENCES
Volume 961

REPARATIVE MEDICINE

GROWING TISSUES AND ORGANS

Edited by Jean D. Sipe, Christine A. Kelley, and
Loré Anne McNicol

The New York Academy of Sciences
New York, New York
2002

Library of Congress Cataloging-in-Publication Data

Reparative medicine : growing tissues and organs /
edited by Jean D. Sipe, Christine A. Kelley, and Loré Anne McNicol.
  p. ; cm. — (Annals of the New York Academy of Sciences ; v.961)
Papers from a conference held by the National Institutes of Health Bioengineering
Consortium (BECON) on June 25–26, 2001 in Bethesda, Maryland.
Includes bibliographical references and index.
ISBN 1-57331-382-3 (cloth : alk. paper) — ISBN 1-57331-383-1 (paper :alk. paper)
1. Biomedical engineering—Congresses. 2. Regeneration (Biology)—Congresses.
3. Organ culture—Congresses. 4. Tissue culture—Congresses. 5. Animal cell
biotechnology—Congresses.
[DNLM: 1. Organ Culture—Congresses. 2. Tissue Engineering—Congresses. QS 530
R425 2002] I. Sipe, Jean D. II. Kelley, Christine A. III. McNicol, Loré Anne. IV. National Institutes of Health (U.S.). Bioengineering Consortium. V. Series.
Q11 .N5 vol. 961
[R856.A2]
500 s—dc21
[610'.]   2002006343

GYAT/B-MP
Printed in the United States of America
ISBN 1-57331-382-3 (cloth)
ISBN 1-57331-383-1 (paper)
ISSN 0077-8923

ANNALS OF THE NEW YORK ACADEMY OF SCIENCES

Volume 961
June 2002

REPARATIVE MEDICINE

GROWING TISSUES AND ORGANS

Editors
JEAN D. SIPE, CHRISTINE A. KELLEY, AND LORÉ ANNE MCNICOL

Extramural Planning Committee

ANTHONY J. ATALA *(Harvard Medical School),*
STEPHEN BADYLAK *(Purdue University),* ELLIOT CHAIKOF *(Emory University),*
MOHAMMED HEIDARAN *(BD Technologies),*
PETER J. JOHNSON *(TissueInformatics),* ALAN P. KORETSKY *(NINDS, NIH),*
DENNIS METZGER *(Albany Medical College),*
LAURA E. NIKLASON *(Duke University),*
ANTHONY RATCLIFFE *(Advanced Tissue Sciences),*
FRED H. GAGE *(Salk Institute),* LINDA G. GRIFFITH *(MIT),*
ROBERT M. NEREM *(Georgia Institute of Technology),* AND
E. HELENE SAGE *(Hope Heart Institute)*

This volume is the result of a symposium entitled **Reparative Medicine: Growing Tissues and Organs** held on June 25–26, 2001 by the National Institutes of Health (NIH) Bioengineering Consortium (BECON) in Bethesda, Maryland.

CONTENTS

Part IV. Tissue Generation

A. Functional Considerations in the Design of Engineered Tissue

B. Bioreactors and Bioprocessing

C. Vascular Assembly in Engineered and Natural Tissues

Financial assistance was received from:

• **NATIONAL EYE INSTITUTE**

• **NATIONAL INSTITUTE OF BIOIMAGING AND BIOMEDICAL
  ENGINEERING**

• **NATIONAL INSTITUTE OF CHILD HEALTH AND HUMAN
  DEVELOPMENT**

NIH Program Planning Committee

MARK S. BROWN, CMP, *Program Manager, MasiMax Resources, Inc.*

ARLENE Y. CHIU, PH.D., *Director, Repair and Plasticity Program, National Institute of Neurological Disorders and Stroke, NIH*

ZOE-ANN COPELAND-SEWELL, *Chief, Administrative Services Branch, Office of Extramural Research, OD, NIH*

DHARAM S. DHINDSA, D.V.M., PH.D., *Scientific Review Administrator and Referral Officer, Center for Scientific Review, NIH*

FRANK EVANS, PH.D., *Research Fellow, National Heart, Lung, and Blood Institute, NIH*

NANCY L. FREEMAN, PH.D., *Scientific Program Director, National Institute on Deafness and Other Communication Disorders, NIH*

CASSANDRA GIBBS, *Administrative Officer, Office of Extramural Research, OD, NIH*

FLORENCE P. HASELTINE, M.D., PH.D, *Director for Center for Population Research, National Institute of Child Health and Human Development, NIH*

WILLIAM J. HEETDERKS, M.D., PH.D., *Program Director, Repair and Plasticity Program, National Institute of Neurological Disorders and Stroke, NIH*

MICHAEL F. HUERTA, PH.D., *Associate Director, Division of Neuroscience and Basic Behavioral Science, National Institute of Mental Health, NIH*

WILLIAM M. JOHNSTON, PH.D., *Program Specialist, Biomaterials, Biomimetics, and Tissue Engineering Branch, National Institute of Dental and Craniofacial Research, NIH*

ERNEST D. MARQUEZ, PH.D., Chief, *Minority Biomedical Research Support Branch, National Institute of General Medical Sciences, NIH*

TERESA NESBITT, D.V.M., PH.D., *Scientific Review Administrator, Surgery and Bioengineering Study Section, Center for Scientific Review, NIH*

TRACY E. ORR, PH.D., *Scientific Review Administrator, Surgery, Radiology and Bioengineering IRG, Center for Scientific Review, NIH*

JAMES S. PANAGIS, M.D., MPH, *Director, Orthopedics Program, National Institute of Arthritis, Musculoskeletal, and Skin Diseases, NIH*

WINIFRED K. ROSSI, M.A., *Health Program Specialist, Genetic Epidemiology and Translational Research Geriatrics Program , National Institute on Aging, NIH*

SHERYL SATO, PH.D., *Director, Cellular Basis of Metabolic Diseases Program, National Institute of Diabetes, Digestive, and Kidney Diseases, NIH*

JEAN D. SIPE, PH.D., *Scientific Review Administrator, Center for Scientific Review, NIH*

MOLLIE SOURWINE, *Special Assistant for Bioengineering, Office of Extramural Research, OD, NIH*

RICHARD E. SWAJA, PH.D., *Senior Advisor for Bioengineering, Office of Extramural Research, OD, NIH*

JOHN T. WATSON, PH.D., *Director, Clinical and Molecular Medicine Program, Division of Heart and Vascular Disease, National Heart, Lung, and Blood Institute, NIH*

MICHAEL WEINRICH, PH.D., *Director, National Center for Medical Rehabilitation Research, National Institute of Child Health and Human Development, NIH*

Preface

Reparative Medicine: Growing Tissues and Organs

JEAN D. SIPE,[a] CHRISTINE A. KELLEY,[b] AND LORÉ ANNE McNICOL[c]

[a]Center for Scientific Review, National Institutes of Health,
Bethesda, Maryland 20892, USA

[b]National Institute of Biomedical Imaging and Bioengineering,
National Institutes of Health, Bethesda, Maryland 20814-9692, USA

[c]National Eye Institute, National Institutes of Health,
Bethesda, Maryland 20892, USA

The fourth National Institutes of Health (NIH) Bioengineering Consortium (BECON) symposium, entitled **Reparative Medicine: Growing Tissues and Organs**, was held in Bethesda, Maryland on June 25–26, 2001. Reparative medicine, for the purposes of the symposium, was defined as the "development or growing of biological substitutes for the body *in vitro* and/or the fostering of tissue regeneration and remodeling *in vivo*, with the purpose being to replace, repair, maintain, or enhance tissue/organ function." A diverse group of stakeholders gathered to assess the current status of reparative medicine and to map out a strategy by which the clinical needs for, and promise of, growing tissues and organs could be achieved. The goals of the conference were (1) to provide a forum in which scientists, engineers, clinicians, administrators and other interested parties could review and integrate existing knowledge and identify hurdles to be overcome, and (2) to stimulate ideas for new experimental approaches with broadened horizons. It was anticipated that clear directions for future research would emerge that would move this field ahead in the next five to ten years.

The program was intended to provide multiple perspectives on reparative medicine from the disciplines of science, engineering, and medicine and from individuals in the public and private sectors. Breakout session reports were written at the meeting to convey the collective thinking of symposium speakers, moderators, panelists, and conferees on major issues that need to be resolved before crucial advances can occur. Poster presentations of ongoing research spanned the multiple elements of reparative medicine from cells to storage and translational issues.

The opinions and statements expressed in this document do not necessarily reflect those of the National Institutes of Health.

sipej@csr.nih.gov
kelleyc@mail.nih.gov
lm271f@nih.gov

Ann. N.Y. Acad. Sci. 961: xiii–xiv (2002). © 2002 New York Academy of Sciences.

We would like to express our gratitude to the conference organizers—the NIH BECON Symposium Program Planning Committee—and to the Extramural Planning Committee. We thank Barbara Goldman, Justine Cullinan, and Sheila Kane of the *Annals* staff of the New York Academy of Sciences for their encouragement, support, and invaluable assistance and advice in the preparation of this volume. We express our sincere appreciation to all of the speakers, discussants, and conference participants who contributed to this volume.

Tissue Engineering and Reparative Medicine

JEAN D. SIPE

Center for Scientific Review, National Institutes of Health,
Bethesda, Maryland 20814, USA

ABSTRACT: Reparative medicine is a critical frontier in biomedical and clinical research. The National Institutes of Health Bioengineering Consortium (BECON) convened a symposium titled "Reparative Medicine: Growing Tissues and Organs," which was held on June 25 and 26, 2001 in Bethesda, Maryland. The relevant realms of cells, molecular signaling, extracellular matrix, engineering design principles, vascular assembly, bioreactors, storage and translation, and host remodeling and the immune response that are essential to tissue engineering were discussed. This overview of the scientific program summarizes the plenary talks, extended poster presentations and breakout session reports with an emphasis on scientific and technical hurdles that must be overcome to achieve the promise of restoring, replacing, or enhancing tissue and organ function that tissue engineering offers.

KEYWORDS: bioreactors; biomaterials, bioscaffolds; cryopreservation; host remodeling

INTRODUCTION

The term reparative medicine is often used to denote the replacement, repair, or functional enhancement of tissues and organs. Reparative medicine has traditionally used materials at hand and the technology of the day to restore or improve function of organs and tissues afflicted with birth defects or the ravages of injury, disease, and age. The driving force for reparative medicine has been clinical need; clinicians, particularly surgeons, have recruited practitioners from multiple disciplines to address complex biological problems that are well beyond a single scientific or engineering discipline. Depending upon the particular organ or tissue at stake, a number of reparative medicine strategies have been employed: (1) substitution of one body part for another (Godbey and Atala report that substitution of one body part for another has been practiced since at least the time of Socrates[1]); (2) repair of the body with non-vital, often synthetic, materials and devices; (3) transplantation of an organ from another individual or from a nonhuman species; (4) use of an external device to augment or substitute for a non-functioning organ; and, (5) use of living cells to restore, maintain, or enhance the function of tissues and organs, that is, what is now known as tissue engineering.

Address for correspondence: Jean D. Sipe, Center for Scientific Review, National Institutes of Health, 6701 Rockledge Drive, Bethesda, MD 20814. Voice: 301-435-1743; fax: 301-480-2644. sipej@csr.nih.gov

Ann. N.Y. Acad. Sci. 961: 1–9 (2002). © 2002 New York Academy of Sciences.

Tissue engineering is the three-dimensional assembly over time of vital tissues/ organs by a process involving cells, signals, and extracellular matrix.[1,2]

The field of tissue engineering shows enormous potential for reparative medicine, if it can be developed and advanced to the point that living tissues and organs can be routinely assembled and reliably integrated into the body to restore, replace, or enhance tissue and organ functions. Thus, the application of tissue engineering to reparative medicine shows great promise for the treatment of a large number of conditions including birth defects, musculoskeletal disorders, Alzheimer's and Parkinson's diseases, diabetes, heart disease, liver and kidney failure, and spinal cord injuries. In addition to its use in reparative medicine, tissue engineering could provide surrogate tissues that could be useful for drug discovery and development and toxicological assessment.[3]

The engineering challenges posed by generation, insertion, and maintenance of functional tissue are enormous. Complex, multidifferentiated, vascular tissues must be generated either in the body or in bioreactors within the "classical engineering constraints of reliability, cost, government regulation, societal acceptance..."(see Griffith,[3] this volume). The dynamics of tissue engineering vary from tissue to tissue according to the hierarchy of tissue or organ function, macroscopic structure, and composition. Initial success with the use of tissue engineering in reparative medicine has been with rather delimited applications, such as skin substitutes that are propagated *in vitro* and which, upon implantation, integrate with host skin or introduction of acellular scaffolding—cornea, bone, heart valve, or cell transplantation into damaged organs, such as heart.[4]

In order to hasten realization of the promise of tissue engineering for reparative medicine, the National Institutes of Health sponsored the two-day symposium "Reparative Medicine: Growing Tissues and Organs." on June 25–26, 2001, in Bethesda, MD. Scientists, engineers, clinicians, administrators, and other interested parties convened to review and integrate existing knowledge and to identify hurdles to be overcome. The essential elements of tissue engineering, (1) cells, (2) signaling, (3) extracellular matrix, (4) design principles, (5) vascular assembly, (6) bioreactors, (7) storage and translation, and (8) host remodeling and immune response, were reviewed in a series of plenary lectures, breakout sessions, and poster presentations. The current status and some of the future issues associated with each of the elements are considered here.

CELLS

The human body comprises on the order of 100 trillion cells, with about 260 different phenotypes, that divide, differentiate and self-assemble over time and space into an integrated system of tissues and organs.[5] Although the source and availability of cells for tissue engineering is critical, up until now, the field has been strongly driven from the perspective of biomaterials, based on the recognition that biomaterials can influence cell function and response.[3–6] Because early attempts with cellular grafts generated outside the body proved problematic, much of the successful work to date, with respect to clinical application, has been acellular.[4,7] With some exceptions, such as the skin substitute Apligraf, which is able to respond to its wound environment and is not rejected, tissue-engineered constructs containing cells have exhibited problems with physical properties, maintenance of cell phenotypes, and the host immune response.

The use of cells in tissue-engineered constructs is hampered not only by the lack of information about how to retain and regulate cellular function in the construct and in the host, but also by problems and limitations with both cell expansion and differentiation.[4] It is a challenge to generate sufficient numbers of a single cell type, to orchestrate the assembly of the needed mixture of multiple cell phenotypes, and to maintain stable phenotypes as needed. To alleviate cell shortages, a number of cell sources are being investigated,[4,8,9] including adult stem cells, adult differentiated cells, embryonic and fetal stem cells, cells generated by nuclear transplantation, and *ex vivo* manipulated cells. A major technical hurdle associated with the use of pluripotent progenitor cells is the lack of specific markers for cells in the pluripotent state.[9]

Cells from the above variety of tissue sources can be (1) autologous (self), which offers the advantage of manipulation with minimal risk of adverse host response and disease transmission[8]; (2) allogeneic (nonself, same species), which offers the advantage of banking prior to need, but is more likely to be complicated by the presence of disease-transmitting viruses; or (3) xenogeneic (animal, other species). Both allogeneic and xenogeneic cells would be more likely to generate an adverse response from the host.[8,9]

The ability to obtain physiologic information on an individual cell basis rather than on the basis of an average response of a population of cells will help to provide needed parameters to design and develop tissue engineering constructs. However, computer software to manage cell images and metabolic data will be required.[10] Cells can also be commandeered within tissue-engineered constructs to serve as cellular bioreactors for the generation of proteins. The generation of proteins at sites where needed offers an advantage over direct protein delivery[11] because the chemical synthesis of proteins greater than 100 amino acids in length can be technically challenging and it can be difficult to deliver and maintain needed protein concentrations at appropriate times.

Cell expansion and differentiation *in vivo* has been effective in some cases, such as regeneration of cardiac tissue.[12,13] However generation of tissues and organs *in vitro* using bioreactors is currently difficult to orchestrate due to problems with supplying oxygen to three-dimensional constructs and the integration of cell expansion, differentiation, and assembly within the extracellular matrix.[14, 15]

SIGNALING

Cells respond to the extracellular environment by sensing a chemical signal or physical stimulus that is transmitted to the nucleus to trigger the expression or repression of genes, the products of which regulate cell division, migration, differentiation, and apoptosis. Much of the signaling information that is currently utilized by tissue engineers has been derived primarily from studies with single cell populations cultured in two dimensions and treated selectively with soluble factors. Recently, the importance of spatial signals during three-dimensional culture has been recognized, and time and force are considered the fourth and fifth dimensions that play crucial roles in tissue engineering.[16] Bottaro et al.[2] emphasize the need for a better understanding of environmental clues given to cells and of how the signals are integrated and assembled into a hierarchy of interactions with cell receptor systems. Also, information is lacking on how pathways specific to cell phenotype are integrated with multimolecular complexes and cellular organelles. Tissue engineering faces the need to switch from two-dimensional to three-dimensional (flexible) cultures and the need to employ design principles in order to become less trial-and-error in approach.[3,15–20] Investigators also recognize that there are important lessons to be learned from embryonic development and the analysis of postnatal master gene expression.[16]

It is recognized that cell division, differentiation, and maintenance of phenotype are influenced by the synergy and interplay between soluble factors, insoluble adhesion molecules within the extracellular matrix, and mechanical forces.[2,18,19] Cellular mechanoregulation refers to the processes by which mechanical forces influence gene expression, metabolic pathways, and tissue patterning and architecture.[19] Cells in tissues are under constant stimulation by mechanical forces, the very minimum of which is gravity. Mechanical forces influence cell shape, which, in turn, affects how a cell responds to the summation of its signals with either a growth response or apoptosis. Thus, study of three-dimensional multicellular model systems is at a crucial forefront of tissue engineering for reparative medicine. Time can be viewed as the fourth dimension of the cellular environment and gravity or forces as the fifth dimension.

EXTRACELLULAR MATRIX (BIOSCAFFOLDS)

It is the insoluble extracellular matrix (ECM) that confers physical, mechanical, and functional properties on tissues and organs, that is, strength of bone, elasticity of skin, et.c[2,3] Throughout the body, ECM is made up of proteoglycans, elastin, fibrillin, and 19 different types of collagen. During development and wound repair, cells synthesize and remodel ECM, and thus all cells spend at least part of their time interacting with the extracellular matrix.[3,6,21] Insoluble signals/factors provided by the ECM interact with soluble signals and mechanical forces to promote adherence, migration, division, and differentiation of cells. There is an intimate link between cell adhesion and cell signaling. Natural polymers, synthetic polymers, and inorganic composites, collectively known as biomaterials, have been used for tissue engineering usually as temporary, surrogate ECM. Natural polymers used for tissue engineering comprise proteoglycans and collagen; collagen is highly prevalent in the extracellular matrix throughout the body and has been shown experimentally to influence cell phenotype, that is, chondrocyte phenotype is maintained in presence of type II, but not type I collagen.[3,6,21]

Griffith[3] reviewed the hierarchy of design scale considerations that apply to scaffolds for tissue engineering: the macroscopic level (on a scale of millimeters to centimeters); an intermediate level (hundreds of microns), involving the topography of pores and channels; and the molecular level, involving surface texture and chemistry (tens of microns). Growing tissues and organs requires that varying numbers of differentiated cells be assembled into a specific architecture in a series of specific events occurring at time intervals ranging from seconds to weeks and months, at dimensions ranging from 0.0001 to 10 cm and involving a range of forces from 3 to 15 orders of magnitude difference.

Initially, scaffolds used for tissue engineering were derived from surgical materials; the tendency to adapt materials in current or prior use for other applications offers advantages from the perspective of regulatory agencies such as the Food and Drug Administration, but does not necessarily promote development of optimal materials with regard to performance characteristics needed for different tissues.[3,17,21] It is desirable that the scaffolding biomaterial can be degraded as cells go through the process of forming their own supportive ECM; the permanent presence of implants almost always can be expected to elicit a foreign-body response. Degradation is influenced by material composition, surface chemistry, and topology.

Bioactive scaffolding materials can be engineered to deliver growth factors/signals, to deliver cells, or to direct the three-dimensional orientation of cells. For example, certain biomaterials can aid hepatocytes to retain epithelial polarization. The number and spatial orientation of cell adhesion ligand moieties is becoming recognized as crucial to cell migration and mechanical signaling and, thus, subsequent differentiation. Scaffolds may also be designed to deliver DNA locally to transduce cells to become bioreactors for production of proteins in situations that require that sufficient local concentrations be produced and that the protein is biologically active.[2,3,6,11,21]

DESIGN PRINCIPLES

When faced with the challenge of replacement of whole organs and tissues, there is the need to understand tissue and organ properties that result from scaled hierarchies, from molecular to cellular to macrosopic organ length, that are currently best understood at the cellular and molecular level.[3,17,20,22] It is considered to be unlikely that the properties of final tissue-engineered constructs will be completely understandable by investigation of the respective components, cell types, and matrix, and vasculature in isolation because of the need to consider properties resulting from issues of hierarchy.[3,22,14]

Tissue engineering, if it is to fulfill its promise for reparative medicine, will require standards and fundamental principles that cross organ-based disciplines and will require a judicious mix of design and iteration.[17,20] It is crucial that rational design be added to studies that have been up until now somewhat trial-and-error in nature. Goldstein[17] and Guilak and copanelists[20] presented design principles for functional tissue engineering that start and end with patients' needs, such as the biomechanics for bone strength and electrical conduction needed for heart contraction. It was noted that the repair of load-bearing structures (tissues that serve a biomechanical function) is problematic and that there is a need to define biomechanical properties of native tissue. The need for early consideration of complexities of design and running clinical trials and for the selection of clinically relevant end points was emphasized. In view of the long lag period from project initiation to clinical application,[1] it is advisable to choose carefully and perhaps limit the numbers of scaffolds and cell sources to be developed.

The fundamental elements of embryonic development may be expected to have direct applicability to functional tissue engineering. As work progresses from animal models to clinical trials, it will be highly desirable to have surrogate markers of function, as the methodology to assess function in the patient must be much less invasive. The continued development of imaging approaches is expected to help greatly in assessment of function.[21,23,24]

VASCULAR ASSEMBLY

A major roadblock to the achievement of three-dimensional tissue-engineered constructs is ability to vascularize the tissues. The need for a vasculature is pervasive throughout the body, as all cells need adequate oxygen levels. Hirschi and coauthors[14] stressed the need to apply an integrative approach to determine the principles that guide tissue and organ formation together with a continued reductionist approach to develop a more sophisticated understanding of vessel formation at the molecular and cellular levels. It appears unlikely that the final vascular structure can be pre-determined in the micro fabrication stage of tissues and that the natural adaptive powers of resident cells will be needed to achieve a vasculature that is sufficient for large, complex tissues. Lessons from embryonic development may be helpful here. Key issues in need of investigation include identification of the various differentiation stages of cells with respect to vasculogenesis, and the mechanism by which differentiation is regulated and cells are supported and directed to sites of injury or tissue repair. Also, there is the need for molecular modeling of the vascular network and study of blood vessel pattern formation using time-lapse microscopy, computational analyses, and computer modeling.[14, 25]

BIOREACTORS

Tissue-engineering efforts must include an early decision as to whether tissues should be generated inside or outside the body. Success has recently been achieved in the area of implantation of cells into diseased tissue (bone marrow stem cells and skeletal muscle to heart).[1,2,13,14,21] In other cases, the approach has been to implant tissues assembled *in vitro* as in the case of bioengineered bladder.[1] The challenges associated with design and implementation of bioreactors, in which functional tissues and cells are produced at the laboratory bench, are enormous, in view of the need for multiple cell phenotypes and an adequate vasculature.[26] A number of nontechnical regulatory issues are associated with bioreactors and bioprocessing, and there is a dilemma with how to establish and evaluate research milestones. In some situations, the bioreactor approach to tissue engineering can be utilized in the form of extracorporeal assist devices such as the bioartificial liver and kidney.[1]

STORAGE AND TRANSLATION

There are a number of translational issues associated with taking a tissue-engineered construct from the laboratory to the clinic. These pertain to cell isolation, cell and tissue culture and differentiation, scale-up of bioreactors, biomaterials, and scaffolds, long-term storage strategies, and safety and regulatory policies.

In order to provide for storage of living cells or tissues, freeze-thawing and vitrification approaches are being considered.[27,28] A great deal of basic information is needed about physicochemical changes associated with intracellular ice formation, how ice is propagated, and about how cryoprotectants protect and interact with cells.

In the vitrification process, cells don't freeze but, instead, become "glassy" in the presence of high concentrations of cryoprotectants. Vitrification is considered to be suitable for preservation of complex three-dimensional structures; a challenge is how to load and remove high concentrations of cryprotectants. Nontoxic sugars have been used at lower concentrations, but cells need to be permeabilized. Gain of an understanding of wound healing at the level of the cell membrane may facilitate cryoprotection by stabilization of cell membranes[29] One desirable goal for reparative medicine would be the direct administration of freeze-thawed cells without the removal of cryoprotectants.

Cell recovery is a major technical hurdle for translation from the laboratory bench to the clinic. It is currently being investigated in this regard whether maintenance of tissues in the dried state may be preferable to freeze-thawing. Desiccation based on principles of anhydrobiosis, which is used in nature by some microorganisms, and which involves a glassy state in which sugars protect membranes, proteins, and supramolecular structures from effects of drying, is being investigated. Here, research is needed on the stability of the dry state as light, oxygen, and other features of the storage environment affect it.[27]

HOST REMODELING AND IMMUNE RESPONSE

Host remodeling is viewed as an inevitable and often beneficial stage of the tissue-engineering process.[21] Remodeling is an essential part of development throughout life, and varies with age, disease state, and species. When organ regeneration is to be achieved in the body, scar-tissue formation is a major barrier. While host remodeling is closely related to the immune response, integration of the tissue engineering and immunology disciplines has been extremely limited.[30,31] Badylak and copanelists[21] commented that "it is likely that the traditional dividing lines that tend to exist between our understanding of the processes of inflammation, immunity, scar tissue formation, developmental biology and wound healing will require rethinking and/or elimination." The need for better methods to track tissue-engineered implants was recognized as was the great need for a registry of human patients in order to establish predictors for success and failure of tissue-engineered products.

Previous work in reparative medicine involving allograft and xenograft transplantation has shown that both innate and adaptive immunity represent formidable barriers. Even autologous materials will undergo remodeling and induce immune reactions.[30] The host immune system may be expected to influence tissue-engineered constructs whether generated in the body or in a bioreactor. In many situations, remodeling may facilitate or be necessary for successful integration of the tissue-engineered construct. One approach that has been used by tissue engineers is encapsulation to isolate the construct from the body's immune response by biomaterial or by a fibrous capsule generated by the body itself. Immunologists[30] consider the body's immune system to comprise the innate immune system, an ancient, "hard wired" system in which macrophages play a prominent role that recognizes molecular patterns found on pathogens, but not higher eukaryotic organisms. In addition to innate immunity, higher organisms possess an adaptive immune system that utilizes T and B lymphocytes to recognize specific antigens that have been processed and presented by the host. Currently, the immunology community is considering a greater role for innate immunity in the immune response, in that the role of the immune system may be to recognize danger to the body, rather than self or non-self, as has been thought for a number of years. A greater understanding of the interplay between remodeling and innate and adaptive immunity is needed to realize the potential of reparative medicine.

CONCLUSION

This volume is a compilation of plenary manuscripts, panelist overviews, breakout session summary reports, and extended poster abstracts covering the range of elements into which tissue engineering can be reduced: cells, signaling, extracellular matrix, design principles, vascular assembly, bioreactors, storage and translation, and host remodeling and immune response. The poster paper manuscripts present current research activities of some of the symposium participants. There was consensus that multidisciplinary efforts involving surgeons, physicians, engineers, physicists, mathematicians, chemists, cell biologists and allied specialties will be needed to achieve the promise of tissue engineering for reparative medicine.[32]

REFERENCES

1. GODBEY, A.T. & A. ATALA. 2002. In vivo and in vitro systems for tissue engineering. Ann. N.Y. Acad. Sci. **961**: this volume.
2. BOTTARO, D.P., A. LIEBMAN-VINSON & M.A. HEIDARAN. 2002. Molecular signaling in bioengineered tissue microenvironments, Ann. N.Y. Acad. Sci. **961**: this volume.
3. GRIFFITH, L. 2002. Emerging design principles in biomaterials and scaffolds for tissue engineering. Ann. N.Y. Acad. Sci. **961**: this volume.
4. PARENTEAU, N.L. 2002. The use of cells in regenerative medicine. Ann. N.Y. Acad. Sci. **961**: this volume.
5. WADE, N. 2002. In tiny cells, glimpses of body's master plan. *New York Times,* December 18, 2001.
6. CHAIKOFF, E.L. *et al.* 2002. Biomaterials and bioscaffolds in reparative medicine. Ann. N.Y. Acad. Sci. **961**: this volume.
7. METZGER, D.W. 2002. Immune responses to tissue-engineered extracellular matrix used as a bioscaffold. Ann. N.Y. Acad. Sci. **961**: this volume.
8. GERMAIN, L. 2002. Engineering human tissues for *in vivo* applications. Ann. N.Y. Acad. Sci. **961**: this volume.
9. FAUSTMAN, D.L. *et al.* 2002. Cells for repair. Ann. N.Y. Acad. Sci. **961**: this volume.
10. KAPUR, R. 2002. Fluorescence imaging and engineered biosensors. Ann. N.Y. Acad. Sci. **961**: this volume.
11. BONADIO, J. & M.L. CUNNINGHAM. 2002. Genetic approaches to craniofacial tissue repair. Ann. N.Y. Acad. Sci. **961**: this volume.
12. SEFTON, M. 2002. Functional considerations in tissue engineering whole organs. Ann. N.Y. Acad. Sci. **961**: this volume.
13. TAYLOR, D.A. 2002. Is *in vivo* remodeling necessary or sufficient for cellular repair of the heart? Ann. N.Y. Acad. Sci. **961**: this volume.
14. HIRSCHI, K.K. *et al.* 2002. Vascular assembly in natural and engineered tissues. Ann. N.Y. Acad. Sci. **961**: this volume.

15. RATCLIFFE, A. & L.E. NIKLASON. 2002. Bioreactors and bioprocessing for tissue engineering. Ann. N.Y. Acad. Sci. **961:** this volume.
16. DUCY, P. 2002. Molecular signaling. Ann. N.Y. Acad. Sci. **961:** this volume.
17. GOLDSTEIN, S.A. 2002. Tissue engineering: functional assessment and clinical outcome. Ann. N.Y. Acad. Sci. **961:** this volume.
18. DAMSKY, C. *et al.* 2002. Molecular signaling. Ann. N.Y. Acad. Sci. **961:** this volume.
19. INGBER, D. 2002. Mechanical signaling. Ann. N.Y. Acad. Sci. **961:** this volume.
20. GUILAK, F. 2002. Functional tissue engineering: the role of biomechanics. Ann. N.Y. Acad. Sci. **961:** this volume.
21. BADYLAK, S.F. *et al.* 2002. *In vivo* remodeling. Ann. N.Y. Acad. Sci. **961:** this volume.
22. YIP, C. 2002. Biomaterials in reparative medicine: biorelevant structure–property analysis. Ann. N.Y. Acad. Sci. **961:** this volume.
23. O'KEEFE, R. 2002. Determination of responsiveness in biological systems is dependent upon highly… Ann. N.Y. Acad. Sci. **961:** this volume.
24. KORETSKY, A.P. 2002. Functional assessment of tissues with magnetic resonance imaging. Ann. N.Y. Acad. Sci. 961: this volume.
25. SKALAK, T.C. *et al.* 2002. Vascular assembly in engineered and natural tissues. Ann. N.Y. Acad. Sci. 961: this volume.
26. NIKLASON, L.E. *et al.* 2002. Bioreactors and bioprocessing.
27. TONER, M. & J. KOCSIS. 2002. Storage and translational issues in reparative medicine. Ann. N.Y. Acad. Sci. **961:** this volume.
28. KOCSIS, J. *et al.* 2002. Storage and translational issues in reparative medicine. Ann. N.Y. Acad. Sci. **961:** this volume.
29. LEE, R. C. 2002. Cytoprotection by stabilization of cell membranes. Ann. N.Y. Acad. Sci. **961:** this volume.
30. HARLAN, D.M. *et al.* 2002. Immunological concerns with bioengineering approaches. Ann. N.Y. Acad. Sci. **961:** this volume.
31. KARP, C.L. 2002. Immunological barriers (opportunities?) to the use of bioengineered tissue. Ann. N.Y. Acad. Sci. **961:** this volume.
32. NAUGHTON, G.K. 2002. From lab bench to market: critical issues in tissue engineering. Ann. N.Y. Acad. Sci. **961:** this volume.

In Vitro Systems for Tissue Engineering

W.T. GODBEY AND A. ATALA

Laboratory for Tissue Engineering, Harvard Medical School/The Children's Hospital, Boston, Massachusetts 02115, USA

ABSTRACT: Tissue engineering, by necessity, encompasses a wide array of experimental directions and scientific disciplines. *In vitro* **tissue engineering involves the manipulation of cells** *in vitro*, **prior to implantation into the** *in vivo* **environment. In contrast,** *in vivo* **tissue engineering relies on the body's natural ability to regenerate over non-cell-seeded biomaterials. Cells, biomaterials, and controlled incubation conditions all play important roles in the construction and use of modern** *in vitro* **systems for tissue engineering. Gene delivery is also an important factor for controlling or supporting the function of engineered cells both** *in vitro* **and post implantation, where appropriate. In this review, systems involved in the context of** *in vitro* **tissue engineering are addressed, including bioreactors, cell-seeded constructs, cell encapsulation, and gene delivery. Emphasis is placed upon investigations that are more directly linked to the treatment of clinical conditions.**

KEYWORDS: tissue engineering; biomaterials; bioreactors

INTRODUCTION

Historical Perspective

The first written record of using one body part for another to meet the needs of the patient dates back to the time of Socrates, and can be referred to as a form of tissue engineering. This concept still exists today, when surgeons in the operating room use tissues from one organ to replace or repair another in the same patient.

The concepts of organ transplantation and tissue regeneration were studied extensively in the 1930s by the surgeon and Nobel Laureate Alexis Carrell and the famous pilot Charles Lindbergh.[1] However, organ transplantation was not possible until the early 1950s, when Joseph Murray performed a non-related kidney transplant from one gentically non-identical patient into another.[2] This transplant, which overcame the immunological barrier, marked a new era in medical therapy and opened the door for using transplantation as a means of therapy for different organ systems. However, the lack of good immunosuppression and the inability to monitor and control rejection, as well as a severe shortage of organ donors, opened the door for other alternatives.

Address for correspondence: Anthony Atala, Laboratory for Tissue Engineering, Harvard Medical School/The Children's Hospital, 300 Longwood Ave., Enders Bldg. #461, Boston, MA 02115 Voice: 617-355-6169; fax: 617-355-6587.
A.Atala@TCH.Harvard.edu

Ann. N.Y. Acad. Sci. 961: 10–26 (2002). © 2002 New York Academy of Sciences.

As times evolved, synthetic materials were introduced in order to replace or re-build diseased tissues or parts in the human body. The advent of new man-made materials such as Teflon and Silicone created new opportunities for human application that involved a wide array of devices. Although these devices could provide for structural replacement, the functional component of the original tissue was not achieved.

Simultaneous with this development was an increased body of knowledge of the biological sciences, which included new techniques for cell harvesting, culture, and expansion. The areas of cell biology, molecular biology, and biochemistry were advancing rapidly. Studies of the extracellular matrix and its interaction with cells as well as with growth factors and their ligands gave way for a further understanding of cell and tissue growth and differentiation.

In the 1960s, a natural evolution occurred wherein researchers started to combine the fields of devices and materials science with cell biology, in effect starting a new field termed "cell transplantation." In an effort to more widely include different fields within the life sciences, the term "tissue engineering" was used in the 1980s. One of its first references was in 1985, when the term was used to describe the creation of prosthetics for corneal replacement.[3] As more scientists from different fields came together with the common goal of tissue replacement, the field of tissue engineering became more formally established. Tissue engineering is now defined as "an interdisciplinary field that applies the principles of engineering and life sciences towards the development of biological substitutes that aim to maintain, restore, or improve tissue function."[4] In the last four decades scientists have attempted to engineer virtually every tissue of the human body.

Today there are several options for tissue and organ replacement. Native tissues can be used within each patient from the same source (skin grafts for burn patients), or from a different source (intestine to replace esophagus, vagina, or bladder). Tissues and organs for transplantation can be either allogeneic (from one human to another), or xenogeneic (from an animal to a human). Cells can also be transplanted in an attempt to normalize deficient function (muscle cells injected for muscular dystrophy). Tissue engineering encompasses many fields and forms.

Strategies for in Vitro Tissue Engineering

Tissue engineering follows the principles of cell transplantation, materials science, and bioengineering towards the development of biological substitutes that would restore and maintain normal function. *In vivo* tissue engineering relies on the body's natural ability to regenerate over non-cell-seeded biomaterials. This approach may require proteins, DNA, or mRNA additives for enhanced regeneration. This technology has been applied to patients for decades. The use of acellular human donor corneas, bone substitutes, and pig-heart valves are common examples.[5,6] In contrast, *in vitro* tissue engineering involves the manipulation of cells *in vitro*, prior to implantation into the *in vivo* environment. Although this technique also relies on the body's ability to regenerate, additional cues are provided with the presence of cells or tissue at the time of implantation. When cells are used for tissue engineering, donor tissue is dissociated into individual cells that are either implanted directly into the host or expanded in culture, attached to a support matrix, and re-implanted after expansion. A basic example of *in vitro* tissue engineering is the expansion of keratinocytes from a small skin biopsy in culture with the subsequent attempt of reorganizing the keratinocytes into a sheet-like application prior to implantation *in vivo*.[7]

BIOMATERIALS FOR CELL-BASED TISSUE ENGINEERING

Biomaterials function as an extracellular matrix (ECM) and elicit biological and mechanical functions of native ECMs found in tissues in the body. Native ECMs brings cells together into tissue, control the tissue structure, and regulate the cell phenotype.[8] Biomaterials facilitate the localization and delivery of cells and/or bioactive factors (e.g., cell adhesion peptides and growth factors) to desired sites in the body, define a three-dimensional space for the formation of new tissues with appropriate structure, and guide the development of new tissues with appropriate function.[9] Direct injection of cell suspensions without biomaterial matrices has been utilized,[10,11] but it is sometimes difficult to control the localization of transplanted cells. In addition, some mammalian cell types are anchorage-dependent and will die if not provided with a cell-adhesion substrate. Biomaterials provide a cell-adhesion substrate and can be used to achieve cell delivery with high loading efficiency to specific sites in the body. The configuration of the biomaterials can guide the structure of an engineered tissue. The biomaterials provide mechanical support against *in vivo* forces, thus maintaining a predefined structure during the process of tissue development. The biomaterials can be loaded with bioactive signals, such as cell-adhesion peptides and growth factors that can regulate cellular function.

Design and Selection of Biomaterials

The design and selection of biomaterials is critical in the development of engineered tissues. The biomaterial must be capable of controlling the structure and function of the engineered tissue in a predesigned manner by interacting with transplanted cells and/or the host cells. Generally, the ideal biomaterial should be biocompatible, promote cellular interaction and tissue development, and possess proper mechanical and physical properties. If the selected biomaterial is biodegradable, the degradation products should not provoke inflammation or toxicity and must be removed from the body via metabolic pathways. The degradation rate and the concentration of degradation products in the tissues surrounding the implant must be at a tolerable level.[12]

The biomaterials should provide an appropriate regulation of cell behavior, such as adhesion, proliferation, migration, and differentiation in order to promote the development of functional new tissue. Cell behavior in engineered tissues is regulated by multiple interactions with the microenvironment, including interactions with cell adhesion ligands[13] and with soluble growth factors.[14] Cell-adhesion-promoting factors (e.g., Arg-Gly-Asp [RGD]) can be presented by the biomaterial itself or be incorporated into the biomaterial in order to control cell behavior through ligand-induced cell receptor signaling processes.[15,16] The biomaterial can also serve as a depot for the local release of growth factors and other bioactive agents that induce tissue-specific gene expression of the cells. The biomaterials should possess appropriate mechanical properties to regenerate tissues with predefined sizes and shapes. The biomaterials should provide temporary mechanical support sufficient to withstand *in vivo* forces exerted by the surrounding tissue and to maintain a potential space for tissue development. The mechanical support of the biomaterials should be maintained until the engineered tissue has sufficient mechanical integrity to support itself.[17] This can be potentially achieved by an appropriate choice of mechanical and degradative properties of the biomaterials.

Types of Biomaterials

Many classes of biomaterials have been used for cell-based tissue engineering. These can be categorized as naturally derived materials (e.g., collagen and alginate), acellular tissue matrices (small intestinal submucosa), and synthetic polymers (e.g., polyglycolic acid [PGA], polylactic acid [PLA], and poly(lactic-co-glycolic acid) [PLGA]). Naturally derived materials and acellular tissue matrices have the potential advantage of biological recognition. An advantage of synthetic polymers is reproducible large-scale production with controlled properties of strength, degradation rate, and microstructure.

Collagen is the most abundant and ubiquitous structural protein in the body, and may be readily purified from both animal and human tissues with an enzyme treatment and salt/acid extraction.[18] Collagen has long been known to exhibit minimal inflammatory and antigenic responses,[19] and has been approved by the United States Food and Drug Administration (FDA) for many types of medical applications, including wound dressings and artificial skin.[20] Collagen implants degrade through a sequential attack by lysosomal enzymes. The *in vivo* resorption rate can be regulated by controlling the density of the implant and the extent of intramolecular crosslinking. The lower the density of the collagen, the greater the interstitial space and generally the larger the pores for cell infiltration, leading to a higher rate of implant degradation. Intramolecular crosslinking reduces the degradation rate by making the collagen molecules less susceptible to an enzymatic attack. Intramolecular crosslinking can be accomplished by various physical (e.g., UV radiation and dehydrothermal treatment) or chemical (e.g., glutaraldehyde, formaldehyde, and carbodiimides) techniques.[18] Collagen contains cell-adhesion domain sequences (e.g., RGD) which exhibit specific cellular interactions. This may assist in retaining the phenotype and activity of many types of cells, including fibroblasts[21] and chondrocytes.[22] Collagen exhibits high tensile strength and flexibility. These mechanical properties can be further enhanced by intramolecular crosslinking. This material can be processed into a wide variety of structures (e.g., sponges, fibers, films).[18,23,24]

Alginate, a polysaccharide isolated from sea weed, has been used as an injectable cell-delivery vehicle[25] as well as a cell-immobilization matrix,[26] owing to its gentle gelling properties in the presence of divalent ions such as calcium. Alginate is relatively biocompatible and is approved by the FDA for human use as wound dressing material. Alginate is a family of copolymers of β-D–mannuronate and α-L–guluronate. The physical and mechanical properties of alginate gel are strongly correlated with the proportion and length of the polyguluronate block in the alginate chains.[27]

Acellular tissue matrices are collagen-rich matrices that are prepared by removing cellular components from tissues. The matrices are often prepared by mechanical and chemical manipulations of a segment of bladder or small intestinal tissue.[5,28] The matrices slowly degrade upon implantation, and are replaced and remodeled by ECM proteins synthesized and secreted by transplanted or ingrowing cells. Acellular

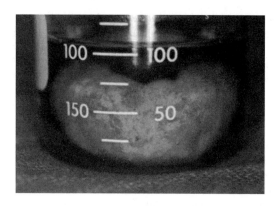

FIGURE 1. A tissue-engineered bladder. An acellular matrix consisting of PLGA-coated PGA was seeded with bladder smooth muscle cells to form the structure pictured. (Photograph by A. Atala.)

tissue matrices have been proven to support cell ingrowth and regeneration of several tissues, including blood vessels, soft tissues, and urethras, with no evidence of immunogenic rejection.[29–31]

Polyesters of naturally occurring α-hydroxy acids, including PGA, PLA, and PLGA, are widely used in tissue engineering (FIG. 1). These polymers have gained FDA approval for human use in a variety of applications, including sutures.[32] The ester bonds in these polymers are hydrolytically labile, and these polymers degrade by nonenzymatic hydrolysis. The degradation products of PGA, PLA, and PLGA are nontoxic, natural metabolites and are eventually eliminated from the body in the forms of carbon dioxide and water.[32] The degradation rates of these polymers can be tailored to require several weeks to several years by altering the crystallinity, initial molecular weight, and, in the case of PLGA, the copolymer ratio of lactic to glycolic acids. Since these polymers are thermoplastics, they can be easily formed into three-dimensional scaffolds with a desired microstructure, gross shape, and dimension by various techniques, including molding, extrusion,[33] solvent casting,[34] phase separation techniques, and gas foaming techniques.[35]

Many applications in tissue engineering often require a scaffold with high porosity and surface-area-to-volume ratio. This requirement has been addressed by processing biomaterials into configurations of fiber meshes and porous sponges using the techniques described above. The mechanical properties of the scaffold can be controlled by the fabrication process. A drawback of synthetic polymers, however, is the lack of biological recognition. As an approach toward incorporating cell-recognition domains into these materials, copolymers with amino acids have been synthesized.[16,36–38] Many other biodegradable synthetic polymers, including poly(anhydrides) and poly(ortho-esters), can also be used to fabricate scaffolds for tissue engineering with controlled properties.[39]

CELL SOURCES AND BIOREACTORS FOR *IN VITRO*
TISSUE ENGINEERING

The fields of cell tranplantation and tissue engineering have been actively studied for several decades, yet there have been relatively few clinical advances. This is mostly related to the inability to grow and expand numerous cell types in large quantities. It is known that 70% of a liver can be surgically resected, and 6 months later it can regrow to its initial volume. Yet, even today, if a biopsy of liver tissue were to be obtained, it would not be possible to grow and expand all the cell types in any appreciable manner. However, scientists have made major advances in the areas of cell growth over the last decade.[39,40] Many different cell lineages have been tried for *in vitro* tissue engineering. These include cells from either autologous or heterologous sources and from allogeneic or xenogeneic lineages. More recently, stem cells have also been applied for tissue engineering, either from embryonic or adult sources.[42,43]

In vitro systems used in tissue engineering span a wide area of subject matter, from pumps and bioreactors to constructs intended for tissue or organ replacement. *Ex vivo* approaches to alter or otherwise control cell behavior are included in this range, and are important for the possible attainment of novel treatments for many human diseases such as diabetes or muscular dystrophy. Bioengineered organs are another goal of tissue engineering, with current approaches aimed at partially constructing and growing the organ in an *in vitro* setting.

As sophisticated as tissue construction *in vitro* can be today, the *in vivo* environment is crucial for the final development of the tissue or organ. Thus, the human body often serves as the "terminal incubator." It has often been stated that cells possess all the genetic information required to reconstruct tissues, whole organs, and even an entire human being. This concept has become even more important today as scientists, governments, and ethicists grapple with the controversies surrounding stem cells and cloning. One concept is certain: that if specific cells are placed in the right environment within a living organism and are given the right cues, then appropriate tissue and organ development may ensue.

The term "bioreactor" refers to a system where conditions are closely controlled to permit or induce a certain behavior in living cells or tissues. The behavior could simply be cell proliferation, or could be as complex as having several sets of cells that sense one or more variable parameters and produce specific chemicals accordingly. The former example could be applied to the common laboratory incubator, while the latter could serve as a guide for a theoretical glucose sensor for the control of insulin and glucagon levels.

The concept of bioreactors is neither new nor restricted to tissue-engineered cells. Bioreactors have been used in the past for wastewater treatment,[44–46] wine production,[47] and even flavor production.[48] The designs of bioreactors intended for tissue engineering are focused upon influencing cell behavior, and incorporate such features as the controlled and steady flow of cell media, pulsatile fluid flow, or culture vessel rotation to influence tissue development. Controlled flow of cell media has been used by several groups for the dynamic seeding of polymer scaffolds with hepatocytes,[49-51] although the approach is not limited to this cell type.[52] Pulsatile fluid flow has been used to develop tissue-engineered heart valves, under the auspices that the mimicry of *in vivo* conditions will yield, in this case, a stronger tissue.[53] Low *et al.* have used the principal of a rotating wall bioreactor to yield relatively large, three-dimensional cultures of neural precursor cells, although the cells had lower growth rates than those of the same type grown in stationary flasks.[54] While numerical modeling has been performed for bioreactors such as the rotating wall bioreactor,[55] one must still keep in mind that changing one parameter of the cellular environment might yield unexpected cellular behaviors due to the complexity of intracellular (and intercellular) molecular interactions.

Another application of *in vitro* bioreactors that has a more direct impact on the *in vivo* setting is the hemodialyzer. This device still fits the definition of an *in vitro* bioreactor in that cellular environments are controllably altered in a test tube setting, but in this case the altered environment (blood) is outside of the body while the cells that will be affected are relatively undisturbed within the patient. Considering acute renal failure, the development of biomaterials with high hydraulic permeability has allowed for the use of continuous hemofiltration as a treatment. A few considerable limitations have been observed with this procedure. Thrombotic occlusion and protein deposition inhibit transport and thereby result in a loss of filtration. Additionally, the need for anticoagulation in the extracorporeal unit results in frequent bleeding. Another limitation of the system stems from the large amounts of fluids that are needed to replace the ultrafiltrate from the filtering unit. Reports on the effects of collagen types I and IV, laminin, and fibronectin on endothelial cell adherence, growth, and differentiation have led to the development of bioartificial hemofilters containing endothelial and other supporting cells.[56,57] In 1995, Humes *et al.* created a bioreactor with seeded endothelial cells.[58] This unit required the use of autologous endothelial cells since they would be in continuous contact with blood. However, these hemofilters were still exposed to the risk of thrombosis. One possible solution to this problem is the transfer of genes that code for the continuous expression of anticoagulant proteins.[59,60] Following a similar approach, commercially available bioreactors have been used as a platform for a bioartificial hemofilter. Woods *et al.* used pronectin-F, a non-degradable substrate for cell attachment, to coat the hollow fibers of the bioreactor.[61] Endothelial cells were then implanted and grown as a monolayer to improve the bioreactor's selectivity to albumin. Although the unit filtered at a low rate, it was successful in its selectivity to albumin and therefore warrants further investigation.

Cell-based bioreactors designed to perform specific functions have also been designed. In an experimental model of a bioartificial tubule,[61] cells were grown within hollow fibers while blood was allowed to flow on the outside of the fibers. The cells were therefore immunoisolated by the synthetic membrane. This device managed to transport salt and water along osmotic gradients.[62] A more recent report described a bioartificial renal tubule in which the lumen of a single hollow fiber was seeded with Madin–Darby canine kidney cells, a renal epithelial cell line.[63] After introducing intraluminally perfused C-inulin,[33] a recovery rate of 98.9% was measured in the cell-lined units. (A control line consisting of hollow fibers without cells obtained less than 7.4% recovery.) Ouabain, an inhibitor of Na^+-K^+ ATPase, and albumin were introduced into the extracapillary space to test the dependency of fluid flux upon transport osmotic and oncotic pressure variances across the tubule. Albumin caused the rate to rise to 4.5 ± 0.4 µl/30 min, whereas ouabain caused the transport rate to fall to nearly the baseline level of 2.1 ± 0.4 µl/30 min. These results demonstrated the functional transport capabilities of an engineered renal epithelial cell confluent monolayer seeded on a support structure.

Bioreactors which replicate microgravity conditions have been designed and used extensively. Chromiak *et al.* examined cellular respiration by using a modified perfusion system that allowed the monitoring of glucose, lactate, and H^+, as well as extracellular protein concentrations as an indication of cellular nitrogen balances.[64] The use of these bioreactors may allow for the discovery of novel cellular functions that may be helpful for the engineering of normal tissues.

BIOMATERIAL–CELL INTERACTIONS

Virtually every tissue in the body has been considered for some form of tissue engineering application. Reactions between the matrix itself and the transplanted cells are important for development of successful tissues. For instance, in studies involving the engineering of liver tissues, chitosan matrices have been modified with collagen, gelatin, or albumin in an effort to enhance hepatocyte attachment.[65] Here it was found that albumin modification of chitosan provided a good surface for cell attachment. Another example is the use of poly(L-lactic acid) sponges as matrices for hepatocyte culture, where it was found that the use of collagen coatings on the sponges or collagen-embedded cells could aid in cell attachment and culture.[66] Others have foregone the use of artificial matrices by using the native architecture of existing livers.[67] Similar findings regarding matrix modification have been noted with various other cell systems. Schwann cells have been grown on cross-linked hyaluronic strands coated with poly(D-lysine) for grafts to aid nerve regeneration.[68] Preadipocytes have been grown on poly(lactic-co-glycolic) acid disks and found to fully mature into adipocytes within the pores of the disks.[69] Work on an artificial salivary gland has been performed using a salivary epithelial cell line seeded on poly(L-lactic acid), poly(glycolic acid), and copolymers of the two materials.[70] Corneal constructs have been manufactured by seeding keratocytes into cross-linked collagen-chondroitin sulfate supports, followed by epithelial and endothelial cell layering below and on top of the substrate, respectively.[71]

Several methods have been proposed for maximizing cell survival after implantation *in vivo*. As an example, keeping corneal cells alive post implantation is still a challenge.[72] A valid approach would be to have the seeded cells express some factor(s) so that the host immune response would be limited after transplantation. Adenovirus-mediated transfection to induce overexpression of interleukin-10 reduced the rate of rejection in a sheep model.[73]

As constructs have become more complex, so have analyses of the construct components. In an attempt to develop a suitable system for soft tissue repair, Eiselt *et al.* examined the effects of different polymer supports, different seeding techniques, and different release kinetics of VEGF-containing microspheres upon seeded smooth muscle cell survival.[74] Their optimal support material tested was poly(L-lactic acid) bonded to poly(glycolic acid), and they found that dynamic seeding under stirred conditions yielded a higher density of seeded cells than did static seeding. Although VEGF delivery was examined, the desired level of angiogenesis was not obtained despite optimization of release kinetics.

The complexity of cell/matrix constructs can vary by application and laboratory. Constructs consisting of a single cell type and one support material have been used successfully in humans clinically, such as in the case study presented in Ref. 75. Here the bony portion of the distal part of the thumb of a patient injured in a machine accident was replaced by a construct consisting of natural coral (hydroxyapetite) seeded with periosteal cells. (Here the patient served as his own bioreactor for the skin covering, as his phalanx was directly attached to his abdomen for 19 days for skin regeneration.) In separate experiments, constructs utilizing two types of cells have been used successfully in a cultured skin application, as for the case of venous ulcer treatment.[76] In this published report, three layers of tissue were achieved first by seeding dermal fibroblasts onto matrices of bovine type I collagen to form a lattice, followed by the seeding of a layer of keratinocytes to serve as the epidermal layer. After the epidermal layer was sufficiently developed, the level of culture medium was lowered to partially expose the keratinocytes to air, allowing for the formation of a stratum corneum. The resulting structure has been used for the treatment of venous ulcers via secondary intention.

Tissue-engineered blood vessels also employ more than one cell type by necessity. One investigation of engineered arteries used smooth muscle cells seeded onto modified PGA matrices (FIG. 2).[77] A bioreactor producing pulsatile media flow was used to induce migration of the smooth muscle cells into the scaffolds. This was eventually followed by seeding a layer of endothelial cells into the lumen of the vessel. A major problem that was addressed by this approach was that of vessel rupture due to poor mechanical properties in the vessel. The observed success of vessel non-rupture was attributed to the smooth muscle cells and the matrix proteins they produced. Another tissue-engineered vascular model used three cell types in its composition.[78] The support for the construct was made from dehydrated fibroblasts wrapped temporarily around an inert tubular support. A sheet of smooth muscle cells was layered around the support, and this was covered with a sheet of fibroblasts to provide an adventitia. After at least 8 additional weeks, a layer of endothelial cells was seeded onto the dehydrated fibroblasts on the luminal side of the constructs. The constructs showed desirable mechanical properties which the authors attributed to the adventitial layer made from the seeded fibroblasts and the matrix they produced. This is a good example of a complex construct that does not rely on an artificial scaffold for its support.

An organ (bladder) has also been engineered using multiple cell types. Autologous bladder muscle and urothelial cells were obtained from a small surgical biopsy (<1 cm^2). The cells were seeded on bladder-shaped pre-configured PGA/PLLA scaffolds in a sequential fashion over a period of a few days. The tissue-engineered bladders implanted in dogs were able to function in a normal fashion, with adequate capacity, filling pressures, and emptying, for the duration of the study (1 year after the initial bladder biopsy).[79]

CELL ENCAPSULATION INTO MICROSPHERES

The principle behind cell encapsulation is to surround cells with a coating that prevents contact between the cells and their surface antigens with host immune components. The coating must be porous to allow for the exchange of nutrients and cel-

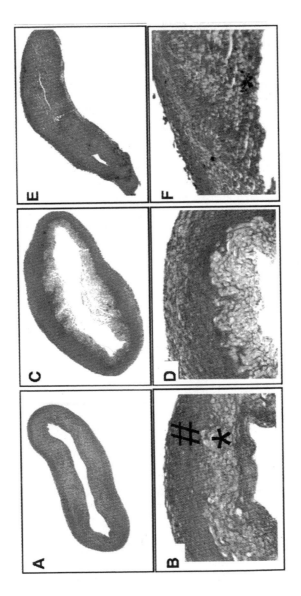

FIGURE 2. Functional tissue-engineered arteries. (**A** and **B**) Vessel grown in pulsed flow for 8 weeks. Verhoff's stain for elastin (**A**); Masson's trichrome stain (blue) for collagen (**B**). # indicates dense cellular region, * indicates polymer region. (**C** and **D**) Non-pulsed vessel grown for 8 weeks. Verhoff's stain (**C**); Masson's stain (**D**). (**E** and **F**) Vessel grown without media supplementation. Verhoff's stain (**E**); Masson's stain (**F**). (From Niklason et al.[77] Reproduced by permission.)

lular products between the interior and exterior of the microcapsules. While the encapsulated cells are used in an *in vivo* setting, their creation is by necessity an *in vitro* process.

Cell encapsulation has been studied on a wide variety of cells for a wide number of conditions. Canine and bovine islets have been investigated for the treatment of diabetes.[80] "Islet-like cell clusters" of porcine origin have been examined in terms of long-term cryopreservation to address the possibility of long-term storage.[81] Characterizations of encapsulation materials in terms of mechanical stability and albumin diffusion have been performed using red blood cells as the encapsulated entity.[82] Parathyroid cells have also been encapsulated in hopes of finding a new treatment for hypoparathyroidism.[83] Additionally, Leydig cell encapsulation has been performed as a method to supplement low testosterone levels.[84]

The level of protein production and the duration of the secretion period can be regulated by modulating the number of engineered cells that are encapsulated per microsphere, as well as the number of microspheres injected. A similar strategy has also been pursued for the genetic engineering of anti-angiogenic factor–secreting cells.[85] The use of non-mitogenic hydrogels has also been employed to control the level of theraputic molecules emanating from cells encapsulated in hollow fibers.[86] The idea in this case was that fewer cells would impose a lesser drain on local nutrient supplies, although alternate methods to combat this problem currently exist.

GENE DELIVERY

A powerful tool that is becoming increasingly useful to the tissue engineer is that of gene delivery. Gene delivery itself is a very large field that is itself going through rapid changes and expansion.

There are many carriers available for gene delivery, and each comes with its own set of advantages and disadvantages.[87] The most common use of gene delivery for tissue engineering is *ex vivo* transfection. The goal of such work is typically either to engineer cells with particular characteristics to repair a defect or replace somatic cells that are lacking in the specific behavior, or to manipulate existing cells to produce a product that will induce the body to respond in a specific way after implantation. An example of the former aim is the creation of an artificial organ from autologous cells of a patient with a chromosomal anomaly, while the latter goal is exemplified by cells implanted to produce VEGF to stimulate angiogenesis in the area of the implant. *Ex vivo* transfection has also been used in investigations of possible cancer vaccines (reviewed in Ref. 88).

Several cell and tissue types have already been used for *ex vivo* gene delivery. Chondrocytes have been transfected for articular cartilage repair,[89] and bone repair has been investigated using bone morphogenic proteins (BMPs) such as BMP-2 transfected onto skeletal muscle cells.[90] Other cell types include hepatocytes, smooth muscle and urothelial cells as a proof of principle,[91, 92] fibroblasts for the repair of anterior cruciate ligaments,[93] and mesothelial cells to combat changes in peritoneal function and structure as a result of peritoneal dialysis.[94] Human skeletal myoblasts have also been retrovirally transduced to produce human growth hormone, a process used to induce cells to fuse to form the myofibrils of a bioartificial muscle.[95]

An approach that has been pursued to increase and stimulate rapid vascularization *in vivo* is to engineer cell lines to secrete high levels of vascular endothelial growth factor (VEGF) through transfection with the appropriate cDNA. The VEGF-secreting cells can then be encapsulated in polymeric microspheres to allow nutrients to reach the cells while the VEGF proteins secreted from the cells diffuse into the surrounding tissues. Soker *et al.* created such a VEGF-secreting cell line in CHO cells in 1996.[96] A similar strategy has also been pursued for the genetic engineering of anti-angiogenic factor–secreting cells.[85]

Gene delivery has also been used in *ex vivo* settings to investigate possible treatments of neurological disorders. Because of the blood–brain barrier, immunological responses to the implantation of genetically engineered cells into the central nervous system are expected to be limited. Transfected fetal human astrocytes have been studied with this in mind as a possible treatment for Parkinson's disease.[97] Others have approached Parkinson's disease in a similar manner, their findings having implications for other neurodegenerative disorders.[98] To combat Huntington's disease, cells have been transfected to produce ciliary neurotrophic factor, followed by encapsulation and implantation.[99] Although performed in an *in vitro* setting, past work with astrocytes by Lin *et al.* has implications for *ex vivo* investigations.[100] Here it was found that astrocytes in the human fetal cortex can be isolated and efficiently infected with an amphotropic retrovirus harboring mouse beta-nerve growth factor (NGF). It was observed that the transduced cells manufactured NGF mRNA, and that NGF was present in the extracellular media of the transfected cells. Such work helps demonstrate the potential importance of gene delivery to *in vitro* tissue engineering, especially for currently untreatable neurological disorders.

CLINICAL APPLICATIONS

In vitro tissue engineering techniques have been used clinically for a variety of organ systems. Like most areas involving biologicals, the translational lag from experimental work to clinical application has varied depending on the complexity involved. In the field of cell transplantation and tissue engineering, there has usually been a 10- to 20-year lag for most tissue types applied clinically to date: skin (7 years),[101,102] pancreas (19 years),[103,104] cartilage (17 years),[11,105] liver (12 years),[106,107] cornea (15 years),[3,108] and bladder (9 years).[109,110] In addition to these specific examples, several other tissue technologies are in the process of being applied clinically. Tissue-engineering techniques using cell-based approaches are already expanding the options for treatment in patients with deficient tissues and organs.

REFERENCES

1. BERG, A.S. 1998. *In* Lindbergh. Putnam. New York.
2. MURRAY, J.E., J.P. MERRILL & J.H. HARRISON. 1955. Renal homotransplantation in identical twins. Surg. Form. **6:** 432–436.
3. WOLTER J.R. & R.F. MEYER. 1985. Sessile macrophages forming clear endothelium-like membrane of successful keratoprosthesis. Trans. Am. Ophthal. Soc. **82:** 187–202.
4. SKALAK R. & C. F. FOX, Eds. 1988. *In* Tissue Engineering. Liss. New York.

5. BADYLAK, S.F., R. TULLIUS, K. KOKINI, *et al.* 1995. The use of xenogeneic small intestinal submucosa as a biomaterial for Achilles tendon repair in a dog model. J. Biomed. Mater. Res. **29:** 977–985.
6. KHAN, S.S., A. TRENTO, M. DEROBERTIS, *et al.* 2001. Twenty-year comparison of tissue and mechanical valve replacement. J. Thorac. Cardiovasc. Surg. **122:** 257–269.
7. RHEINWALD, J.G. & H. GREEN. 1975. Serial cultivation of strains of human epidermal keratinocytes: the formation of keratinizing colonies from single cells. Cell **6:** 331–343.
8. ALBERTS, B., D. BRAY, J. LEWIS, *et al.* 1994. *In* Molecular Biology of the Cell. :971–995. Garland. New York.
9. KIM, B.S. & D.J. MOONEY. 1998. Development of biocompatible synthetic extracellular matrices for tissue engineering. Trends Biotechnol. **16:** 224–230.
10. PONDER, K.P., S. GUPTA, F. LELAND, *et al.* 1991. Mouse hepatocytes migrate to liver parenchyma and function idenfinitely after intrasplenic transplantatin. Proc. Natl. Acad. Sci. USA **88:** 1217–1221.
11. BRITTBERG, M., A. LINDAHL, A. NILSSON., *et al.* 1994. Treatment of deep cartilage defects in the knee with autologous chondrocyte transplantation. N. Engl. J. Med. **331:** 889–895.
12. BERGSMA, J.E., F.R. ROZEMA, R.R.M. BOS, *et al.* 1995. Biocompatibility and degradatin mechanism of predegraded and non-degraded poly(lactide) implants: an animal study. Mater. Med. **6:** 715–724.
13. HYNES, R.O. 1992. Integrins: versatility, modulation and siignaling in cell adhesion. Cell **69:** 11–25.
14. DEUEL, T.F. 1997. Growth factors. *In* Principles of Tissue Engineering. R. P. Lanza, R. Langer & W. L. Chick, Eds.: 133–149. Academic Press. New York.
15. BARRERA, D.A., E. ZYLSTRA, P.T. LANSBURY, *et al.* 1993. Synthesis and RGD peptide modification of a new biodegradable copolymer poly (lactic acid-co-lysine) J. Am. Chem. Soc. **115:** 11010–11011.
16. COOK, A.D., J.S. HRKACH, N.N. GAO, *et al.* 1997. Characterization and development of RGD-peptide-modified poly(lactic acid-co-lysine) as an interactive, resorbable biomaterial. J. Biomed. Mater. Res. **35:** 513–523.
17. ATALA, A. 1998. Autologous cell transplantation for urologic reconstruction. J. Urol. **159:** 2–3.
18. LI, S.T. 1995. Biologic biomaterials: tissue-derived biomaterials (collagen). *In* The Biomedical Engineering Handbook. J.D. Brozino, Ed. :627–647. CRS Press. Boca Raton, FL.
19. FURTHMAYR, H. & R. TIMPL. 1976. Immunochemistry of collagens and procollagens. Int. Rev. Connect. Tiss. Res. **7:** 61–99.
20. PACHENCE, J.M. 1996. Collagen-based devices for soft tissue repair. J. Biomed. Mater. Res. (Appl. Biomater.) **33:** 35–40.
21. SILVER, F. H. & G. PINS. 1992. Cell growth on collagen: a review of tissue engineering using scaffolds containing extracellular matrix. J. Long-term Effects Med. Implants **2:** 67–80.
22. SAM, A.E. & A.J. NIXON. 1995. Chondrocyte-laden collagen scaffolds for resurfacing extensive articular cartilage defects. Osteoarthritis & Cartilage **3:** 47–59.
23. YANNAS, I.V. & J.F. BURKE. 1980. Design of an artificial skin. I. Basic design principles. J. Biomed. Mater. Res. **14:** 65–81.
24. CAVALLARO J.F., P.D. KEMP & K. KRAU. 1994. Collagen fabrics as biomaterials. Biotechnol. Bioeng. **43:** 781–791.
25. ATALA, A., L.G. CIMA, W. KIM, *et al.* 1993. Injectable alginate seeded with chondrocytes as a potential treatment for vesicoureteral reflux. J. Urol. **150:** 745–747.
26. LIM, F. & A.M. SUN. 1980. Microencapsulated islets as bioartificial endocrine pancreas. Science **210:** 908–910.
27. SMIDSRØD, O. & G. SKJÅK-BRÆK. 1990. Alginate as an immobilization matrix for cells. Trends Biotechnol. **8:** 71–78.
28. YOO, J.J., J. MENG, F. OBERPENNING, *et al.* 1998. Bladder augmentation using allogenic bladder submucosa seeded with cells. Urology **51:** 221–225.

29. SOLAN, A., V. PRABHAKAR & L. NIKLASON. 2001. Engineered vessels: importance of the extracellular matrix. Transplant. Proc. **33:** 66–68.
30. KIM, B.S., J. NIKOLOVSKI, J. BONADIO, et al. 1999. Engineered smooth muscle tissues: regulating cell phenotype with the scaffold. Exp. Cell Res. **251:** 318–328.
31. CHEN, F., J.J. YOO & A. ATALA. 1999. Acellular collagen matrix as a possible "off the shelf" biomaterial for urethral repair. Urology **54:** 407–410.
32. GILDING, D.K. 1981. Biodegradable polymers. *In* Biocompatibility of Clinical Implant Materials. D. F. Williams, Ed.: 209–232. CRC Press. Boca Raton, FL.
33. FREED, L.E., G. VUNJAK-NOVAKOVIC, R.J. BIRON, et al. 1994. Biodegradable polymer scaffolds for tissue engineering. Biotechnology **12:** 689–693.
34. MIKOS, A.G., A.J. THORSEN, L.A. CZERWONKA, et al. 1994. Preparation and characterization of poly(L-lactic acid) foams. Polymer **35:** 1068–1077.
35. HARRIS, L.D., B.S. KIM & D.J. MOONEY. 1998. Open pore biodegradable matrices formed with gas foaming. J. Biomed. Mater. Res. **42:** 396–402.
36. BARRERA, D.A., E. ZYLSTRA, P.T. LANSBURY, et al. 1993. Synthesis and RGD peptide modification of a new biodegradable copolymer poly (lactic acid-co-lysine) J. Am. Chem. Soc. **115:** 11010–11011.
37. INTVELD, P.J.A., Z.R. SHEN, G.A.J. TAKENS, et al. 1994. Glycine glycolic acid based copolymers. J. Polym. Sci. Polym. Chem. **32:** 1063–1069.
38. BARRERA, D.A., E. ZYLSTRA, P.T. LANSBURY, et al. 1995. Copolymerization and degradation of poly (lactic acid-co-lysine). Macromolecules **28:** 425–432.
39. PEPPAS, N.A. & R. LANGER. 1994. New challenges in biomaterials. Science **263:** 1715–1720.
40. CILENTO, B.G., M.R. FREEMAN, F.X. SCHNECK, et al. 1994. Phenotypic and cytogenetic characterization of human bladder urothelia expanded in vitro. J. Urol. **152:** 665–670.
41. ATALA, A. & R. LANZA. 2001. *In* Methods of Tissue Engineering. Academic Press. San Diego, CA.
42. BARTSCH, G, J. YOO, B. KIM, et al. 2000. Stem cells in tissue engineering applications for incontinence. J. Urol. **1009S:** 227.
43. YOO, J.J., R. LANZA, J.B. CIBELLI, et al. 2001. Embryonic stem cells as a source for urologic tissue reconstruction. J. Urol. **165:** 33.
44. LIM, B.R., X. HUANG, H.Y. HU, et al. 2001. Effects of temperature on biodegradation characteristics of organic pollutants and microbial community in a solid phase aerobic bioreactor treating high strength organic wastewater. Water Sci. Technol. **43:** 131–137.
45. LAPARA, T.M., C.H. NAKATSU, L. PANTEA, et al. 2000. Nucleotide phylogenetic analysis of bacterial communities in mesophilic and thermophilic bioreactors treating pharmaceutical wastewater. Appl. Environ. Microbiol. **66:** 3951–3959.
46. PAMPEL, L.W. & A.G. LIVINGSTON. 1998. Anaerobic dechlorination of perchloroethene in an extractive membrane bioreactor. Appl. Microbiol. Biotechnol. **50:** 303–308.
47. ICONOMOU, L., M. KANELLAKI, S. VOLIOTIS, et al. 1996. Continuous wine making by delignified cellulosic materials supported biocatalyst: an attractive process for industrial applications. Appl. Biochem. Biotechnol. **60:** 303–313.
48. VAN DER SLUIS C., C.J. STOFFELEN, S.J. CASTELEIN, et al. 2001. Immobilized salt-tolerant yeasts: application of a new polyethylene-oxide support in a continuous stirred-tank reactor for flavour production. J. Biotechnol. **88:** 129–139.
49. TOROK, E., J.M. POLLOK, P.X. MA, et al. 2001. Optimization of hepatocyte spheroid formation for hepatic tissue engineering on three-dimensional biodegradable polymer within a flow bioreactor prior to implantation. Cells Tissues Organs **169:** 34–41.
50. KIM S.S., C.A. SUNDBACK, S. KAIHARA, et al. 2000. Dynamic seeding and *in vitro* culture of hepatocytes in a flow perfusion system. Tissue Eng. **6:** 39–44.
51. POLLOK, J.M., D. KLUTH, R.A. CUSICK, et al. 1998. Formation of spheroidal aggregates of hepatocytes on biodegradable polymers under continuous-flow bioreactor conditions. Eur. J. Pediatr. Surg. **8:** 195–199.
52. WU, F., N. DUNKELMAN, A. PETERSON, et al. 1999. Bioreactor development for tissue-engineered cartilage. Ann. N.Y. Acad. Sci. **875:** 405–411.

53. HOERSTRUP S.P., R. SODIAN, J.S. SPERLING, *et al.* 2000. New pulsatile bioreactor for *in vitro* formation of tissue engineered heart valves. Tissue Eng. **6:** 75–79.
54. LOW, H.P., T.M. SAVARESE & W.J. SCHWARTZ. 2001. Neural precursor cells form rudimentary tissue-like structures in a rotating-wall vessel bioreactor. In vitro Cell Dev. Biol. Anim. **37:** 141–147.
55. POLLACK, S.R., D.F. MEANEY, E.M. LEVINE, *et al.* 2000. Numerical model and experimental validation of microcarrier motion in a rotating bioreactor. Tissue Eng. **6:** 519–530.
56. CARLEY, W.W., A.J. MILICI & J.A. MADRI. 1988. Extracellular matrix specificity for the differentiation of capillary endothelial cells. Exp. Cell Res. **178:** 426–434.
57. MILICI, A.J., M.B. FURIE & W.W. CARLEY. 1985. The formation of fenestrations and channels by capillary endothelium *in vitro*. Proc. Natl. Acad. Sci. USA **82:** 6181–6185.
58. HUMES, D.H., D.A. CEILESKI & A.J. FUNKE. 1995. Cell therapy for erythropoietin (EPO) deficient anemias. J. Am. Soc. Nephrol. **6:** 535.
59. WILSON, J.M., L.K. BIRINYI, R.N. SALOMON, *et al.* 1989. Implantation of vascular grafts lined with genetically modified endothelial cells. Science **244:** 1344–1346.
60. ZWEIBEL, J.A., S.M. FREEMAN, P.W. KANTOFF, *et al.* 1989. High-level recombinant gene expression in rabbit endothelial cells transduced by retroviral vectors. Science **243:** 220–222.
61. WOODS, J.D. & D.H. HUMES. 1997. Prospects for a bioartificial kidney. Semin. Nephrol. **17:** 381–386.
62. NIKOLOVSKI, J., S. POIRIER, A.J. FUNKE, *et al.* 1996. Development of a bioartificial renal tubule for the treatment of acute renal failure. J. Am. Soc. Nephrol. [abstr.] **7:** 1376.
63. MACKAY, S.M., A.J. FUNKE, D.A. BUFFINGTON, *et al.* 1998. Tissue engineering of a bioartificial tubule. ASAIO J. **44:** 179–183.
64. CHROMIAK, J.A., J. SHANSKY, C. PERRONE, *et al.* 1998. Bioreactor perfusion system for the long-term maintenance of tissue-engineered skeletal muscle organoids. In vitro Cell Dev. Biol. Animal **34:** 694–703.
65. ELCIN, Y.M., V. DIXIT & G. GITNICK. 1998. Hepatocyte attachment on biodegradable modified chitosan membranes: *in vitro* evaluation for the development of liver organoids. Artif. Organs **22:** 837–846.
66. KAUFMANN, P.M., S. HEIMRATH, B.S. KIM, *et al.* 1997. Highly porous polymer matrices as a three-dimensional culture system for hepatocytes. Cell Transplant. **6:** 463–468.
67. TAKEZAWA, T., M. INOUE, S. AOKI, *et al.* 2000. Concept for organ engineering: a reconstruction method of rat liver for *in vitro* culture. Tissue Eng. **6:** 641–650.
68. HU, M., E.E. SABELMAN, C. TSAI, *et al.* 2000. Improvement of Schwann cell attachment and proliferation on modified hyaluronic acid strands by polylysine. Tissue Eng. **6:** 585–593.
69. PATRICK, C.W., JR, P.B. CHAUVIN, J. HOBLEY, *et al.* 1999. Preadipocyte seeded PLGA scaffolds for adipose tissue engineering. Tissue Eng. **5:** 139–151.
70. AFRAMIAN, D., J.E. CUKIERMAN, J. NIKOLOVSKI, *et al.* 2000. The growth and morphological behavior of salivary epithelial cells on matrix protein-coated biodegradable substrata. Tissue Eng. **6:** 209–216.
71. GRIFFITH, M., R. OSBORNE, R. MUNGER, *et al.* 1999. Functional human corneal equivalents constructed from cell lines. Science **286:** 2169–2172.
72. ORWIN, E.J. & A. HUBEL. 2000. *In vitro* culture characteristics of corneal epithelial, endothelial, and keratocyte cells in a native collagen matrix. Tissue Eng. **6:** 307–319.
73. KLEBE, S., P.J. SYKES, D.J. COSTER, *et al.* 2001. Prolongation of sheep corneal allograft survival by *ex vivo* transfer of the gene encoding interleukin-10. Transplantation **71:** 1214–1220.
74. EISELT, P., B.S. KIM, B. CHAKO, *et al.* 1998. Development of technologies aiding large-tissue engineering. Biotechnol. Prog. **14:** 134–140.
75. VACANTI, C.A., L.J. BONASSAR, M.P. VACANTI, *et al.* 2001. Replacement of an avulsed phalanx with tissue-engineered bone. N. Engl. J. Med. **344:** 1511–1514.
76. SABOLINSKI, M.L., O. ALVAREZ, M. AULETTA, *et al.* 1996. Cultured skin as a "smart material" for healing wounds: experience in venous ulcers. Biomaterials **17:** 311–320.

77. NIKLASON, L.E., J. GAO, W.M. ABBOTT, *et al.* 1999. Functional arteries grown *in vitro*. Science **284:** 489–493.
78. L'HEUREUX, N., S. PAQUET, R. LABBE, *et al.* 1998. A completely biological tissue-engineered human blood vessel. FASEB J. **12:** 47–56.
79. OBERPENNING, F.O., J. MENG, J. YOO, *et al.* 1999. De novo reconstitution of a functional urinary bladder by tissue engineering. Nature Biotech. **17:** 149–155.
80. LANZA, R.P., D.M. ECKER, W.M. KUHTREIBER, *et al.* 1999. Transplantation of islets using microencapsulation: studies in diabetic rodents and dogs. J. Mol. Med. **77:** 206–210.
81. MURAKAMI, M., H. SATOU, T. KIMURA, *et al.* 2000. Effects of micro-encapsulation on morphology and endocrine function of cryopreserved neonatal porcine islet-like cell clusters. Transplantation **70:** 1143–1148.
82. CHANDY, T., D.L. MOORADIAN, G.H. RAO. 1999. Evaluation of modified alginate-chitosan-polyethylene glycol microcapsules for cell encapsulation. Artif. Organs **23:** 894–903.
83. PICARIELLO, L., S. BENVENUTI, R. RECENTI, *et al.* 2001. Microencapsulation of human parathyroid cells: an *"in vitro"* study. J. Surg. Res. **96:** 81–89.
84. MACHLUF, M., S. BOORJIAN, J. CAFFARATTI, *et al.* 1998. Microencapsulation of Leydig cells: a new system for the therapeutic delivery of testosterone. Pediatrics **102S:** 32.
85. JOKI, T., M. MACHLUF, A. ATALA, *et al.* 2001. Continuous release of endostatin from microencapsulated engineered cells for tumor therapy. Nat. Biotechnol. **19:** 35–39.
86. LI, R.H., S. WILLIAMS, M. WHITE, *et al.* 1999. Dose control with cell lines used for encapsulated cell therapy. Tissue Eng. **5:** 453–466.
87. GODBEY, W.T. & A.G. MIKOS. 2001. Recent progress in gene delivery using non-viral transfer complexes. J. Control. Release **72:** 115–125.
88. VAN TENDELOO, V.F., C. VAN BROECKHOVEN & Z.N. BERNEMAN. 2001. Gene-based cancer vaccines: an *ex vivo* approach. Leukemia **15:** 545–558.
89. GOOMER, R.S., L.J. DEFTOS, R. TERKELTAUB, *et al.* 2001. High-efficiency non-viral transfection of primary chondrocytes and perichondrial cells for *ex-vivo* gene therapy to repair articular cartilage defects. Osteoarthritis Cartilage **9:** 248–256.
90. MUSGRAVE, D.S., R. PRUCHNIC, V. WRIGHT, *et al.* 2001. The effect of bone morphogenetic protein-2 expression on the early fate of skeletal muscle-derived cells. Bone **28:** 499–506.
91. ANDREOLETTI, M., N. LOUX, C. VONS, *et al.* 2001. Engraftment of autologous retrovirally transduced hepatocytes after intraportal transplantation into nonhuman primates: implication for *ex vivo* gene therapy. Hum. Gene Ther. **12:** 169–179.
92. YOO, J.J. & A. ATALA. 1997. A novel gene delivery system using urothelial tissue engineered neo-organs. J. Urol. **158:** 1066–1070.
93. MENETREY, J., C. KASEMKIJWATTANA, C.S. DAY, *et al.* 1999. Direct-, fibroblast- and myoblast-mediated gene transfer to the anterior cruciate ligament. Tissue Eng. **5:** 435–442.
94. HOFF, C.M. 2001. *Ex vivo* and *in vivo* gene transfer to the peritoneal membrane in a rat model. Nephrol. Dial. Transplant. **16:** 666–668.
95. POWELL, C., J. SHANSKY, M. DEL TATTO, *et al.* 1999. Tissue-engineered human bioartificial muscles expressing a foreign recombinant protein for gene therapy. Hum. Gene Ther. **10:** 565–577.
96. SOKER, S., H. FIDDER, G. NEUFELD, *et al.* 1996. Characterization of novel vascular endothelial growth factor (VEGF) receptors on tumor cells that bind VEGF165 via its exon 7-encoded domain. J. Biol. Chem. **271:** 5761–5767.
97. YADID, G., N. FITOUSSI, N. KINOR, *et al.* 1999. Astrocyte line SVG-TH grafted in a rat model of Parkinson's disease. Prog. Neurobiol. **59:** 635–661.
98. ISHIDA, A. & F. YASUZUMI. 2000. Approach to *ex vivo* gene therapy in the treatment of Parkinson's disease. Brain Dev. **22 Suppl 1:** S143–S147.
99. BACHOUD-LEVI, A.C., N. DEGLON, J.P. NGUYEN, *et al.* 2000. Neuroprotective gene therapy for Huntington's disease using a polymer encapsulated BHK cell line engineered to secrete human CNTF. Hum. Gene Ther. **11:** 1723–1729.
100. LIN, Q., L.A. CUNNINGHAM, L.G. EPSTEIN, *et al.* 1997. Human fetal astrocytes as an *ex vivo* gene therapy vehicle for delivering biologically active nerve growth factor. Hum. Gene Ther. **8:** 331–339.

101. RHEINWALD, J.G. & H. GREEN. 1974. Growth of cultured mammalian cells on secondary glucose sources. Cell **2:** 287–293.
102. BURKE, J.F., I.V. YANNAS, W.C. QUINBY, JR., *et al.* 1981. Successful use of a physiologically acceptable artificial skin in the treatment of extensive burn injury. Ann. Surg. **194:** 413–428.
103. CHICK, W.L., A. LIKE & V. LAURIS. 1975. Beta cell culture on synthetic capillaries: an artificial endocrine pancreas. Science **187:** 847–849.
104. SOON-SHIOG, P., R.E. HEINTS, N. MERIDETH, *et al.* 1994. Insulin independence in a type 1 diabetic patient after encapsulated islets transplantation. Lancet **143:** 950–951.
105. GREEN, W.T., JR. 1977. Articular cartilage repair: behavior of rabbit chondrocytes during tissue culture and subsequent allografting. Clin. Orthop. **124:** 237–250.
106. DEMETRIOU, A.A., J.F. WHITING, D. FELDMAN, *et al.* 1986. Replacement of liver function in rats by transplantation of microcarrier-attached hepatocytes. Science **233:** 1190–1192.
107. FOX, I.J., J.R. CHOWDHURY, S.S. KAUFMAN, *et al.* 1998. Treatment of the Crigler-Najjar syndrome type I with hepatocyte transplantation. N. Engl. J. Med. **338:** 1422–1426.
108. SCHWAB, I.R. & R.R. ISSEROFF. 2000. Bioengineered corneas—the promise and the challenge. N. Engl. J. Med. **343:** 136–138.
109. ATALA, A., J.P. VACANTI, C.A. PETERS, *et al.* 1992. Formation of urothelial structures in vivo from dissociated cells attached to biodegradable polymer scaffolds in vitro. J. Urol. **148:** 658–662.
110. ATALA, A. 2001. Bladder regeneration by tissue engineering. Br. J. Urol. In press.

The Use of Cells in Reparative Medicine

NANCY L. PARENTEAU AND JANET HARDIN YOUNG

Organogenesis Inc., Canton, Massachusetts 02021, USA

ABSTRACT: Cells are the functional elements of reparative medicine and tissue engineering. The use of living cells as a therapy presents several challenges. These include identification of a suitable source, development of adequate methods, and proof of safety and efficacy. We are now well aware that stem or pluripotent cells offer an exciting potential source for a host of functional cell types. Their true potential will only be realized through continued effort to increase basic scientific understanding at all levels, the development of adequate methods to achieve a functional phenotype, and attention to safety issues associated with adequate control of cell localization, proliferation, and differentiation. There is also new understanding regarding the immunology of parenchymal cells and new promising approaches to immune modulation, which will open the door to broader therapies using allogeneic cell sources without prohibitive immune suppression. Control of cell growth and phenotypic expression does not end in the culture vessel, but goes beyond to the patient. A living therapy is not static but dynamic, as is the host response. The cells or tissue construct in most cases will not behave as a whole-organ transplant. It is therefore important that we understand a cell or tissue therapy's ability to react and interact within the host since clinical effectiveness has proven to be one of the most difficult milestones to achieve. A living cell therapy offers great potential to alter the human condition, encompassing alteration of the current biological state of a targeted tissue or organ, augmentation of depleted or lost function, or absolute functional tissue replacement. The extent to which we are able to achieve effective cell therapies will depend on assimilating a rapidly developing base of scientific knowledge with the practical considerations of design, delivery, and host response.

KEYWORDS: living cell therapy; immune modulation; tissue engineering

THE USE OF CELLS IN REPARATIVE MEDICINE: WHO WILL LEAD THE WAY?

Fundamental knowledge is integral to effective design in tissue engineering; whether the goal is the development of a novel scaffold material to promote tissue regeneration, or a living cellular implant. Without it, technology is developed in the dark, using an iterative approach often lacking the dimension and understanding to produce a successful, predictable outcome in a timely manner. Unless the appropriate knowledge and experience are put to the problem, the potential of living therapy will not be realized.

Address for correspondence: Nancy L. Parenteau, Amaranth Bio, Inc., 5 Mainstone Road, Wayland, MA 01778. Voice and fax: 508-650-1665.
amaranth@ix.netcom.com

Ann. N.Y. Acad. Sci. 961: 27–39 (2002). © 2002 New York Academy of Sciences.

The need for cell therapy is well recognized even by the nonscientist, as evidenced by the perceived need for some form of stem cell research. Yet the complex nature of dealing with actual cells themselves to achieve an outcome is still not fully realized. While there is burgeoning information on the genetics front, advances in technology for rapid proteomic analysis, rapidly growing information on factors that effect cell lineage,[1,2] identification of transcription factors involved in the development of tissue structure and control of morphogenesis,[3] and advancing preclinical and clinical research on cell implantation, there is a very important need to bring all aspects together. Engineers and physician scientists have been instrumental in leading the way in academic tissue engineering research, although they desperately need the participation of workers in other disciplines, such as molecular biologists and cell biologists, to fill the important gaps in understanding between them. We must not be naïve.

How do we stimulate the interest in critical areas such as applied research in cell biology and foster interdisciplinary collaboration? How do we provide academic recognition for being an important *part* of a significant achievement? Academic laboratories must be given an incentive to work on common goals. Multidisciplinary research is more prevalent in industry, out of necessity. Industrial research is usually team-oriented and collaborative, often having a common product goal. But this environment also must be managed carefully so as not to devalue the importance of individual excellence, contribution, and responsibility. New paradigms and metrics must be established both in academics and industry as we delve into complex biological problems that are well beyond a single scientific or engineering discipline. Academic institutions, granting agencies, and professional societies must do their part to foster, recognize, and reward the accomplishments of great teams and the accomplished individuals within them, without letting "entrepreneurial spirit" and "product" focus overshadow the value of critical knowledge.

Academic research must lead the way in establishing the knowledge base that will lead to product opportunities. In the biological disciplines at least, we cannot ask or expect industry to create, research, *and* develop, for there are too many inputs needed for any one group to take on the entire job. In addition, we cannot expect industry support to correctly target and fund academic research for the broad advancement of the field. Academic research must be where things begin, helping to create the opportunities and the paths to those opportunities through basic experimentation and the acquisition of fundamental understanding of biological processes and cell behavior. Industry is expected to take the discovery, knowledge, and opportunity and bring it to meaningful use. This will require innovation, technology development, and ongoing discovery. Universities and industry must work in partnership to bring the technology forward, because without industry participation, products of scientific discovery and knowledge are unlikely to get to the patient in an effective way. Biotechnology will be the proving ground for many of these opportunities, but basic research must be there to continually provide support of theories, scientific validation, and additional knowledge.

The elements involved in delivering any new therapy to the patient are depicted in FIGURE 1. Carrying out this scenario has been particularly difficult in dealing with a living cell therapy because of the novelty, and the multidisciplinary challenges. To date, success has been slow in coming; yet we recognize the great potential of using

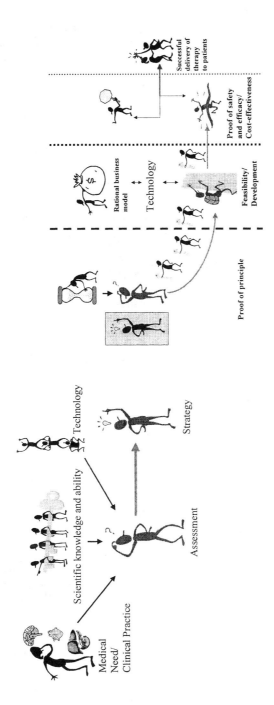

FIGURE 1. Development of a successful therapy requires strong research, technology, and management.

cells for repair and therapeutic use. The assessment of opportunity and the strategy of bringing forward any new technology relies heavily on the basic scientific state of the art, which is still rudimentary for most cell therapy applications. The remainder of this report discusses a few of the considerations in delivering an effective cell therapy to the patient.

CELLS–THERE IS NO ESCAPING THEM

Repair of the body implies a re-growth, a re-formation of tissue, and a re-establishment of function. Cells are the functional elements of reparative medicine, no matter whether we are concerned with creating a noncellular implant that will guide tissue formation or are concerned with the growth and manipulation of cells outside the body to form living tissues for transplantation. For example, much of tissue engineering over the last ten years has been focused in the area of biomaterials to control cell response.[4] To use an example from my company's own work, we have been concerned with the development of a small-caliber vascular graft for some time. This interest first began more than 15 years ago with the work of Weinberg and Bell,[5] who used a cellular construct of collagen and fibrin. The concept of a cellular graft has seen renewed interest in academic research more recently.[6–8] Our own efforts in a cellular approach were abandoned due to issues concerning physical strength and suturability, the immunogenicity of certain cell types, and the activation of undesirable properties in certain cell types. Some of these issues are now being addressed by academic laboratories.[9–12] We moved to the concept of using an acellular scaffold to promote neovessel formation. The idea of using processed native tissues for repair was not new and showed promise for vascular applications.[13–15] However, their ease of use, clinical feasibility, practicality, and the ability to control both the material and the outcome were hurdles in bringing the technology forward for a small-diameter graft application. This has required new technology for tissue processing,[16] engineering to enable entubation of the collagen material to meet physical requirements, matrix biochemistry to enable modification of the collagen surface,[17] and chemistry to ensure the delivery of a reproducible, nonthrombogenic lumen.[18] This scaffold approach has shown promising results in a rabbit carotid model,[19] demonstrating both gross function in the form of long-term patency and favorable cellular infiltration and remodeling in the form of an endothelialized luminal surface and a vasoactive response. This work demonstrated early scientific and technical feasibility. The graft has been further refined and key chemical and physical parameters are being defined as part of the development process; in preparation are large animal studies in both the carotid and coronary positions as the team proceeds to the next step, preclinical efficacy. As the graft gets closer to human use, the team of engineers, biologists, chemists, and clinicians now must not only investigate gross measures of success such as graft patency, but also analyze more dynamic biological parameters such as rate of cell infiltration, cell phenotype, behavior, physiological function, tissue development, and physical properties over time. Knowledge of vascular cell biology, vascular physiology, cell–matrix interaction, and mechanisms of inflammation and vascular disease progression are all key elements that must be brought to the analysis of graft performance and likely outcome. Only with such an analysis, with each discipline involved, will we be able to recognize issues that might mean the success or failure of the construct in the clinic. A sophisticated supporting technology such as bioimaging could greatly facilitate assessment of progress and outcome, preclinically and clinically.

This example illustrates that a program that began heavily focused on biomaterials, matrix biochemistry, and engineering now adds a level of advanced biological (cellular) considerations going forward. This is what distinguishes tissue engineering from a more traditional medical device approach.

CELLS FOR THERAPY AND FUNCTIONAL IMPLANTS

Effective use of cells for therapeutic purposes hinges on our ability to:

(1) accurately predict cell response;
(2) acquire the appropriate cells, either through recruitment *in situ*, expansion *ex vivo* from self or non-self, or manipulation of novel sources such as from different species or pluripotent embryonic or adult cells; and
(3) direct and control cell response toward the desired phenotype to achieve the desired outcome, level of function, and assurance of safety.

The recent stem cell debate in scientific circles made it clear that lack of knowledge in these areas makes it difficult to ensure progress or success of any one path. We are only now beginning to understand cell response *in vitro*, a highly artificial environment, with limited dimension. We can only begin to appreciate that cells implanted in a living human being, often under varying conditions, are an even bigger challenge, not "automatic" in outcome, but not insurmountable either. The use of cells and living tissues is dynamic, not static. This dynamic property on the surface may appear to create numerous questions and hurdles in the development of reparative therapies, but this same dynamic nature is also the property that makes the use of living cells so powerful and promising. Cell therapy is a give and take between donor and recipient; they are not organ transplants. There will be different immunological considerations, differing donor–host responses and a dynamic interaction, as the living tissues react, respond, and repair. One of the greatest challenges is delivering cells, tissues, or organ "equivalents" in a form that will be biologically meaningful. Our experience with the living skin substitute, Apligraf®,[a] illustrates this point.

Apligraf is a bilayered skin construct of human skin reconstructed *in vitro* through the use of organotypic culture techniques.[20] It is made through the large-scale cultivation of both normal human dermal fibroblasts and keratinocytes, obtained from human infant foreskin tissue.[21] The cells are combined with purified bovine collagen and cultivated *in vitro* to form a bilayered, organotypic culture of skin (FIG. 2). The skin equivalent possesses many features of normal human skin: biochemical,[22] structural,[23] and functional. Apligraf was designed with the intent to provide properties of a living skin graft, something clinicians knew was of clinical benefit.[24]

Apligraf is able to respond to its environment, go through the biological and structural aspects of wound healing *in vitro,* and take routinely as a skin graft lasting

---

[a]Apligraf is a registered trademark of Novartis Pharmaceuticals.

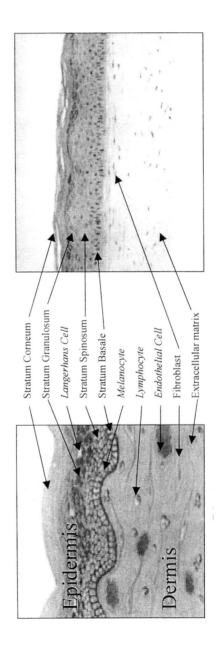

FIGURE 2. The basic components of skin and engineered skin equivalent. *Left:* Diagram of human skin showing the major cell types and their organization. Note that stratified keratinocytes make up the epidermis and display distinct morphologic phenotypes. *Right:* Histologic section of Apligraf human skin equivalent. Original magnification ×142; hematoxylin and eosin stain. Italics indicate cell types presesnt in real skin, but not in the engineered skin equivalent.

at least one year on immune-compromised mice. There is no evidence of rejection in humans.[25] It became the first manufactured living product to be approved for two indications through the traditional U.S. Food and Drug Administration Pre-Market Approval route, having demonstrated clinical efficacy in the treatment of both venous leg ulcers and, later, diabetic foot ulcers. It is currently under study for other indications, but is used quite broadly by physicians in the field.

Given our scientific, clinical, and practical experience, it is evident that Apligraf has both similarities and differences in its response to different wounds *in vivo*. Away from the controlled environment of the athymic or SCID mouse, Apligraf appears to respond to the patient's wound in multiple ways to provide its benefit.[26] We are allowed at least ready gross observation of this response since Apligraf is applied to the body surface. Even with that, it is sometimes difficult to ascertain the exact nature of the response. Clinical research still continues in order to gain added insight.[27–30] While the seeming lack of control we have on the exact behavior of the skin construct may be perceived to be a difficulty, it is actually a benefit, since the normal biological responses and interactions between the living construct and the recipient proceeds in a beneficial way, even without us knowing how. Therefore, the dynamic nature of the living cell response is the key element to its success. As understanding grows, through observation and clinical experience, clinical outcomes become increasingly predictable even though the mechanisms may not yet be well understood. Continued scientific study will provide this understanding. Further analysis is also important in setting a course of future development as we must consider ways to improve the technology. Given what might have been a relatively straightforward scenario with skin, what does this alert us to in the development of other cellular constructs and therapies where access to the graft, assessment of outcome, and pathways to success may be even less forthcoming?

SCIENTIFIC AND TECHNOLOGICAL ADVANCES IN OTHER AREAS OF SCIENCE WILL PLAY KEY ENABLING ROLES

There are a number of ways in which basic scientific understanding of cell genetics and cell processes will enable the successful use of cells for repair. Accurate prediction of cell response is important, not only in enabling us to control proliferation and differentiation pathways *in vitro*, but also in understanding how the responses and factors controlling them will influence their action, reaction, persistence, and safety *in vivo*. This fundamental knowledge is integral to the direct manipulation of cells, and also to the design and development of the enabling technologies surrounding them; such as the development of novel scaffolds, new biomaterials, immunotherapy, microinvasive surgery, and bioimaging.

Immunology

Clinical and basic research on the immunology of the allogeneic fibroblasts and keratinocytes in Apligraf has demonstrated that at least certain allogeneic cells of a nonprofessional nature may be transplanted without immune recognition and rejection.[31,32] Basic research indicates that this is in part due to lack of an operable co-stimulatory pathway,[32] necessary for T-cell activation. Further work continues to help us to understand other biological factors affecting the ability of the graft to persist in different types of wounds. This information not only helps us to understand the skin construct, but also provides important understanding that we can apply to other cell therapies as well.

One important, but neglected aspect governing cell and/or functional persistence is the cell or tissue response to inflammation. This will be a critical area of research going forward, as will immunological research in the areas of immune tolerance, which is expected to play a key role in allowing the transplantation of many cell types and tissues.[33–35] We must appreciate that lack of persistence is due to multiple factors, only one of which may be the recognition of non-self, and that allogenicity does not always equate with rejection.

Genomics and Proteomics

The genomics era in some ways, competes for biotech and pharmaceutical industry interest and resources, but the technology itself can play an important role in enabling cell therapy to advance. For example, DNA arrays will be used at first to identify similarities and differences in cell phenotypic expression *in vitro*, serving to guide our discovery and development of conditions to permit the desired expression. Is an insulin-producing cell *in vitro* always a beta cell? Probably not. As we find novel ways to use stem cells, progenitor cells, or even propagate more committed parenchymal cells, there may be multiple ways to promote insulin production,[36–42] but only a few ways to produce a stable, clinically meaningful, phenotype. Genomic analysis will provide an important tool to recognize key elements of cell differentiation amongst a backdrop of cell plasticity and adaptation, particularly in an extremely artificial *in vitro* environment.[43] Genomics will help provide a roadmap toward the optimum functional phenotype and provide a measure for quality assurance and assessment as the applications approach clinical and commercial use. It will also serve as a vital link connecting knowledge gained from basic research in cell biology and cancer research with the goals of developing stable, effective, and safe cell and tissue therapies.

Proteomics is another important tool for the cell biologist/tissue engineer in several ways. First, it helps define the environment that we are creating and to which the cells are contributing in a dynamic way. It will help identify important factors regulating cell response. Second, proteomics can aid in the rapid identification of known substances of interest on a small scale. This is particularly important as we must analyze many parameters to understand cell response and optimize conditions for maximum cell response. Both genomics and proteomics have the potential to fuel advances in the use of cells for reparative medicine. Methods of cultivation and organotypic culture technologies, i.e., methods that foster and support maximum phenotypic expression, will advance using this knowledge. These technologies will help not only the development of technology, but our ability to predict outcome.

Both genomics and proteomics will help foster advances in cell culture methodology as well. This, combined with innovative bioreactor designs that foster cell–cell interaction and allow dimension for tissue growth, such as the microgravity rotary wall vessel originally developed by NASA (Synthecon Inc., Houston, TX), will play important roles in achieving the cell development and control needed for the generation of functional tissue implants. We are rapidly gaining insights into transcription factors that are important in certain aspects of tissue development.[44] Innovative developmental biological research is illustrating the plasticity of cell phenotype and the potential of the cell to exhibit pluripotency.[45,46] But for the correct enabling *in vitro* environment, for example, a skin construct might be made containing not only epidermis and dermis, but also appendages, such as glands and hair, derived from a single epithelial cell source. Technology must advance to support this scientific potential.

Issues of Cell Sourcing

The choice of cell source has an impact on a technology beyond immunology and safety. A more in-depth discussion of cell sourcing issues can be found elsewhere,[47] but a few points deserve mention. The use of autologous cells often implies an assumption of minimal manipulation and an inherent safety in the use of one's own cells. This is not entirely correct, as culture processes and reagents can alter cells, regardless of the origin. The use of allogeneic (non-self) and xenogeneic (animal) sources presents additional, but different, immunological and safety considerations. Allogeneic cell screening and testing is needed to prevent the transmission of human pathogens, and immunological issues must be evaluated. However, the allogeneic cells should be *biologically* no different from similarly sourced autologous cells. As mentioned in the earlier discussion on Apligraf, it is increasingly evident that certain cell types may be transplanted without rejection. However, our ability to culture only the desired cell types is important for the implementation of such allogeneic cell therapies since professional antigen-presenting cells and possibly some non-professional antigen presenting cells, will elicit a T-cell response. The culture of allogeneic cells allows for a defined human cell source that can be adequately tested, not only for safety, but also for functionality. It also removes the hurdles of person-to-person variability that must be addressed in the processing and manipulation of autologous cell sources. The potential existence of pluripotent stem cells, both from adult humans and from embryonic stem cells, broadens the potential sources for allogeneic cells.

The use of xenogeneic cells has been viewed as an important alternative to the problem of cell sourcing. One approach to dealing with the immunological hurdles of xenogeneic therapy has been to genetically alter the animal source.[48] While some envision this as an alternative source for whole-organ transplants, the animals might also serve as cell sources for engineered cells and tissues.

The choice of cell source can have an impact upon the research strategy, the design requirements, the clinical strategy, and the potential target market. For example, an extracorporeal liver-assist device containing porcine cells will have certain research and development issues, such as a controlled source, reproducible methods of isolation, assurances of batch-to-batch quality and physiological function, immune isolation, etc. This type of product will most likely be limited to severe cases where the expense of processing, the risk of immune sensitization, and risk of exposure to xenogeneic tissue is justified by the life-saving benefit of the device. The ability to incorporate human hepatocytes into the device opens the door to broader use in the population of patients with chronic disease, expanding market potential significantly. This approach has very different research and development hurdles and priorities, such as learning how to cultivate large quantities of human hepatocytes, parenchy-

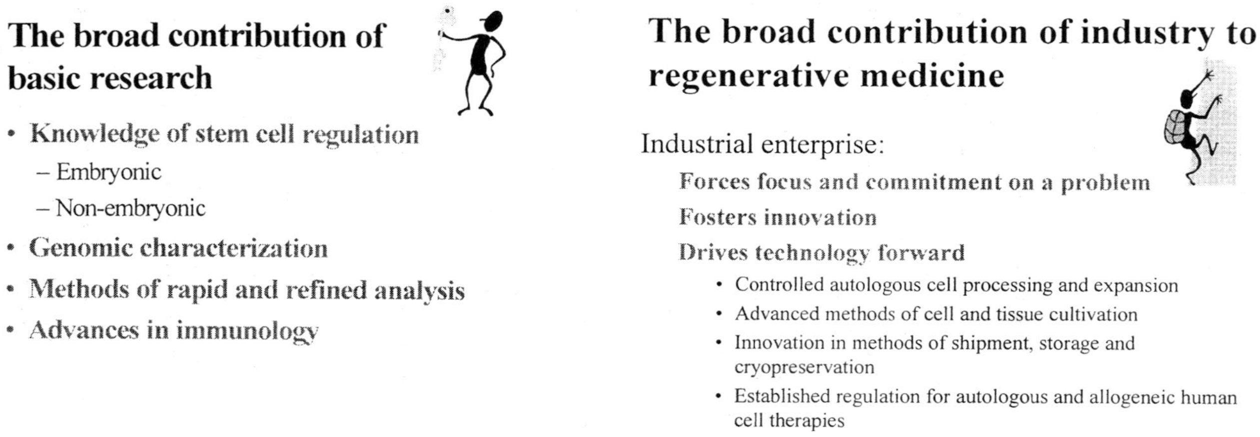

The broad contribution of industry to regenerative medicine

Industrial enterprise:

Forces focus and commitment on a problem

Fosters innovation

Drives technology forward

- Controlled autologous cell processing and expansion
- Advanced methods of cell and tissue cultivation
- Innovation in methods of shipment, storage and cryopreservation
- Established regulation for autologous and allogeneic human cell therapies

The broad contribution of basic research

- **Knowledge of stem cell regulation**
  - Embryonic
  - Non-embryonic
- **Genomic characterization**
- **Methods of rapid and refined analysis**
- **Advances in immunology**

FIGURE 3. The contributions of both basic and applied sciences will be necessary to realize successful cell and tissue therapy.

mal or progenitor, and developing conditions to achieve or maintain adequate physiological function. In both cases, the hard-device engineering could be similar, although cost of the device will be more of a consideration in its acceptance for the chronic patient and may require additional innovation with respect to methods of storage, delivery, and cost of materials. Companion technology such as cryopreservation and cold-storage technologies become important for this application and for many other cell therapies.

CONCLUSION

We have used a few examples from our own experience to illustrate the need for multidisciplinary approaches, open minds, and new tools. We have also attempted to highlight some of the important complexities cells bring to the task. The outlines presented in FIGURE 3 summarize the contributions both basic research and industry can make to cellular reparative medicine. Awareness, focus, and support must be built into many areas to adequately address the complex tasks ahead. The successful delivery of a living cell or tissue therapy to the patient will require the participation of all of us.

REFERENCES

1. SCHULDINER, M., O. YANUKA, J. ITSKOVITZ-ELDOR, et al. 2000. Effects of eight growth factors on the differentiation of cells derived from human embryonic stem cells. Proc. Natl. Acad. Sci. USA **97**: 11307–11312.
2. SHEN, C.N., J.M. SLACK & D. TOSH. 2000. Molecular basis of transdifferentiation of pancreas to liver [in process citation]. Nat. Cell Biol. **2**: 879–887.
3. MERRILL, B.J., U. GAT, R. DASGUPTA, et al. 2001. Tcf3 and Lef1 regulate lineage differentiation of multipotent stem cells in skin. Genes Dev. **15**: 1688–1705.
4. HUBBELL, J.A. 1995. Biomaterials in tissue engineering. Biotechnology **13**: 565–576.
5. WEINBERG, C.B. & E. BELL. 1986. A blood vessel model constructed from collagen and cultured vascular cells. Science **231**: 397–400.
6. L'HEUREUX, N., L. GERMAIN, R. LABBE, et al. 1993. In vitro construction of a human blood vessel from cultured vascular cells: A morphologic study. J. Vasc. Surg. **17**: 499–509.
7. L'HEUREUX, N., S. PAQUET, R. LABBE, et al. 1998. A completely biological tissue-engineered human blood vessel. FASEB J. **12**: 47–56.
8. NIKLASON, L.E., J. GAO, W.M. ABBOTT, et al. 1999. Functional arteries grown in vitro. Science **284**: 489–493.
9. NIKLASON, L.E., W. ABBOTT, J. GAO, et al. 2001. Morphologic and mechanical characteristics of engineered bovine arteries. J. Vasc. Surg. **33**: 628–368.
10. BAROCAS, V.H., T.S. GIRTON & R.T. TRANQUILLO. 1998. Engineered alignment in media equivalents: magnetic prealignment and mandrel compaction. J. Biomech. Eng. **120**: 660–666.
11. GIRTON, T.S., T.R. OEGEMA & R.T. TRANQUILLO. 1999. Exploiting glycation to stiffen and strengthen tissue equivalents for tissue engineering. J. Biomed. Mater. Res. **46**: 87–92.
12. KAUSHAL, S., G.E. AMIEL, K.J. GULESERIAN, et al. 2001. Functional small-diameter neovessels created using endothelial progenitor cells expanded ex vivo. Nat. Med. **7**: 1035–1040.
13. MATSUMOTO, T., R.H. HOLMES, C.O. BURDICK, et al. 1966. Replacement of large veins with free inverted segments of small bowel: autografts of submucosal membrane in dogs and clinical use. Ann. Surg. **164**: 845–848.

14. LAWLER, M.R., J.H. FOSTER & H.W. SCOTT. 1971. Evaluation of canine intestinal submucosa as a vascular substitute. Am. J. Surg. **122:** 517–519.
15. BADYLAK, S.F., G.C. LANTZ, A. COFFEY, et al. 1989. Small intestinal submucosa as a large diameter vascular graft in the dog. J. Surg. Res. **47:** 74–80.
16. ABRAHAM, G.A., R.M. CARR, P.D. KEMP, et al., inventors; Organogenesis, assignee. 1999. U.S Patent Number: 5,993,844. Date of Application: May 8, 1997.
17. KEMP, P.D., R.M.J. CARR & J.G. MARESH, inventors; Organogenesis, assignee. 1993. U.S. Patent 5,256,418. Date of application: April 6, 1990.
18. SULLIVAN, S.J. & K.G.M. BROCKBANK. 2000. Small-diameter vascular grafts. *In* Principles of Tissue Engineering. R. P. Lanza, R. Langer & J. Vacanti, Eds. :447–454. Academic Press. San Diego, CA.
19. HUYNH, T., G. ABRAHAM, J. MURRAY, et al. 1999. Remodeling of an acellular collagen graft into a physiologically responsive neovessel. Nat. Biotechnol. **17:** 1083–1086.
20. PARENTEAU, N.L., P. BILBO, C.J. NOLTE, et al. 1992. The organotypic culture of human skin keratinocytes and fibroblasts to achieve form and function. Cytotechnology **9:** 163–171.
21. PARENTEAU, N. 1994. Skin equivalents. *In* Keratinocyte Methods. I. Leigh & F. Watt, Eds.: 45–54. Cambridge University Press. London.
22. BILBO, P.R., C.J.M. NOLTE, M.A. OLESON, et al. 1993. Skin in complex culture: the transition from "culture" phenotype to organotypic phenotype. J. Toxicol. Cut. Ocular Toxicol. **12:** 183–196.
23. NOLTE, C.J., M.A. OLESON, P.R. BILBO, et al. 1993. Development of a stratum corneum and barrier function in an organotypic skin culture. Arch. Dermatol. Res. **285:** 466–474.
24. WILKINS, L.M., S.R. WATSON, S.J. PROSKY, et al. 1994. Development of a bilayered living skin construct for clinical applications. Biotechnol. Bioeng. **43:** 747–756.
25. FALANGA, V., D. MARGOLIS, O. ALVAREZ, et al. 1998. Rapid healing of venous ulcers and lack of clinical rejection with an allogeneic cultured human skin equivalent. Human Skin Equivalent Investigators Group. Arch. Dermatol. **134:** 293–300.
26. SABOLINSKI, M.L., O. ALVAREZ, M. AULETTA, et al. 1996. Cultured skin as a "smart material" for healing wounds: experience in venous ulcers. Biomaterials **17:** 311–320.
27. FALANGA, V. & M. SABOLINSKI. 1999. A bilayered living skin construct (APLIGRAF®) accelerates complete closure of hard-to-heal venous ulcers. Wound Repair Regen, **7:** 201–207.
28. CHANG, D.W., L.A. SANCHEZ, F.J. VEITH, et al. 2000. Can a tissue-engineered skin graft improve healing of lower extremity foot wounds after revascularization? Ann. Vasc. Surg. **14:** 44–49.
29. WAYMACK, P., R.G. DUFF & M. SABOLINSKI. 2000. The effect of a tissue engineered bilayered living skin analog over meshed split-thickness autografts on the healing of excised burn wounds [in process citation]. Burns **26:** 609–619.
30. BREM, H., J. BALLEDUX, T. BLOOM, et al. 2000. Healing of diabetic foot ulcers and pressure ulcers with human skin equivalent: a new paradigm in wound healing. Arch. Surg. **135:** 627–634.
31. BRISCOE, D.M., V.R. DHARNIDHARKA, C. ISAACS, et al. 1999. The allogeneic response to cultured human skin equivalent in the hu-PBL-SCID mouse model of skin rejection. Transplantation **67:** 1590–1599.
32. LANING, J.C., J.E. DELUCA, C.M. ISAACS, et al. 2001. In vitro analysis of CD40-CD154 and CD28-CD80/86 interactions in the primary T cell response to allogeneic "nonprofessional" antigen presenting cells. Transplantation **71:** 1467–1474.
33. DURHAM, M.M., A.W. BINGAMAN, A.B. ADAMS, et al. 2000. Cutting edge: administration of anti-CD40 ligand and donor bone marrow leads to hemopoietic chimerism and donor-specific tolerance without cytoreductive conditioning. J. Immunol. **165:** 1–4.
34. GRACA, L., K. HONEY, E. ADAMS, et al. 2000. Cutting edge: anti-CD154 therapeutic antibodies induce infectious transplantation tolerance [in process citation]. J. Immunol. **165:** 4783–4786.
35. YOO-OTT, K.-A., H. SCHILLER, F. FANDRICH, et al. 2000. Co-transplantation of donor-derived hepatocytes induces long-term tolerance to cardiac allografts in a rat model. Transplantation **69:** 2538–2546.

36. RAMIYA, V.K., M. MARAIST, K.E. ARFORS, *et al.* 2000. Reversal of insulin-dependent diabetes using islets generated in vitro from pancreatic stem cells. Nat. Med. **6:** 278–282.
37. SORIA, B., E. ROCHE, G. BERN, *et al.* 2000. Insulin-secreting cells derived fromembryonic stem cells normalize glycemia in streptozotocin-induced diabetic mice. Diabetes **49:** 157–162.
38. SHAMBLOTT, M.J., J. AXELMAN, J.W. LITTLEFIELD, *et al.* 2001. Human embryonic germ cell derivatives express a broad range of developmentally distinct markers and proliferate extensively *in vitro*. Proc. Natl. Acad. Sci. USA **98:** 113–118.
39. ASSADY, S., G. MAOR, M. AMIT, *et al.* 2001. Insulin production by human embryonic stem cells. Diabetes **50:** 1691–1697.
40. LUMELSKY, N., O. BLONDEL, P. LAENG, *et al.* 2001. Differentiation of embryonic stem cells to insulin-secreting structures similar to pancreatic islets. Science **292:** 1389–1394.
41. TOMA, J.G., M. AKHAVAN, K.J. FERNANDES, *et al.* 2001. Isolation of multipotent adult stem cells from the dermis of mammalian skin. Nat. Cell Biol. **3:** 778–784.
42. GMYR, V., J. KERR-CONTE, B. VANDEWALLE, *et al.* 2001. Human pancreatic ductal cells: large-scale isolation and expansion. Cell Transplant. **10:** 109–121.
43. PARENTEAU, N.L. 2000. Cell differentiation, animal. *In* Encyclopedia of Cell Technology. R. E. Spier, Ed. :365–377. John Wiley. New York.
44. FUCHS, E., B.J. MERRILL, C. JAMORA, *et al.* 2001. At the roots of a never ending cycle. Devel. Cell **1:** 13–25.
45. TAYLOR, G., M.S. LEHRER, P.J. JENSEN, *et al.* 2000. Involvement of follicular stem cells in forming not only the follicle but also the epidermis. Cell **102:** 451–461.
46. OSHIMA, H., A. ROCHAT, C. KEDZIA, *et al.* 2001. Morphogenesis and renewal of hair follicles from adult multipotent stem cells. Cell **104:** 233–245.
47. HARDIN-YOUNG, J., J. TEUMER, R. ROSS, *et al.* 2000. Approaches to transplanting engineered cells and tissues. *In* Principles of Tissue Engineering. R.P. Lanza, R. Langer & J. Vacanti, Eds.: 281–291. Academic Press. San Diego.
48. GREENSTEIN, J.L. & D.H. SACHS. 1997. The use of tolerance for transplantation across xenogeneic barriers. Nat. Biotechnol. **15:** 235–238.

Reversal of Established Autoimmune Diabetes by *in Situ* β-Cell Regeneration

DENISE L. FAUSTMAN

Harvard Medical School, Charlestown, Massachusetts 02129, USA

KEYWORDS: autoimmune diabetes; B-cell regeneration

In the NOD mouse, a model of autoimmune diabetes, various immunomodulatory interventions prevent progression to diabetes. However, after hyperglycemia is established, such interventions rarely alter the course of disease or sustain engraftment of islet transplants. This situation is mirrored in humans with type 1 diabetes, where allogeneic islet and pancreas transplants are susceptible to recurrent disease and islet transplants rarely result in sustained normoglycemia even with the concurrent use of high-dose immunosuppression.

Recently published data suggest that stem cells could exist in pancreatic islets or ducts and might provide a solution to inadequate islet tissue for diabetic treatments with the added benefit of heightened or more long-lasting regenerative survival *in vivo*. Importantly, it is recognized that the *in vitro* demonstration of insulin production from stem cells does not guarantee *in vivo* regulation or function or guarantee long-term lineage commitment without tumors.

In the NOD mouse, we are investigating novel ways to accomplish *in situ* islet regeneration after severe hyperglycemia is fully established. These new methods restore endogenous islet regrowth in up to 75% of severely diabetic mice after discontinuation of treatment after a brief 40-day intervention. A drug therapy aimed at the underlying disease in autoimmune diabetes is thus able to effect an apparent long-term cure in established type 1 diabetes in the NOD mouse by promoting spontaneous *in situ* islet regeneration.

REFERENCE

1. RYU, S., S. KODAMA, K. RYU, *et al.* 2001. Reversal of established autoimmune diabetes by restoration of endogenous beta cell function. J. Clin. Invest. **108:** 63–72.

Address for correspondence: Denise Faustman, M.D., Ph.D., Immunobiology Laboratory, Massachusetts General Hospital, Building 149, Thirteenth Street, Room 3601, Charlestown, MA 02129. Voice: 617-726-4084; fax: 617-726-4095.
faustman@helix.mgh.harvard.edu

Ann. N.Y. Acad. Sci. 961: 40 (2002). © 2002 New York Academy of Sciences.

Pancreatic Islet Cell Replacement

Successes and Opportunities

SEUNG K. KIM

Stanford University, Stanford, California 94305, USA

KEYWORDS: islet cell replacement; diabetes mellitus

Diabetes mellitus, whether from autoimmune β-cell destruction (type 1) or from β-cell failure and insulin resistance (type 2), results from inadequate insulin production and secretion.[1] Clinical trials demonstrate that tight glucose regulation can prevent the development of diabetic complications,[2] but attempts to achieve this regulation by exogenous insulin administration are only partially successful. Recent evidence suggests that islet cell transplantation with improved systemic immunosuppression may provide a long-term remission in insulin requirements in type I diabetics.[3,4] However, there are limitations with this approach, including (1) a shortage of functioning donor islets; (2) the lack of demonstrated self-renewal by these allografts with the attendant need for possible multiple transplantations; and (3) the need for some form of immunosuppressive therapy.

EXPANDING AND CHARACTERIZING POPULATIONS OF CELLS THAT MAY DIFFERENTIATE INTO ISLET CELLS

In Vitro Studies

Work by a number of groups, including that reported by Bonner-Weir *et al.*,[5] Wang *et al.*,[6] and Zulewski *et al.*[7] suggests that adult pancreatic tissue may harbor multipotent cells with the potential to be manipulated to generate insulin-producing cells *in vitro*. Thus one potential (renewable) source of engraftable material would be derived from cells expanded from adult pancreatic cell cultures. A future challenge will be to develop conditions under which regulated insulin secretion by these cells can be increased and maintained.

Recent studies on embryonic stem (ES) cells[8] and tissue-specific stem cell populations[9] suggested that pancreatic islet cells may be derived from ES or pancreas stem cells. Lumelsky *et al.*[10] showed that islet-like structures could be generated

Address for correspondence: Seung K. Kim, Stanford University, Beckman Center, Room B300, Mail Stop 5329, 279 Campus Drive, Stanford, CA 94305. Voice: 650-723-6230; fax: 650-725-7739.

seungkim@cmgm.stanford.edu

Ann. N.Y. Acad. Sci. 961: 41–43 (2002). © 2002 New York Academy of Sciences.

from ES cells *in vitro*, and that the stimulus-secretion coupling by these structures resembled that of normal islets of Langerhans. Analogous to the experience of Bonner-Weir *et al.* and Zulewski *et al.*, sustained production of insulin by these *in vitro*-derived differentiated cells was problematic, and rescue from diabetes-related lethality in animal models has yet to be achieved. Nevertheless, these reports demonstrate the feasibility and promise of *in vitro* approaches to islet cell regeneration.

In Vivo Studies

Expansion of functioning ß-cell mass has been the goal of recent studies on the genetic regulation of ß-cell growth and differentiation (Refs. 11–17 and reviewed in Ref. 18.) These studies provide evidence that soluble factors may be used to stimulate or permit compensatory ß-cell growth and function. As an alternative, recent gene therapy strategies suggest that *in vivo* transduction of surrogate cells, like hepatocytes, may provoke sufficient insulin production to rescue hyperglycemia and its complications in specific diabetes animal models.[19, 20]

CHALLENGES AND OPPORTUNITIES

Our understanding of islet cell growth and differentiation has grown rapidly in the past several years, provoking new questions, ambitions, and public expectations: (1) The ability to direct the differentiation of self-renewing multipotent cells with growth factors, regardless of initial source, will greatly enhance the feasibility of islet transplantation as a practical method for insulin replacement. (2) In addition to addressing issues of practicability and safety, studies to develop islet cells that are "universally" histocompatible will be important and may aid cell replacement strategies for other tissues. Such studies may include the development of cell-lines lacking "self" antigens, and the development of genetic or bioengineering approaches to evading cell-mediated immune responses. (3) Until long-term self-renewal is achieved, improving methods for (repeated) engraftment of islets may be useful.

REFERENCES

1. POLONSKY, K.S. *et al.* 1996. Seminars in Medicine of the Beth Israel Hospital, Boston. Non-insulin-dependent diabetes mellitus: a genetically programmed failure of the beta cell to compensate for insulin resistance. N. Engl. J. Med. **334:** 777–783.
2. MAZZE, R.S. *et al.* 1993. The effect of intensive treatment of diabetes on the development and progression of long-term complications in insulin-dependent diabetes mellitus. The Diabetes Control and Complications Trial Research Group. N. Engl. J. Med. **329:** 977–986.
3. SHAPIRO, A.M.J. *et al.* 2000. Islet transplantation in seven patients with type 1 diabetes mellitus using a glucocorticoid-free immunosuppressive regimen. N. Engl. J. Med. **343:** 230–238.
4. RYAN, E.A. *et al.* 2001. Clinical outcomes and insulin secretion after islet transplantation with the Edmonton protocol. Diabetes. **50:** 710–719.
5. BONNER-WEIR, S. *et al*, 2000. *In vitro* cultivation of human islets from expanded ductal tissue. Proc. Natl. Acad. Sci. USA **97:** 7999–8004.
6. WANG, X. *et al.* 2001. Liver repopulation and correction of metabolic liver disease by transplanted adult mouse pancreatic cells. Am. J. Pathol. **158:** 571–579.

7. ZULEWSKI, H. *et al.* 2001. Multipotential nestin-positive stem cells isolated from adult pancreatic islets differentiate *ex vivo* into pancreatic endocrine, exocrine, and hepatic phenotypes. Diabetes **50:** 521–533.
8. SCHULDINER, M. *et al.* 2000. From the cover: effects of eight growth factors on the differentiation of cells derived from human embryonic stem cells. Proc. Natl. Acad. Sci. USA **97:** 11307–11312.
9. BRAZELTON, T.R. *et al.* 2001. From marrow to brain: expression of neuronal phenotypes in adult mice. Science **290:** 1775–1779.
10. LUMELSKY, N. *et al.* 2001. Differentiation of embryonic stem cells to insulin-secreting structures similar to pancreatic islets. Science **292:** 1389–1394.
11. WITHERS, D.J. *et al.* 1998. Disruption of IRS-2 causes type 2 diabetes in mice. Nature **391:** 900–904.
12. KULKARNI, R.N. *et al.* 1999. Tissue-specific knockout of the insulin receptor in pancreatic beta cells creates an insulin secretory defect similar to that in type 2 diabetes. Cell **96:** 329–339.
13. XU, G. *et al.* 1999. Exendin-4 stimulates both beta-cell replication and neogenesis, resulting in increased beta-cell mass and improved glucose tolerance in diabetic rats. Diabetes **48:** 2270–2276.
14. RANE, S.G. *et al.* 1999. Loss of Cdk4 expression causes insulin-deficient diabetes and Cdk4 activation results in beta-islet cell hyperplasia. Nat. Genet. **22:** 44–52.
15. STOFFERS, D.A. *et al.* 2000. Insulinotropic glucagon-like peptide-1 agonists stimulate expression of homeodomain protein IDX-1 and increase islet size in mouse pancreas. Diabetes **49:** 741–748.
16. DEL ZOTTO, H. *et al.* 2000. Possible relationship between changes in islet neogenesis and islet neogenesis-associated protein-positive cell mass induced by sucrose administration to normal hamsters. J. Endocrinol. **165:** 725–733.
17. HART, A.W. *et al.* 2000. Attenuation of FGF signalling in mouse beta-cells leads to diabetes. Nature **408:** 864–868.
18. KIM, S.K. & M. HEBROK. 2001. Intercellular signals regulating pancreas development and function. Genes Dev. **15:** 111–127.
19. LEE, H.C. *et al.* 2000. Remission in models of type 1 diabetes by gene therapy using a single-chain insulin analogue. Nature **408:** 483–488.
20. FERBER, S. *et al.* 2000. Pancreatic and duodenal homeobox gene 1 induces expression of insulin genes in liver and ameliorates streptozotocin-induced hyperglycemia. Nat. Med. **6:** 568–572.

Building Animals from Stem Cells

RON McKAY

National Institute of Neurological Disorders and Stroke, National Institutes of Health, Bethesda, Maryland 20892, USA

KEYWORDS: neurotrophins; glia; neural stem cells

The identification of stem cells in both fetal and adult tissues has many scientific and clinical consequences. A critical feature of stem cell biology is the rapid and specific response to external signals. As nervous stem cells can be obtained in large numbers, they provide ideal systems with which to analyze how extracellular signals act to control stem cell differentiation. Analysis of CNTF/LIF- and BMP-activated mechanisms in fetal and adult CNS stem cells will be discussed. Novel pathways will be described that control precursor cell identity, proliferation, and differentiation.

It is important to determine whether stem cells give rise to functional progeny. A signal loop between neurons and glia that activates neurotrophin release and controls the early steps in synaptic differentiation has been defined. Distinct neurotrophins control the balance between glutamatergic and GABA-ergic synaptic transmission. Neurotrophins have the same effects on synapse activation on primary hippocampal neurons and neurons derived from hippocampal stem cells. These results suggest that hippocampal stem cells generate appropriate functional neuron types .

Experiments in tissue culture and in animal models will be used to illustrate how control of the origins of neuronal and glial cells may give new insight into Parkinson's, Alzheimer's, and demyelinating disease. New evidence will be presented suggesting that stem cell technology will also play a significant role in cardiac and endocrine disease. Manipulating stem cells will allow us to build organs and move towards understanding the complex molecular basis of cellular function in health and disease.

SELECTED READING

1. MASKOS, U *et al.* 2001. Long-term survival, migration, and differentiation of neural cells without functional NMDA receptors *in vivo*. Devel. Biol. **231:** 103–112.

Address for correspondence: Ron McKay, National Institute of Neurological Disorders and Stroke, NIH, Building 36, Room 5A29, 36 Center Drive, Bethesda, MD 20892. Voice: 301-496-9110; fax: 301-402-0528.
mckay@codon.nih.gov

Ann. N.Y. Acad. Sci. 961: 44 (2002). © 2002 New York Academy of Sciences.

Cells for Repair

Breakout Session Summary

Moderators

DENISE L. FAUSTMAN, *Harvard Medical School*

ROGER L. PEDERSEN, *University of California, San Francisco*

Panelists

SEUNG K. KIM, *Stanford University School of Medicine*

IHOR R. LEMISCHKA, *Princeton University*

RONALD D. McKAY, *National Institute of Neurological Disorders and Stroke, NIH*

Rapporteur

DENISE L. FAUSTMAN, *Harvard Medical School*

BROAD STATEMENT

Cell sourcing is a major issue in the development of all *in vitro* tissue-engineered products and *in vivo* cell therapies to repair or replace the loss of tissue function due to damage, disease, age, or other complications. The cell sources currently available are autologous, allogeneic, or xenogeneic and include adult differentiated cells, adult stem cells, embryonic and fetal stem cells, *ex vivo* manipulated cells, and cells generated by nuclear transfer. The generation of any optimal cell source for a particular application will depend on a rigorous characterization of these various cell sources in regard to their plasticity, propagation, and control of differentiation both *in vitro* and *in vivo*. Basic research that will elucidate the properties of stem cells must then be integrated with the multidisciplinary aspects of tissue engineering to successfully deliver the appropriate number and types of cells and promote a positive outcome: formation of new, functional tissue. The strategy of cell sourcing and use of cells for tissue engineering and cellular therapy is a central and important topic of basic and applied sciences.

VISION

Recent advances in stem cell biology hold great promise for regenerative medicine in humans. The identification of stem/multipotent progenitor cells in fetal and adult tissues and organs has many scientific and clinical consequences. It is hoped that stem/progenitor cells could be manipulated, both *in vitro* and *in vivo*, to differentiate into many different types of cells, therefore overcoming supply problems of traditional organ transplantation.

Ann. N.Y. Acad. Sci. 961: 45–47 (2002). © 2002 New York Academy of Sciences.

OBJECTIVES

In order to produce *in vitro* or enable *in vivo* immune disguised cells that can differentiate into diverse targets, maintain a stable phenotype, and restore *in vivo* function, a number of approaches are needed, including:

(A) A broad range of studies in basic cell and developmental biology of stem cells. This is particularly needed to rigorously characterize the properties of all types of candidate stem cells with respect to the following criteria: (1) Estimated *in vitro* and *in vivo* proliferative capacity; (2) precise molecular pathways to control *in vitro* and/or *in vivo* differentiation; (3) capacity for clonal growth (to distinguish between bona fide pluripotent stem cells and mixtures of specialized progenitors of terminally differentiated cells; and (4) immune compatibility of reparative cells and the host.

(B) Derivation of diverse sources of donor cells that can ultimately yield a desired phenotype and provide a functional cellular replacement in diverse targets.

(C) Determination of the parameters that allow an indefinite supply of cells but also result in desired *in vivo* phenotypes and function.

(D) Molecular and biochemical analysis of the host environmental factors that: (1) aid exogenous supplied cells or endogenous cells to home to the correct site and 2) aid long-term survival and function of exogenous supplied or introduced cells.

(E) Rigorous testing of novel cells for repair in spontaneous small and large animal models of disease to understand the host disease (diabetes, neurological, orthopedic, etc.) as well as to understand *in vivo* regeneration. Animal models will be essential in understanding and optimizing the host environment in order to promote the regeneration of internal organs and promote survival and correct targeting of introduced cells.

OBSTACLES AND CHALLENGES

The limitations of conventional organ and cell transplantation cannot be overstated, that is, supply and demand, recurrent disease, rejection, and long-term immunosuppression, for example. The recent flurry of research events in stem cell biology is exciting and could have a significant impact on future disease treatments. Diseases such as diabetes, cardiac failure, strokes, CNS degeneration, and hematopoietic disorders could be amenable to restorative therapies that validate the therapeutic effectiveness of these novel modalities.

There is an accumulating body of evidence that self-renewal of tissues and organs is possible from adult bone marrow as well as stem cells residing within a regenerating organ. There has been a recent explosion of literature demonstrating that adult stem cells from one tissue or organ can be made to differentiate into cells of other organs, both *in vivo* and *in vitro*. Demonstrations of these concepts include differentiation of bone marrow cells into muscle and brain and liver; adult neural cells into blood and heart; and, perhaps, skin cells to brain cells. Although many of these studies are in preliminary stages of investigation, the present cells could serve as a potential source of cells for replacement therapy, and therefore warrant further study.

Although the therapeutic potential of stem cells shows promise, more basic research is needed to overcome the obstacles that exist in translating these findings into clinical treatments. For instance, many adult-origin stem cells that are cultured *in vitro* and subsequently reintroduced *in vivo* fail to differentiate into the correct cell type and indeed in many cases demonstrate metastatic potential. The prevalence and location of multipotent stem/progenitor cells in various adult tissues is not clear. The expansion of purified clonal populations of cells in conjunction with the characterization of these cells using appropriate clonogenic and/or functional assays has not been achieved. In most cases, the signals/signaling pathways needed to promote a particular cell lineage are not well understood. The degree of host engraftment may not be robust or properly sustained for functional phenotypes to persist and confer phenotypes that permanently reverse disease.

Pluripotent embryonic stem (ES) cells derived from mouse or human can give rise to many different cell types in culture, and are capable of self-renewal. If ES cells are to be used therapeutically to generate tissue for transplantation, then information on how to manipulate undifferentiated cells efficiently down different developmental pathways is critical. The derivation of additional human pluripotent embryonic stem cell lines and their characterization and direct comparison to stem/progenitor cells derived from human adult tissues is needed.

The known challenges to this field include:

(1) defining the molecular and cellular features, location, and lineage-specific gene programs of stem cells in developing and adult organs and tissues; (2) understanding the molecular signals/signaling pathways that regulate self-renewal, cell lineage, proliferation, and differentiation of stem/progenitor cells; (3) understanding the complex interplay between "environmental" factors and "intrinsic" factors that regulate self-renewal and differentiation; (4) understanding the mechanism of how stem/progenitor cells continue to maintain residency, respond to insult, and regenerate during adult life and in different disease states; (5) understanding cell migration and homing of stem cells/progenitors to their environment; (6) understanding the plasticity of hematopoietic, neural, mesenchymal, and other tissue-specific stem cells; (7) designing strategies for the propagation and differentiation of embryonic stem cells or multipotential progenitor cells to specific cell types *in vitro*; (8) developing quantitative clonogenic assays that are reliable and convenient for characterizing stem/progenitor cells and determining whether the cells are capable of giving rise to functional progeny; (9) studying stem cell tumorigenicity after *in vitro* culture and expansion; and (10) determining the function, reliability, long-term survival, and safety of isolated cells upon reintroduction into the host.

ACKNOWLEDGMENT

The Cells for Repair Breakout Session was coordinated by Dr. Sheryl Sato (NIDDK, NIH) and Dr. Florence Haseltine (NICHD, NIH).

Genetic Approaches to Craniofacial Tissue Repair

JEFFREY BONADIO[a] AND MICHAEL L. CUNNINGHAM[b,c,d]

Departments of [a]Bioengineering and [b]Pediatrics, [c]Biological Structure, and [d]Oral Biology, University of Washington, Seattle, Washington 98195, USA

ABSTRACT: This review discusses in some detail the opportunities and challenges of applying gene therapy to the important clinical problem of wound repair and regeneration.

KEYWORDS: craniofacial skeletal malformation; gene therapy; tissue engineering

INTRODUCTION

DNA-based therapies for tissue regeneration blend the technologies of gene therapy and tissue engineering.[1] Tissue engineering technology is based on the generation of living constructs that consist of a biomaterial scaffold and seeded cells. Three tissue engineering strategies are currently in use. Conductive strategies use matrices either as passive three-dimensional scaffolds or as barriers that control cell access to defect sites.[2] In contrast, inductive strategies attempt to control cell migration and behavior, often by delivering growth factors to surrounding tissues. For example, Boyne et al.[3] used an inductive strategy in clinical trials to deliver bone morphogenetic protein (BMP) and successfully regenerate periodontal bone. The third strategy involves cell transplantation. Here, cells are isolated, expanded in vitro, and seeded on biomaterial scaffolds. Eventually, a vital construct is formed that can restore, maintain, or improve tissue function.[4] Autologous, allogeneic, and xenogeneic cell sources as well as stem and progenitor cells have all been tried with considerable success. As regards the latter, Hollinger and colleagues recently established a line of osteoprecursor cells that will respond to BMP, grow on tissue-engineered scaffolds,[5] and restore critical-sized defects in rat calvarium.[6]

On the other hand, genetic approaches to tissue repair are based on the delivery of DNA (gene therapy) to the wound bed. Typically, recombinant vectors encoding therapeutic molecules are formulated with porous biomaterial carriers/scaffolds.[7] The biomaterial fills the wound bed, holding the DNA vector in situ until endogenous repair cells arrive. As these cells migrate within the scaffold, they are transfected/transduced, essentially becoming local in vivo bioreactors that produce the therapeutic factor encoded by the DNA. Thus, genetic approaches to tissue repair in-

Address for correspondence: Jeffrey Bonadio, M.D., Department of Bioengineering, Box 351720, University of Washington, Seattle, WA 98195. Voice: 206-221-5876; fax: 206-616-9763.

jbonadio@u.washington.edu

Ann. N.Y. Acad. Sci. 961: 48–57 (2002). © 2002 New York Academy of Sciences.

volve *in situ* migration of wound healing cells on biomaterial scaffolds and local gene delivery and expression. The targeting of repair cells for transfection/transduction can be characterized as a passive process, that is, target cells encounter the DNA as they migrate within the biomaterial scaffold.

VECTORS

Initially, plasmid vectors were used for tissue repair applications (such as bone fracture repair) because of the many perceived advantages in formulation, cost, and safety conferred by plasmid DNA.[7] Thus, plasmids present a flexible chemistry, have no theoretical subcloning limit (e.g., Ref. 8), show a broad targeting specificity, and transfect cells as episomes (therefore presenting a low level of risk for insertional mutagenesis). On the other hand, plasmid vectors are inefficient, and, consequently, high doses typically are required to achieve clinical utility. High-dose administration may be limited, however, by pricing concerns (even though the manufacture of plasmid DNA is relatively inexpensive), and by motifs in the vector backbone that are capable of modulating the immune system.[9–13] Both immune stimulation and immune suppression have been observed, and, in some instances, immune modulation has led to significant tissue toxicity (and even death) in preclinical animal models.

Formulation of plasmids with cationic agents results in nanometer-in-size particles (synthetic polyplexes and lipoplexes) that show improved efficiency relative to naked plasmids.[14] To enhance efficiency even further, one may use polyethylene glycol to control surface properties of synthetic complexes, or incorporate fusogenic peptides or pH-responsive polymers that enhance escape from endosomes/lysosomes. While this is an active and fast-moving area of technology development, the gain in gene-transfer efficiency associated with synthetic complexes (which is generally less than an order of magnitude in *in vitro* assays) must be balanced against the general lack of stability of polyplex and lipoplex vectors *in vivo*. In addition, the tendency for locally delivered cationic agents to cause tissue necrosis (e.g., in the wound bed) can be dramatic.

Viruses are natural vectors for the transfer of recombinant DNA into cells, and this attribute has led to engineered recombinant viral vectors for gene therapy.[15] Retroviral vectors, for example, integrate efficiently into the chromatin of target cells. Integration does not, however, guarantee stable expression of the transduced gene, and clinical applications must be chosen with care since disruption of the nuclear membrane is absolutely required for the pre-integration complex to gain access to chromatin, and a productive transduction by retroviral vectors is strictly dependent on target cell mitosis. In contrast to retroviral vectors, replication-defective adenoviral (Ad) vectors efficiently transduce both mitotically active and quiescent cells *in vivo*. Since they integrate at low frequency, Ad vectors present a greatly reduced risk of insertional mutagenesis relative to retroviral vectors. Again, however, clinical applications must be chosen with care, since Ad vectors may induce an intense inflammatory/immune response that silences gene expression. For example, Rivera *et al.*[16] recently studied the feasibility of Ad gene delivery after intramuscular injection in mice. The investigators employed an Ad vector encoding human growth hormone (hGH) as a marker gene, and gene expression was controlled by rapamycin, a cell-permeant small molecule. In initial experiments using immune-deficient mice, a single i.p. injection of rapamycin (5.0 mg/kg) resulted in a 100-fold increase in the plasma hGH level. Levels then diminished to baseline over the next 14 days. Similar induction profiles were noted after five subsequent injections (administered periodically over 6 months), and a direct relationship was observed between the peak hGH level and the amount of rapamycin administered. However, in immune-competent animals, peak hGH levels were 50-fold lower, and no induction was observed after the first administration of rapamycin. These results were attributed to destructive cellular and humoral immune responses to the Ad vector. On the other hand, acute diseases (i.e., diseases that must be treated relatively soon after onset) may represent a good match between clinical need and the inherent advantages of Ad vectors.[17]

Promising results have been obtained recently with recombinant adeno-associated viral (rAAV) vectors. rAAV vectors efficiently transduce both dividing and non-dividing cells, and the rAAV genome persists as integrated tandem repeats in chromosomal DNA.[15] Upon coinfection with helper virus, AAV also transduces cells as an episome. Elimination of AAV *rep* and *cap* coding sequences from rAAV prevents the generation of wild-type helper virus. Transgene expression *in vivo* typically reaches a steady state after a gradual 2- to 10-week rise. Together, host chromosome integration and the absence of a cytotoxic T-lymphocyte response provide a viable mechanism for long-term transgene expression, as demonstrated in skeletal muscle and brain of immunocompetent animals, and in skeletal muscle of human subjects. It is important that small-scale procedures allow the efficient manufacture of sterile rAAV stocks at titers of 10^{11}–10^{12} vector genomes/mL,[18] which should facilitate a thorough exploration of the utility of this vector system. While these attributes offer many apparent advantages, strategies to optimize the tropism of rAAV remains to be determined for many cell types. Moreover, a concern was raised recently about the potential for inadvertent germline transmission of rAAV, although the current consensus is that the risk is minimal (e.g., Ref. 19).

CLINICAL APPLICATIONS

Gene therapy is being applied to a variety of tissue engineering applications, and several reviews have been published recently that discuss these efforts.[7,20] As a consequence, the remainder of this review will depart slightly from the mandate to survey the current status of the field. Instead, the review will discuss in some detail the opportunities and challenges of applying gene therapy to an important clinical problem, namely, craniofacial bone regeneration. The discussion will, it is hoped, touch on several themes that relate broadly to other tissue engineering applications and to the field of wound healing at large.

CRANIOFACIAL BONE DEFECTS

Defects of the craniofacial complex are heterogeneous and may result from: dental/oral/craniofacial trauma, age-related bone loss, heritable/developmental disorders, caries and periodontal disease, and iatrogenic causes (e.g., oncologic surgery). Regardless of pathogenesis, defects in the craniofacial complex must be repaired

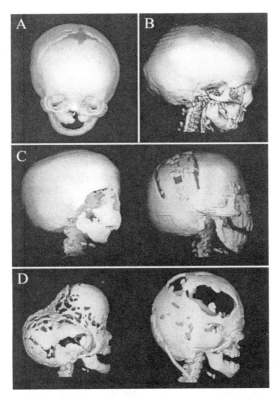

FIGURE 1. Heritable defects of the craniofacial complex. This figure demonstrates four examples of defects in the craniofacial skeleton represented by 3-D computerized tomography (3-D CT) of the skull. (**A**) 3-D CT image of a child with a cleft of the primary palate (cleft of the alveolar bone). (**B**) 3-D CT image shows micrognathia and a zygomatic arch cleft in a child with hemifacial microsomia. (**C**) 3-D CT image shows absent zygomatic arch, lateral orbital clefting, and a malformed mandible in a child with severe Treacher-Collins syndrome. 3-D CT image on right was obtained after several bone grafting procedures to reconstruct the orbit, zygoma, zygomatic arch, and mandible. The image shows harvest areas in the calvaria for the periorbital reconstruction. (**D**) 3-D CT image on left shows cloverleaf skull deformity due to FGFR3 mutation in transmembrane domain of the receptor. Severe increased intracranial pressure has led to multiple bony erosions (shown here in patient at 2 months of age). 3-D CT image on right was obtained after expansion cranioplasty was performed. After successful reconstruction, huge skull defects remain due to the relative deficiency of autologous bone available for transplantation.

whenever they cause debilitating functional loss and/or significant disfigurement. Despite intense efforts for more than a decade, general agreement exists that new strategies for craniofacial reconstruction are urgently needed. The unmet need is two-fold: (*a*) to accelerate the process of craniofacial bone regeneration and (*b*) to completely heal craniofacial osseous defects regardless of size.

Problems with complete healing may arise in association with mandibular segmental defects,[20] zygomatic arch clefts,[21] maxillary defects,[22] and with the need to reconstruct the atrophic alveolar ridge (e.g., before surgical placement of a dental implant[23]). A consideration of congenital defects also underscores the unmet clinical need.[24] As shown in FIGURE 1, congenital defects may be extensive, and thus require an extraordinary amount of bone for reconstruction. In these cases multiple surgical harvest procedures typically are performed in order to reconstruct the skeleton. Most would agree that the current patient care pathway is inefficient, expensive, and painful, and therefore deserves to be improved.

BONE GRAFTING AND BONE GRAFT SUBSTITUTES: CURRENT STANDARD OF CARE

Although autogenous bone grafting remains the treatment of choice for craniofacial reconstruction, efficacy is limited by donor morbidity, the lack of tissue resource availability, and the undesirable alterations in volume and geometry that may occur to the graft post implantation.[25] Unfortunately, we also lack a robust bone graft substitute technology. For example, while initial studies showed that various allogeneic bone preps supported alveolar bone regeneration, more recent studies in orthotopic models have provided little, if any, evidence of benefit.[26] In addition, while xenogeneic bone-mineral preparations are well incorporated into bone, slow resorption rates may alter bone quality,[27] and potential adverse effects may result from immunologic reactions or infectious disease transmission. Furthermore, osteoconductive calcium-based ceramics, bioactive glass, and synthetic polymers have inherent mechanical integrity and are capable of maintaining a void volume for skeletal reconstruction,[28] but in selecting these materials clinicians must carefully consider whether early resorption will interfere with bone formation, or, alternatively, if late resorption will compromise bone quality. Finally, we note an enabling technology (guided tissue regeneration) that employs passive membrane devices that separate tissues during healing.[2] Although widely accepted in periodontics and implant dentistry, clinical application is restricted to stable defects because compromised wound closure or early mechanical wound failure, which exposes the membrane to the oral cavity, may restrict outcomes.

While recombinant bone morphogenetic proteins (BMPs) are now in final approval stages by the FDA, optimal conditions for human use have yet to be defined. During the past decade, BMP-7/OP-1 demonstrated considerable clinical utility and safety in more than 500 patients worldwide. Specifically, the outcome of patients with tibial nonunion in a randomized, prospective clinical trial involving rOP-1 administration and intramedullary fixation were equivalent to the outcomes of control patients treated with the same surgical protocol and bone autograft.[29] In a second study (in human fibular defects), demineralized bone matrix controls and rOP-1 induced similar amounts of new bone.[30] To achieve their promise, however, most investigators agree that BMPs will need to show results that exceed the current standard of care. A recently realized hurdle may be the role of BMPs in osteoclast formation.[31] Because BMPs stimulate the release of factors that drive osteoclast production, pharmacologic doses may lead to BMP-induced bone resorption, which may limit efficacy (e.g., particularly when rigid fixation is required). Bisphosphonates prevent osteoclast formation and activation, but clinical trials designed to test a BMP-bisphosphonate treatment regime have yet to be performed. Lack of an appropriate delivery vehicle has also limited the success of BMP treatments. Ideal delivery vehicles should be biocompatible, biodegradable, and sufficiently porous to allow cell infiltration.[32] Issues of carrier chemistry and local/systemic bioavailability remain unresolved despite years of effort.

GENE THERAPY FOR CRANIOFACIAL REGENERATION: OPPORTUNITIES AND CHALLENGES

While osteogenic gene therapy has been studied in some detail,[7,33,34] it is reasonable to suggest that much remains to be learned. For example, the mechanism of gene delivery at a molecular level is complex and remains largely unknown, and many control parameters for gene delivery have yet to be quantified. The wound healing sequence in mammalian bone is highly evolved and highly conserved.[35] Cells that participate in wound healing include platelets, acute inflammatory cells, macrophages, fibroblasts, endothelial cells, pericytes, and tissue-specific progenitor cells. To coordinate the cellular response, cytokines and growth factors act locally through wound-specific signal transduction cascades. The typical response to acute tissue injury initially focuses on controlling hemorrhage and removing damaged cells. Repair then begins with the formation of granulation tissue and ends with bone regeneration or scar formation. (The latter is referred to as nonunion/malunion, a debilitating and painful condition.) This certainly is a dynamic environment for gene delivery when compared to quiescent skeletal muscle, and the pharmacokinetics and pharmacodynamics of the vector-encoded protein production are expected to reflect this dynamic complexity. In this regard, it would be helpful to define gene transfer efficiency using direct measures (rather than surrogates) of both DNA persistence and RNA/protein expression. To be meaningful, efficiency must also be defined *in vivo* over reasonably long time frames.

It also may be argued that certain of the factors that govern the safety of osteogenic gene therapy are largely unknown. For example, we have little information regarding the consequences of vector binding and internalization by cells in the acute wound bed. Neither immune suppression nor stimulation presents an acceptable risk to human patients with non-healing wounds, and we must understand more about the response of the immune system to the administration of gene therapy vectors, which, after all, retain certain of the signals of invading microbial pathogens.

Finally, it is hard to imagine a gene-therapy product for skeletal repair that does not involve regulated therapeutic gene expression as a way to avoid toxicity and still respond to the evolving nature of the regenerative process. The sustained expression of rAAV vectors ("constitutive gene therapy") raises important potential safety concerns that must be addressed with care. (For example, the relevant concerns associated with the sustained local expression of an osteogenic factor include: exuberant bone overgrowth, dystrophic calcification of orthotopic and ectopic soft tissues, and tumor formation.) Breakthroughs in controllable gene therapy have allowed therapeutic transgene expression to be regulated with precision over a period of months to years. Ideally, this type of technology will feature low baseline transgene expression, a high induction ratio, and small-molecule control. These breakthroughs offer a number of potentially exciting clinical applications that are unique to gene therapy—that is, similar levels of drug stability and control simply do not exist for more traditional drug substances (small molecules, peptides, and proteins).

CONTROLLABLE GENE THERAPY

Several gene-regulatory systems are available, including those in which transgene expression is induced by tetracycline, antiprogestins, and ecdysone.[36] In this regard, promising technology employs a heterologous transcription factor that selectively binds and activates transcription in response to a cell-permeant controller molecule (e.g., Refs. 37–39). Activation is achieved by reconstitution of a transcription factor complex that consists of independently expressed protein chimeras.[40,41] One transcription factor chain consists of a unique DNA-binding domain genetically fused to FKBP, while the other consists of the activation domain (of the p65 subunit of NFkB), fused with the rapamycin-binding domain of FRAP (FRB). Elegant feasibility studies have been reported using rAAV2 vectors (e.g., Ref. 42). Packaging limits required that the regulated gene therapy be incorporated into two vectors, that is, one vector expressed both transcription factors from a single transcriptional unit, while the other vector expressed the therapeutic gene driven by a promoter recognized by the ZFHD1 DNA-binding domain. Infection of permissive human cells in vitro with equal quantities of the two AAV vectors at a high multiplicity of infection resulted in full reconstitution of the regulated system, with at least a 100-fold induction after exposure to rapamycin. The feasibility of reconstituting the regulated system in vivo has also been determined.[42] Here, a controllable rAAV vector cocktail (2×10^8 infectious particles, with rAAV vectors at a 1:1 ratio) was injected into skeletal muscle of immune-competent mice. Administration of rapamycin resulted in 200-fold induction of plasma erythropoietin. Stable engraftment of this humanized system was achieved for 6 months, with similar results for 3 months in an immune-competent rhesus model.

The "transcriptional-switch" technology described above features a therapeutic protein induction-decay response on a time-scale of days: transgene-encoded protein in blood typically peaks at about 24 hr and then decreases to background over 4 to 14 days. On the one hand, this kinetic profile probably reflects an "early-point" of transgene regulation as well as the pharmacokinetics and pharmacodynamics of rapamycin,[43] the dynamic process of rAAV transduction, and the bioavailability of the therapeutic protein. Such prolonged kinetics may be appropriate for certain proteins (e.g., erythropoietin) that govern relatively slow physiological processes, but they may not always be as appropriate. To address this potential limitation, Rivera et al.[44] recently developed technology that allows protein secretion from the endoplasmic reticulum (ER) to be rapidly regulated. Therapeutic proteins are expressed as fusions with a conditional aggregation domain (CAD). CADs self-interact, and fusion proteins therefore form an aggregate in the ER that is far too large to be transported. Rivera et al. showed that the addition of cell-permeant ligand ("disaggregator") to transfected cells dissolves the aggregates and permits the rapid transport of therapeutic proteins from the ER via the constitutive secretory pathway.

To produce bioactive proteins, CAD moieties must be removed. A furin cleavage site was interposed between the coding sequence for therapeutic protein and the CAD repeats in order to solve this problem. In one example, Rivera et al.[44] demonstrated that a natural version of human growth hormone (hGH) could be secreted in a controllable fashion using disaggregator technology. Recombinant hGH was generated via a cDNA construct consisting of a CMV promoter, signal sequence, four CAD motifs, a furin cleavage signal, and growth hormone (proinsulin was also used). Vectors were stably transfected into HT1080 cells and fluorescence microscopy was used to demonstrate ER retention of both insulin and growth hormone in the absence of disaggregator. Cells expressing fusion proteins were then treated with increasing concentrations of disaggregator for 2 hr. The authors showed that accumulated protein was released by disaggregator administration, and the rate of release was controllable over an ~20-fold dose range. In the absence of the ligand, fusion proteins were found only in cell lysate samples, whereas 2 hours after addition of the ligand, fusion proteins were cleaved appropriately and secreted (as determined by Western analysis). Finally, myoblast transfer was used to demonstrate feasibility of the system in animal models. Engineered cells were implanted into mice made diabetic by treatment with streptozotocin. Administration of vehicle failed to normalize serum glucose concentrations. However, insulin was detected in serum within 15 min, and peaked by 2 hr, after intravenous administration of ligand. Indeed, 2hr after administration of a 10.0 mg/kg dose of ligand, the circulating insulin concentration increased to greater than 200.0 pM, and serum glucose decreased concomitantly to normal.

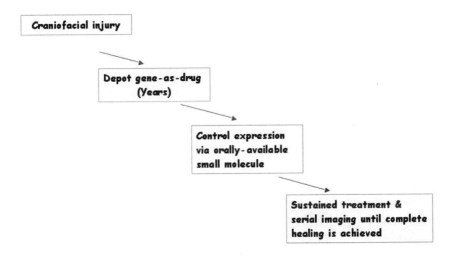

FIGURE 2. Proposed patient-care pathway based on controllable gene delivery for craniofacial bone regeneration. A patient-care pathway is envisioned in which gene delivery to the craniofacial defect is followed by a therapeutic regimen that involves the ingestion of a cell-permeant small molecule. The extent of healing is monitored by serial, non-destructive imaging techniques. Oral therapy is continued until an appropriate amount of new bone is formed and then stopped. In this way, a mechanism exists to avoid toxicity in the individual patient and still allow gene therapy to be responsive to the evolving nature of the regenerative response, all but eliminating the need for surgical revision.

POTENTIAL APPLICATION OF CONTROLLABLE GENE THERAPY TO THE TREATMENT OF CRANIOFACIAL DEFECTS

With a controllable osteogenic gene therapy, one may envision a patient-care pathway in which gene delivery to the craniofacial skeletal defect is followed by a therapeutic regimen that involves the ingestion of a cell-permeant small molecule (FIG. 2). The extent of healing is monitored by serial, non-destructive imaging techniques. Oral therapy is continued until an appropriate amount of new bone is formed and then stopped. In this way, a mechanism exists to avoid toxicity in the individual patient and still allow gene therapy to be responsive to the evolving nature of the regenerative response, all but eliminating the need for revision surgery.

The ability to personalize bone regeneration therapy represents a paradigm shift for this technology and for patient care. Bone graft substitutes based on a controllable gene-delivery strategy may be applied to craniofacial pathology that arises from traumatic injury, age-related bone loss (e.g., osteoporosis), heritable/developmental defects (e.g., clefts), chronic inflammatory disease (e.g., caries, periodontal disease), chronic degenerative disease (TMJ complex), and iatrogenic causes (e.g., tumor surgery).

REFERENCES

1. LAUFFENBURGER, D.A. & D.V. SCHAFFER. 1999. Nat. Med. 5: 733–734.
2. BUSER, D. et al. 1994. Guided Bone Regeneration in Implant Dentistry. Quintessence Books. Berlin.
3. BOYNE, P.J. et al. 1997. A feasibility study evaluating rhBMP-2/absorbable collagen sponge for maxillary sinus floor augmentation. Int. J. Periodontics Restorative Dent. 17: 11–25.
4. STOCK, U.A. & J.P. VACANTI. 2001. Tissue engineering: current state and prospects. Annu. Rev. Med. 52: 443–451.
5. WINN, S.R. et al. 1999. Establishing an immortalized human osteoprecursor cell line. J. Bone Miner. Res. 14: 1721–1733.
6. WINN, S.R. et al. 1999. A tissue engineered bone biomimetic to regenerate calvarial critical-sized defects in athymic rats. J. Biomed. Mater. Res. 45: 414–421.
7. BONADIO, J. 2000. Tissue engineering via local gene delivery: update and future prospects for enhancing the technology. Adv. Drug Del. Rev. 44: 185–194.
8. CAMPEAU, P. et al. 2001. Transfection of large plasmids in primary human myoblasts. Gene Ther. 8: 1387–1394.
9. MACCOLL, G. et al. 2001. Intramuscular plasmid DNA injection can accelerate autoimmune responses. Gene Ther. 8: 1354–1356.
10. BALLAS, Z.K. et al. 1996. Induction of NK activity in murine and human cells by CpG motifs in oligodeoxynucleotides and bacterial DNA. J. Immunol. 157: 1840–1845.
11. KLINMAN, D.M. et al. 1996. CpG motifs present in bacterial DNA rapidly induce lymphocytes to secrete interleukin 6, interleukin 12, and interferon gamma. Proc. Natl. Acad. Sci. USA 93: 2879–2883.
12. DENG, G-M. et al. 1999. Intra-articularly localized bacterial DNA containing CpG motifs induces arthritis. Nat. Med. 5: 702–705.
13. PAYETTE, P.J. et al. 2001. Immune-mediated destruction of transfected myocytes following DNA vaccination occurs via multiple mechanisms. Gene Ther. 8: 1395–1400.
14. STAYTON, P.S. et al. 2000. Molecular engineering of proteins and polymers for targeting and intracellular delivery of therapeutics. J. Controlled Release 65: 203–220.
15. KAY, M.A. et al. 2001. Viral vectors for gene therapy: the art of turning infectious agents into vehicles of therapeutics. Nat. Med. 7: 33–40.

16. RIVERA, V.M. *et al.* 1999. Long-term regulated expression of growth hormone in mice after intramuscular gene transfer. Proc. Natl. Acad. Sci. USA **96**: 8657–8662.
17. FACTOR, P. 2002. Gene therapy for acute diseases. Mol. Ther. **4**: 515–524.
18. AURICCHIO, A. *et al.* 2001. Isolation of highly infectious and pure adeno-associated virus type 2 vectors with a single-step gravity-flow column. Hum. Gene Ther. **12**: 71–76.
19. FAVRE, D. *et al.* 2001. Immediate and long-term safety of recombinant adeno-associated virus injection into the nonhuman primate muscle. Mol Ther. **4**: 559–566.
20. ALSBERG, E. *et al.* 2001. Craniofacial tissue engineering. Crit. Rev. Oral Biol. Med. **12**: 64–75.
21. YUDELL, R.M. & M.S. BLOCK. 2000. Bone gap healing in the dog using recombinant human bone morphogenetic protein-2. J. Oral Maxillofac. Surg. **58**: 761–766.
22. BOYNE, P.J. *et al.* 1998. Human recombinant BMP-2 in osseous reconstruction of simulated cleft palate defects. Br. J. Oral Maxillofac. Surg. **36**: 84–90.
23. SIMION, M. *et al.* 2001. Long-term evaluation of osseointegrated implants inserted at the time or after vertical ridge augmentation: a retrospective study on 123 implants with 1-5 year follow-up. Clin. Oral Implants Res. **12**: 35–45.
24. CUNNINGHAM, M.L. 2002. Craniofacial malformations and syndromes. *In* Rudolph's Pediatrics, 21st ed. M. Rudolph, G. Lister & M.K. Hostetter, Eds. McGraw-Hill. Columbus, Ohio. In press.
25. MARX, R.E. & M.J. MORALES. 1988. Morbidity from bone harvest in major jaw reconstruction: a randomized trial comparing the lateral anterior and posterior approaches to the ilium. J. Oral Maxillofac. Surg. **46**: 196–203.
26. ASPENBERG, P. *et al.* 1988. Failure of bone induction by bone matrix in adult monkeys. J. Bone Joint Surg. Br. **70**: 625–627.
27. SKOGLUND, A. *et al.* 1997. A clinical and histologic examination in humans of the osseous response to implanted natural bone mineral. Int. J. Oral Maxillofac. Implants **12**: 194–199.
28. OUHAYOUN, J.R. 1997. Bone grafts and biomaterials used as bone grafts substitutes. *In* Proceedings of the 2nd European Workshop on Periodontology: Chemicals in Periodontics. N.P. Lang, T. Karring & J. Lindhe, Eds. :313–358. Quintessence Books. Berlin.
29. FRIEDLAENDER, G.E. *et al.* 2001. Osteogenic protein-1 (bone morphogenetic protein-7) in the treatment of tibial nonunions. J. Bone Joint Surg. Am. **83A** (Suppl 1;Pt 2): S151–S158.
30. GEESINK, R.G. *et al.* 1999. Osteogenic activity of OP-1 bone morphogenetic protein (BMP-7) in a human fibular defect. J. Bone Joint Surg. Br. **81**: 710–718.
31. ITOH, K. *et al.* 2001. Bone morphogenetic protein 2 stimulates osteoclast differentiation and survival supported by receptor activator of nuclear factor-kappa-β ligand. Endocrinology **142**: 3656–3662.
32. AGHA-MOHAMMADI, S. & M.T. LOTZE. 2000. Regulatable systems: applications in gene therapy and replicating viruses. J. Clin. Invest. **105**: 1177–1183.
33. RIVERA, V.M. *et al.* 1996. A humanized system for pharmacologic control of gene expression. Nat. Med. **2**: 1028–1032.
34. MAGARI, S.R. *et al.* 1997. Pharmacologic control of a humanized gene therapy system implanted into nude mice. J. Clin. Invest. **100**: 2865–2872.
35. POLLOCK, R. *et al.* 2000. Delivery of a stringent dimerizer-regulated gene expression system in a single retroviral vector. Proc. Natl. Acad. Sci. USA **97**: 13221–13226.
36. BROWN, E.J. *et al.* 1994. A mammalian protein targeted by G1-arresting rapamycin-receptor complex. Nature **369**: 756–758.
37. STANDAERT, R.F. *et al.* 1990. Molecular cloning and overexpression of the human FK506-binding protein FKBP. Nature **346**: 671–674.
38. YE, X. *et al.* 1999. Regulated delivery of therapeutic proteins after *in vivo* somatic cell gene transfer. Science **283**: 88–91.
39. MAHALATI, K. & B.D. KAHAN. 2001. Clinical pharmacokinetics of sirolimus. Clin. Pharmacokinet. **40**: 573–585.
40. RIVERA, V.M. *et al.* 2000. Regulation of protein secretion through controlled aggregation in the endoplasmic reticulum. Science **287**: 826–830.

Genetic Approaches to Tissue Repair

JEFFREY BONADIO

Department of Bioengineering, University of Washington,
Seattle, Washington 98195-1720, USA

KEYWORDS: tissue engineering; gene therapy

Genetic approaches to tissue repair are based on delivery of DNA to the wound bed. The delivery technology oftentimes is a blend of gene therapy and tissue engineering in that genes are formulated with porous biomaterial scaffolds, a form of guided tissue regeneration, rather than formulated in a buffer solution. The scaffold fills the wound bed, holding the DNA vector *in situ* until endogenous repair cells arrive. These cells migrate along the scaffold, and, once transfected, essentially become local *in vivo* bioreactors that produce the therapeutic factor encoded by the DNA. Thus, gene delivery and expression are local phenomena rather than systemic, and cell targeting is a passive process rather than an active one.

Typically, the therapeutic goal is to induce new tissue formation in a nonhealing wound or to inhibit an exaggerated tissue-repair response. Compared to other biologics, DNA offers the unique feature of sustained low-level expression by repair cells (which may optimize growth factor expression). The most experience has been with plasmid DNA as a vector system. Plasmid DNA (especially as a vaccine) has an established safety record in humans, which is an important consideration, since tissue engineering applications have inherently low risk–benefit profiles. However, for tissue regeneration, large amounts of DNA have been required in preclinical animal models, and high-dose administration in humans may be limited by CpG motifs that activate an immune response after ligation with Toll-like receptors. In this regard, plasmid DNA administration to animals has generated anti-DNA antibodies, leading to renal disease and premature death. Therefore, potency and safety probably both need to be enhanced before plasmid vectors could be realistically used for human therapy.

Viral vectors (derived from retrovirus, lentivirus, adenovirus [Ad], and herpes simplex virus) maximize gene-transfer efficiency and may be formulated with biomaterial scaffolds. However, efficient vector preparation and purification methods have not yet been established, and unwanted recombination between vector and helper virus may occur. Additionally, Ad vectors induce unacceptable immune responses that decrease efficacy and increase toxicity. Among a multiplicity of alter-

Address for correspondence: Jeffrey Bonadio, University of Washington, Box 351720, Seattle, WA 98195-1720. Voice: 206-221-5876; fax: 206-616-9763.

jbonadio@u.Washington.edu

Ann. N.Y. Acad. Sci. 961: 58–60 (2002). © 2002 New York Academy of Sciences.

natives, promising viral vector technology includes the recombinant adeno-associated viral (rAAV) vectors. Small-scale procedures are available that allow manufacture of pure, sterile preparations at titers of 10^{12-13} particles/ml. rAAV efficiently transduces both dividing and nondividing cells, and the vector genome persists as integrated tandem repeats in chromosomal DNA. (The rAAV vectors also transduce cells as an episome.) Elimination of AAV coding sequences from the vector effectively prevents an immune response to viral gene products and generation of wild-type helper virus. Together, host chromosome integration and the absence of a cytotoxic T-lymphocyte response provide viable mechanisms for long-term transgene expression (demonstrated in immunocompetent animals and human subjects). In head-to-head comparisons in animal models, rAAV2-*LacZ* appears to be several orders-of-magnitude more effective than plasmid vectors. While rAAV vectors appear at this early stage to have a safety/efficacy profile suitable for tissue-repair applications, potential advantages must be balanced against potential disadvantages associated with vector integration (i.e., random insertional mutagenesis), and the fact that ~80% of the U.S. population is immunized against certain parental AAV strains.

Cationic agents (dendrimers, peptides, proteins and lipids) bind noncovalently to plasmid DNA and, when properly formulated, result in synthetic polyplexes and lipoplexes with improved efficiency. Alternatively, porous biomaterial scaffolds contribute to gene transfer efficiency by providing the surface on which cells and DNA vectors interact. Type I collagen from animals, which is easily formulated with DNA vectors, is a proven biomaterial for human applications. On the other hand, use of animal collagens presents a risk of disease transmission, and the controlled bioavailability of DNA vectors from collagen is poor. The advent of recombinant human collagens with normal structure and function effectively eliminates the possibility of disease transmission and offers novel collagen scaffolds that actively promote repair. Among synthetic polymer alternatives, three-dimensional PLGA scaffolds allow for controlled DNA release, and these scaffolds can be pre-designed according to tissue-engineering principles (for both *in vitro* and *in vivo* application). PLGA also may be formulated with DNA vectors as novel medical-device coatings.

Factors outside the field of tissue engineering should have also have a significant impact: (1) Genome sequencing will, of course, increase the number of rational gene-drug candidates for tissue engineering applications. (2) The growing number of mouse strains with targeted mutations allows one to develop local gene transfer formulations in defined genetic backgrounds, a powerful scientific advantage. (3) The regulated marketplace will likely favor the development of cost-effective gene delivery technology (i.e., technology that is simple, stable, and cheap; capable of producing therapeutic protein in a dose-responsive, controllable manner; and easily managed by physician and patient). (4) Integrity and open disclosure of scientific activities will be required to engender public trust in gene-delivery technology.

While gene therapy has been intensively studied, there is much that remains in need of rigorous investigation. For example, the mechanism of gene delivery remains largely unknown at the molecular level, and many of the control parameters for gene delivery have yet to be quantified. These control parameters include gene transfer efficiency and the pharmacokinetics and pharmacodynamics of the vector-encoded protein. Efficiency must be defined using direct measures (rather than surrogates) in terms of both DNA persistence and RNA/protein expression. Moreover, efficiency must be defined over a reasonably long time course in order to be meaningful. Quantitative models of gene delivery and expression should be developed.

It also may be argued that the factors that govern safety are largely unknown. For example, almost no information exists regarding the consequences of vector binding and internalization by targeted cells. (Is function significantly altered in mammalian cells that take-up and express gene therapy vectors?) We also must understand more about the response of the immune system to the administration of gene therapy vectors, which, after all, retain signals of invading microbial pathogens. Neither immune suppression nor stimulation presents an acceptable risk to patients with non-healing wounds. Finally, we must develop more insight into the potential for adverse interaction between local tissues and gene-therapy formulations. Necropsy and biodistribution studies are the best way to evaluate potential tissue toxicity.

Finally, it is hard to imagine a gene-therapy product that does not involve regulated therapeutic gene expression as a way to avoid toxicity (e.g., neoplasia; catastrophic tissue failure) and still respond to the evolving nature of disease. To qualify, a system conferring regulated gene expression should feature low baseline transgene expression, a high induction ratio, and tight control by a small-molecule drug.

REFERENCES

1. MARTIN, P. 1997. Wound healing: aiming for perfect skin. Science **276**: 75–81.
2. BONADIO, J. 2000 Tissue engineering via local gene delivery: update and future prospects for enhancing the technology. Adv. Drug Del. Rev. **44:** 185–194.
3. KAY, M.A *et al.* 2001. Viral vectors for gene therapy: the art of turning infectious agents into vehicles of therapeutics. Nat. Med. **7:** 33–40.
4. SCHAFFER, D.V. & D.A. LAUFFENBURGER. 2000. Targeted synthetic gene delivery vectors. Curr. Opin. Mol. Ther. **2:** 155–161.
5. LARSON, P.J. & K.A. HIGH. 2001. Gene therapy for hemophilia B:AAV-mediated transfer of the gene for coagulation factor IX to human muscle. Adv. Exp. Med. Biol. **489:** 45.
6. AGHA-MOHAMMADI, S. & M.T. LOTZE. 2000. Regulatable systems:applications in gene therapy and replicating viruses. J. Clin. Invest. **105:** 1173–1176..
7. POLLOCK, R. & V.M. RIVERA. 1999. Regulation of gene expression with synthetic dimerizers. Meth. Enzymol. **306:** 263–281.
8. OZAWA, C.R *et al.* 2000. A novel means of drug delivery: myoblast-mediated gene therapy and regulatable retroviral vectors. Annu. Rev. Pharmacol. Toxicol. **40:** 295–317.
9. WOO, S.L. 2000 Gene therapy researchers react to field's pitfalls and promises. FDA Consum. **34:** 40.
10. WOO, S.L. 2001. Clinical gene transfer: education is the key. Mol.Ther. **4:** 92.

Gene Transfer in Tissue Regeneration and Reparative Medicine

SAVIO L. C. WOO

The Carl C. Icahn Institute for Gene Therapy and Molecular Medicine, Mount Sinai School of Medicine, New York, New York 10029, USA

KEYWORDS: gene therapy; tissue regeneration

Gene therapy is the use of genes as medicines to prevent disease or to alter the clinical course of an existing disease. Over the past decade, dramatic progress has been made in the development and refinement of technologies for delivery of genes to various cells and organs, both in animal models and human. In several instances, significant treatment benefits achieved in laboratory animal models of human disease have been observed in clinical studies as well. An example is Hemophilia B, which is caused by a deficiency of clotting factor IX in the blood. Normal blood clotting times have been restored for extended durations of time after a single administration of the gene to genetically affected mice and dogs. Encouraging results have also been reported from patients in early-phase clinical studies after intramuscular delivery of a recombinant adeno-associated virus expressing the human factor IX gene. Another example is X-Linked Severe Combined Immunodeficiency Syndrome secondary to a deficiency of the gamma chain of cytokine receptors on T cells. Autologous transplantation of $CD34^+$ cells transduced with a recombinant retroviral vector expressing the normal human gene has resulted in the reconstitution of T-cell counts and immune functions in several affected children for up to one year; this has allowed them to live normally with their families without further treatment. The recent technological advances that permitted these achievements may be expected to lead not only to better treatment of patients with heritable disorders, but also to the treatment of complex and acquired disorders such as cardiovascular disease, diabetes, neurodegenerative disorders, and cancer.

Another major area in which gene medicines may be expected to have application is in induction of local tolerance to transplanted tissues in immune-competent subjects, which will permit allogeneic transplantation without life-long immunosuppression. Finally, an improved capacity for stem cell isolation and propagation will permit their genetic reconstitution prior to autologous transplantation for tissue regeneration in patients.

Address for correspondence: Savio L. C. Woo, Mount Sinai School of Medicine, P.O. Box 1496, New York, NY 10029. Voice: 212-659-8260; fax: 212-849-2572.
savio.woo@mssm.edu

Ann. N.Y. Acad. Sci. 961: 61–62 (2002). © 2002 New York Academy of Sciences.

The development of such novel modalities for the treatment of diseases by gene transfer and tissue repair will no doubt be driven by scientific discoveries, technological innovations, validation in preclinical animal models of disease, and clinical translational research. Strategically, the following areas need particular attention:

(1) gene transfer vector development, including viral and nonviral vectors as well as targeting vectors;

(2) tissue-specific transgene expression and regulation, including the enhancement of transgene expression while retaining tissue specificity;

(3) preclinical efficacy and pharmacology/toxicology studies in relevant animal modals of disease, including the development of such animal models in the first place;

(4) innate and adaptive immunological responses to the vectors and the transgene products in immune-competent recipients;

(5) adult tissue stem cells, including their identification, molecular characterization, *in vitro* expansion, vector transduction, and organ formation; and

(6) systematic development of clinical translational research, including the recruitment and training of clinical investigators who are familiar with clinical research, the science of gene therapy and federal regulations, and guidelines governing clinical gene transfer studies.

SELECTED REFERENCE

1. AGUILAR, L.K. *et al.* 2000. A prescription for gene therapy. Mol. Ther. **1:** 385–388.

Gene-Based Approaches to the Treatment of Hemophilia

KATHERINE HIGH

The Children's Hospital of Philadelphia, Philadelphia, Pennsylvania 19104, USA

KEYWORDS: gene therapy; hemophilia

Several examples of novel gene- and cell-based approaches to the treatment of hemophilia were reviewed. These included *in vivo* gene transfer strategies in which an adenoassociated virus (AAV) vector is used to introduce the factor IX gene into hepatocytes or skeletal muscle, as well as *ex vivo* approaches in which the gene for factor VIII is introduced into autologous fibroblasts or endothelial cells. Although progress in these areas has been rapid and some approaches are already under evaluation in clinical studies, many problems remain to be solved.

Vector technology must be further developed. There is a need for analysis of determinants of sustained expression. For inherited diseases like hemophilia, long-term expression of the donated gene is the goal. Development of gene targeting technology is needed. For many diseases, expression of the donated gene must occur in a specific cell type. Current efforts are addressed at defining cell-type tropisms for different vectors and at engineering vectors so that they can transduce specific cell types. "Switch" systems that allow regulated expression of a transgene are in need of refinement.

A number of safety issues are critical for human applications. Compilation of data on long-term effects of therapy with integrating gene sequences in humans will be needed, including evaluation of safety and efficacy of *in vivo* and *ex vivo* approaches and analysis of the molecular state of vector sequences in recipient cells. The molecular consequences of integration will need to be determined, including the analysis of immunological consequences of *de novo* endogenous synthesis of proteins to which the recipient may not be tolerant.

The use of autologous cells for tissue replacement is in need of further investigation. Nuclear reprogramming may permit the use of mature cells to give rise to multipotent precursors. The ability to effect stem cell redirection will have many applications. The role of specific peptide factors in determination of cell fate needs to be better understood. Finally, all stakeholders need to pay attention to educating the lay public about possibilities afforded by, and limitations of, gene transfer/tissue engineering strategies.[1-9]

Address for correspondence: Katherine High, The Children's Hospital of Philadelphia, 3516 Civic Center Boulevard, Philadelphia, PA 19104. Voice: 215-590-4521; fax: 215-590-3660.
high@email.chop.edu

Ann. N.Y. Acad. Sci. 961: 63–64 (2002). © 2002 New York Academy of Sciences.

REFERENCES

1. ROTH, D.A., *et al.* FOR THE FACTOR VIII TRANSKARYOTIC THERAPY STUDY GROUP. 2001. Nonviral transfer of the gene encoding coagulation factor VIII in patients with severe hemophilia A. N. Engl. J. Med. **344:** 1735–1742.

2. HIGH, K.A. 2000. Gene therapy in haematology and oncology. Lancet **356**(suppl.)**:** s8.

3. KAY, M.A., *et al.* 2000. Evidence for gene transfer and expression of blood coagulation factor IX in patients with severe hemophilia B treated with an AAV vector. Nat. Genet. **24:** 257–261.

4. SOMIA, N. & I.M. VERMA. 2000. Gene therapy: trials and tribulations. Nat. Rev. Genet. **1:** 91–99.

5. WILLIAMS, D.A. & F.O. SMITH. 2000. Progress in the use of gene transfer methods to treat genetic blood diseases. Hum. Gene Ther. **11:** 2059–2066.

6. BARON, U. & H. BUJARD. 2000. Tet-repressor-based system for regulated gene expression in eukaryotic cells: principles and advances. Methods Enzymol. **327:** 402–421.

7. HERZOG, R.W., *et al.* 1999. Long-term correction of canine hemophilia B by AAV-mediated gene transfer of blood coagulation factor IX. Nat. Med. **5:** 56–63.

8. KAY, M.A. & K. HIGH. 1999. Gene therapy for the hemophilias [commentary]. Proc. Natl. Acad. Sci. U.S.A. **96:** 9973–9975.

9. SNYDER, R.O. *et al.* 1999. Correction of hemophilia B in canine and murine models using recombinant adeno-associated viral vectors. Nat. Med. **5:** 64–70.

Genetic Approaches to the Repair of Connective Tissues

STEVEN C. GHIVIZZANI

Center for Molecular Orthopedics, Harvard Medical School,
Boston, Massachusetts 02115, USA

KEYWORDS: gene transfer therapy; connective tissue repair

Through advances in molecular and cellular biology, numerous proteins have been identified with activities that could benefit the treatment of orthopedic injuries or diseases. Certain proteins are known to stimulate such processes as angiogenesis, cellular mitosis and differentiation, and synthesis of extracellular matrices, while others have anti-inflammatory or immunomodulatory properties. Harnessing the reparative and therapeutic potential of these molecules has, in general, remained problematic. Proteins are often unstable and are therefore difficult to administer effectively. They can also be difficult and expensive to manufacture. Furthermore, many growth factors are pleiotropic and have stimulatory properties that could be undesirable in collateral or nontarget tissues.

Gene transfer technology is being adapted as a means to overcome problems encountered by conventional methods of protein delivery, such as injection or oral administration. By delivering cDNAs encoding the protein(s) of interest to cells at sites of injury or disease, the modified cells become local factories for drug production. This process would permit persistent presentation of the specific peptide agent, concentrated at the site of need. Through regulated delivery of the appropriate factors to the proper cell types in this manner, it should be possible to provide improved treatment of specific orthopedic conditions or to affect enhanced tissue repair.

Initial application of this premise was directed toward the treatment of rheumatoid arthritis. For this, the cDNA for interleukin-1 receptor antagonist (IL-1Ra) was delivered in an *ex vivo* method to the synovial lining of joints. The procedure involved surgical removal of synovial tissue, the isolation and genetic modification of synoviocytes using a replication-defective retrovirus, and then the autologous transplant of the cells to the joint where the IL-1Ra gene product was expressed and secreted into the joint space. The strategy proved to be successful in treating different animal models of arthritis and has proven safe and feasible in human trials.

Address for correspondence: Steven C. Ghivizzani, Center for Molecular Orthopedics, Harvard Medical School, 221 Longwood Avenue, BLI-152, Boston, MA 02115. Voice: 617-732-8607; fax: 617-730-2846.

sghivizzani@rics.bwh.harvard.edu

Ann. N.Y. Acad. Sci. 961: 65–67 (2002). © 2002 New York Academy of Sciences.

Further experimentation in animals has shown that relevant levels of gene expression can also be achieved using direct methods of gene transfer, where certain recombinant, replication-defective, viral vectors based on adenovirus, adeno-associated virus, or herpes simplex virus are injected directly into the joint space to infect cells of the synovial lining *in situ*. Based on the success of gene-based treatment strategies for RA, experimentation has been initiated to explore the feasibility of using gene transfer for other applications such as healing of bone, ligament, and tendon, as well as for cartilage repair/regeneration.

Development of a gene-based therapy in humans generally requires four integrated stages. First, it is necessary to demonstrate sufficient levels of gene transfer to the target tissue. This is generally performed using different marker genes, such as luciferase or alkaline phosphatase, that enable quantitation of gene expression or visual identification of the location and numbers of genetically modified cells. The second step is to determine if the gene product of interest has the predicted biological effect in target cells *in vitro* or *in vivo*. This is achieved by delivering the gene to specific cells and measuring the ensuing biological response stimulated by the gene product, such as increased synthesis of specific collagens or proteoglycans in the case of cartilage repair. The third step is to demonstrate efficacy and safety in a relevant animal model of disease. Does delivery and expression of the gene product cause effective or improved healing or repair? Following success at the earlier stages, the fourth step is then the evaluation of the strategy in humans with disease in both safety and efficacy trials.

Thus far, it has been demonstrated that exogenous marker genes can be delivered and expressed in the cells of many connective tissues, such as bone, ligament, tendon, meniscus, intervertebral disk, and cartilage. This has been achieved by both *ex vivo* delivery and direct injection of recombinant vectors. An interesting observation arising from these studies is that the expression of nonnative, xenogenic proteins via adenovirus and herpes simplex virus vectors has been noted to persist for uncharacteristically long periods of time, in some cases for greater than a year. The dense matrix surrounding the cells resident in these tissues and their limited vascularity likely limit the exposure of the modified cells to components of the immune system.

A number of groups have also shown that gene transfer and expression of certain cytokines, such as TGF-β1, IGF-1, BMP-2, or BMP-7, can stimulate extracellular matrix synthesis in chondrocytes, mesenchymal stem cells, and meniscal cells, among others. Data are also beginning to emerge that demonstrate efficacy for bone healing and spinal fusion via delivery of BMP-2 or LIM mineralization protein-1 (LMP-1), as well as cartilage repair in animal models by implantation of chondroprogenitors modified to express BMP-7.

The outlook for the timely development of enhanced methods for treating and repairing connective tissue using gene-based strategies appears particularly bright. It is possible to deliver and express exogenous genes within the cells of these tissues. Overexpression of certain gene products can stimulate certain cell types toward desired differentiation or biosynthetic pathways. Expression of certain transgenes appears to enhance or accelerate tissue repair in some animal models, suggesting that the strategy is feasible and efficacious. In contrast to genetic or chronic disease conditions, it is probable that expression of a specific transgene product(s) would be required only for a limited period, sufficient to adequately stimulate cells toward synthesis of repair tissue. Because transient expression is desirable, if not a requirement, for these types of applications, current generations of vectors may be sufficient. Further, unlike other gene-based treatments for diseases such as cancer, the standard for successful and useful application is considerably lower. It is only necessary to augment or accelerate the repair of damaged tissue.

However, given the elective nature of gene therapy applications for repair of connective tissues, thorough evaluation and candid representation of findings relative to the safety of the various methods are of particular importance. Many of the cytokines proposed for use in these applications can have detrimental side effects if overexpressed or expressed in nontarget organs such as the heart, lung, or kidney. Thus, it is necessary to explore methods of gene transfer that will afford maximum control over expression with limited dissemination of vectors, genetically altered cells, or transgene products.[1-4]

REFERENCES

1. GHIVIZZANI, S.C., *et al.* 2001. Direct gene delivery strategies for the treatment of rheumatoid arthritis. Drug Discovery Today **6**(5): 259–267.
2. EVANS, C.H., *et al.* 2000. Using gene therapy to protect and restore cartilage. Clin. Orthop. **379**(suppl.): S214–S219.
3. KANG, R., *et al.* 2000. Young Investigator Award—Orthopaedic applications of gene therapy: from concept to clinic. Clin. Orthop. **375**: 324–337.
4. BALTZER, A.W. *et al.* 2000. Genetic enhancement of fracture repair: healing of an experimental segmental defect by adenoviral transfer of the BMP-2 gene. Gene Ther. **7**(9): 734–739.

Genetic Approaches to Tissue Repair

DAVID V. SCHAFFER

University of California, Berkeley, Berkeley, California 94720-1462, USA

KEYWORDS: gene therapy; tissue engineering; stem cell replacement

Genetic approaches for tissue repair and regeneration lie at the interface of two exciting fields—gene therapy and tissue engineering. As these two fields advance towards maturity, combining their successes may yield new and powerful therapeutic avenues. However, a number of challenges must still be overcome before this promise can translate into a therapeutic reality.

Gene therapy for tissue regeneration is currently faced with a number of questions, and further research is required to elucidate the correct solutions for specific therapeutic applications. First, when tissue undergoes injury due to either disease or trauma, a choice must be made between rescuing the damaged cells or replacing them, and each path offers particular challenges. For cell rescue, the choices of the therapeutic gene and the mode of delivery will determine the outcome. For cell replacement, a new supply of cells, such as stem cells, must be identified, and the basic biology of their function and behavior must be sufficiently understood before we can most effectively coax them into integrating with and regenerating damaged tissue.

The first major challenge is then the development of improved gene delivery vehicles. Earlier in the development of this field, gene delivery was most often conducted *ex vivo*. For example, delivery to the nervous system, our laboratory's target tissue, often involved retroviral gene delivery of growth or neurotrophic factors to autologous fibroblasts or myoblasts, followed by grafting to the site of injury. Although this approach has recently entered clinical trials for Alzheimer's disease, the development of efficient direct gene delivery vehicles is generally more attractive than the *ex vivo* use of these cell types due to reduced complexity and potentially equal chances for success. A number of vehicles, both viral and synthetic, are under development for direct gene transfer, although they vary widely in a number of properties including efficiency, immunogenicity, and sustained and regulatable expression of the transgene cargo. Even with the more successful adeno-associated and lentiviral vehicles, improvements in efficiency are still required, and careful quanti-

Address for correspondence: David V. Schaffer, University of California, Berkeley, 201 Gilman Hall, Mail Stop 1462, Berkeley, CA 94720-1462.

Ann. N.Y. Acad. Sci. 961: 68–70 (2002). © 2002 New York Academy of Sciences.

tative analysis of the gene delivery pathway can provide the information needed for further vector engineering. Furthermore, the directed evolution of vectors is a promising, emerging approach that may lead to enhanced performance and mechanistic understanding of vehicles.

In addition to enhancing the capabilities of the vehicle, we must improve our understanding of the cargo. Specifically, gaining a deeper grasp of the molecular mechanisms of cell injury and death will provide us with data to guide our choice of a therapeutic gene. Apoptosis has emerged as a ubiquitous mode for cell death in many types of tissue injury, and the elucidation of signaling pathways involved in this process has led to identification of molecular targets for its inhibition. For example, the use of growth factors and antiapoptotic genes in models of CNS injury has led to cell rescue in animal models. However, further careful analysis of the branching of the apoptotic signaling cascade may lead to the identification of optimal targets for intervention since inhibiting or halting the process at different stages of its progress could make the difference between cell death, nonfunctional cell survival, or fully functional recovery.

Finally, we must deepen our understanding of the most attractive cellular targets for gene delivery. Stem cells have recently emerged as exciting targets for tissue regeneration, and it is particularly in this exciting area that the advances of gene therapy can combine with those of the field of tissue engineering. The gene therapy field recognizes that, in contrast to many differentiated cell phenotypes, stem cells are potentially permanent residents of the body, and transducing them will therefore potentially have the most lasting therapeutic effects. Tissue engineers view stem cells as promising renewable cell sources for the development of artificial tissues and organs. However, genetic modification of these cells may aid in their ability to differentiate and integrate into functional tissue, and this approach represents a combination of the most powerful aspects of both fields. However, we must understand stem cell biology at a more basic level before such efforts can be generally successful. In particular, the factors and signaling pathways that control stem cell survival, proliferation, and differentiation must be elucidated. As development has taught us, the quantitative nature of these instructive signals may be a key determinant of the resulting cell responses.

Progress has been made in utilizing principles from engineering product design to enhance the properties of vectors for *in vivo* gene delivery to the nervous system. In addition, signal transduction in adult neural stem cells has been analyzed, at a quantitative and mechanistic level, to gain a better understanding of how to control and harness these cells for neural regeneration.

Genetic enhancement of stem cells for tissue engineering and regeneration can potentially combine many of the advantages of the approaches described above. If stem cells can be instructed to differentiate towards a particular lineage and functionally integrate into a tissue, they can replace cells that have been lost in a patient. In addition, genetic modification not only may assist in controlling their behavior, but may also revive *ex vivo* gene delivery of growth or other therapeutic factors to regenerate the surrounding tissue. Quantitative and mechanistic analysis of gene delivery, stem cell biology, and the molecular pathology of disease will provide valuable data to enable such efforts and advances in the converging fields of gene therapy and tissue engineering.[1,2]

REFERENCES

1. SCHAFFER, D.V. & D.A. LAUFFENBURGER. 2000. Targeted synthetic gene delivery vectors. Curr. Opin. Mol. Ther. **2:** 155–161.
2. LAUFFENBURGER, D.A. & D.V. SCHAFFER. 1999. The matrix delivers. Nat. Med. **5:** 733–734.

Genetic Approaches to Tissue Engineering

Breakout Session Summary

Moderators

JEFFREY BONADIO, *University of Washington*

SAVIO L.C. WOO, *Mount Sinai School of Medicine*

Panelists

KATHERINE HIGH, *University of Pennsylvania*

STEVEN C. GHIVIZZANI, *Harvard Medical School*

DAVID SCHAFFER, *University of California, Berkeley*

Rapporteur

JEFFREY BONADIO, *University of Washington*

BROAD STATEMENT

Genetic approaches for tissue repair and regeneration lie at the interface of two exciting fields, gene therapy and tissue engineering. Typically, the therapeutic goal is to induce new tissue formation in an injured site or to inhibit an exaggerated tissue-repair response.

Researchers involved in tissue engineering strive to repair lost tissue function through transplantation of living cells grown on bioresorbable scaffolds. Exciting advances have been made in the regeneration of bone, skin, and blood vessel, yet significant challenges remain.

With the advent of modern gene transfer technologies, damaged cells and tissues can be repaired through somatic gene delivery and expression. Gene therapy is the use of genes as medicines for the purpose of preventing, ameliorating, or curing disease.

VISION

Genes may be transferred to cells cultured on bioresorbable scaffolds in order to: (1) improve the manufacture of tissue-engineered constructs; (2) promote the biocompatibility of these constructs once implanted into patients; and (3) deliver therapeutic genes to surrounding tissues. It is anticipated that gene transfer will augment current tissue engineering technology, resulting in improved treatment outcomes for patients with diseases of the cardiovascular, renal, endocrine, musculoskeletal, nervous and other organ systems.

Ann. N.Y. Acad. Sci. 961: 71–72 (2002). © 2002 New York Academy of Sciences.

OBJECTIVES

The modern medical marketplace will favor the development of cost-effective gene delivery technology that will be: (1) easily managed by physician and patient; (2) simple, safe, stable, and relatively inexpensive to manufacture; and (3) capable of producing therapeutic protein in a controllable manner.

OBSTACLES AND CHALLENGES

The field must have safe, effective, and acceptable gene delivery technology at its disposal. Toward this end, the scientific community needs to:

(1) develop efficient and safe gene transfer vectors, and rigorously investigate the basic mechanisms of gene transfer;

(2) establish and quantify the control parameters (e.g., biodistribution, pharmacokinetics, and pharmacodynamics) of gene delivery;

(3) develop targeting vectors capable of tissues-specific gene transfer and expression;

(4) develop regulated gene expression as a way to achieve desired therapeutic outcome while minimizing or avoiding potential toxicities. (A system conferring regulated gene expression should feature low baseline transgene expression, a high induction ratio, and tight control by a small-molecule drug.);

(5) develop vector systems for genetic modification of stem/progenitor cells, and rigorously investigate the functional consequences of vector transduction into these cells;

(6) develop vectors that do not induce innate/adaptive immune responses and those that induce local immune tolerance to the grafts;

(7) establish multidisciplinary approaches toward clinical translational research, so that novel therapeutics in tissue engineering can be validated in the clinical trials while maximally protecting patient safety;

(8) carefully and meticulously establish database and tissue repository in support of future clinical translational studies; and

(9) engage the public in earnest about the real promises and challenges in tissue engineering and regenerative medicine.

Integrity and open disclosure of scientific and clinical translational activities will be required to gain the public trust in gene-delivery technology.

ACKNOWLEDGMENTS

The Genetic Approaches to Tissue Engineering Breakout Session was coordinated by Dr. William Johnston (NIDCR, NIH) and Ms. Winifred Rossi (NIA, NIH).

Different Osteochondral Potential of Clonal Cell Lines Derived from Adult Human Trabecular Bone

ANNA M. OSYCZKA, ULRICH NÖTH,[a] KEITH G. DANIELSON, AND ROCKY S. TUAN[b]

Department of Orthopaedic Surgery, Thomas Jefferson University, Philadelphia, Pennsylvania 19107, USA

[a]*Department of Orthopaedic Surgery, König-Ludwig-Haus, Julius-Maximilians-University, Wurzburg, Germany*

[b]*Cartilage Biology and Orthopaedics Branch, National Institute of Arthritis, Musculoskeletal and Skin Disorders, National Institutes of Health, Bethesda, Maryland*

ABSTRACT: Cells derived from human trabecular bones have been shown to have multipotential differentiation ability along osteogenic, chondrogenic, and adipogenic lineages. In this study, we have derived two clonal sublines of human trabecular bone cells by means of stable transduction with human papilloma virus E6/E7 genes. Our results showed that these clonal sublines differ in their osteochondral potential, but are equally adipogenic, indicative of the heterogeneous nature of the parental cell population. The availability of these cell lines should be useful for the analysis of the mechanisms regulating the differentiation of adult mesenchymal progenitor cells.

KEYWORDS: adult human; trabecular bone; cell lines; osteoblasts; osteogenesis; chondrogenesis; adipogenesis; mesenchymal progenitors

INTRODUCTION

Adult human trabecular bone explant cultures give rise to osteoblast-like cells, which reproducibly undergo osteogenic differentiation and matrix mineralization.[1] These cells have also been reported to differentiate *in vitro* into adipocytes,[2] as well as undergo chondrogenic differentiation *in vivo*,[3] suggesting the presence of pluripotent mesenchymal progenitors within this population. We have recently shown that these adult human trabecular bone cells (hTBC) are also capable of chondrogenic differentiation *in vitro*.[4] In view of these findings, the goal of this investigation was to immortalize these trabecular bone–derived pluripotent cells and to characterize clonal populations with regard to their mesenchymal multilineage differentiation

Address for correspondence: Rocky S. Tuan, Ph.D., Cartilage Biology and Orthopaedics Branch, National Institute of Arthritis, Musculoskeletal and Skin Diseases, National Institutes of Health, 50 South Drive, Building 50, Room 1503, Bethesda, MD 20892-8022. Voice: 301-451-6854; fax: 301-402-2724.

tuanr@mail.nih.gov

Ann. N.Y. Acad. Sci. 961: 73–77 (2002). © 2002 New York Academy of Sciences.

potential. We report here the initial characterization of two clonal sublines derived from adult human trabecular bone cells, stably transduced with human papilloma virus type 16 (HPV-16) E6/E7 genes, that differ in their osteochondral potential, but are equally capable of adipogenic differentiation.

MATERIALS AND METHODS

Initiation of adult human trabecular bone explant cultures and retroviral transduction of primary cells with HPV-16 E6/E7 genes were carried out as described previously.[1,5,6] Stable transduced cells were subjected to ring cloning, and two clonal cell populations, hTBC-1 and hTBC-2, were selected for further study on the basis of their weak and strong alkaline phosphatase (ALP) staining (Sigma cat. no. 86-C), respectively. Postconfluent monolayer cultures of hTBC-1 and hTBC-2, grown in DMEM-F12K supplemented with 10% fetal bovine serum, were stained for ALP (control) and analyzed by RT-PCR for the expression of osteoblast-related transcripts using primers designed for ALP,[7] collagen type I (Col I),[8] osteopontin (OP, custom primers), and osteocalcin (OC),[8] with glyceraldehyde-3-phosphate dehydrogenase (GAPDH)[8] as a housekeeping gene. Cell monolayers were also treated with osteogenic or adipogenic supplements[7,9–11] and then stained for ALP (14-day cultures), matrix mineralization (28-day cultures, alizarin red S[12]), or accumulation of intracellular lipid droplets (21-day cultures, oil red O[7]), respectively. Furthermore, cells were cultured as chondrogenic high-density aggregates in a serum-free medium supplemented with TGF-β1,[7,13–15] and then sectioned and stained for cartilaginous extracellular matrix (3-week cultures, Alcian blue[16,17]).

RESULTS AND DISCUSSION

RT-PCR analysis of the hTBC-1 and hTBC-2 clonal populations (FIG. 1) revealed that they both share phenotypic markers of osteoblasts, as evidenced by mRNA

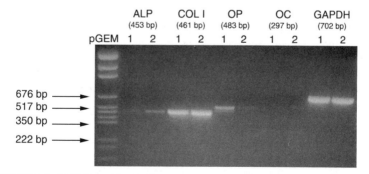

FIGURE 1. RT-PCR analysis of hTBC-1 (1) and hTBC-2 (2) cells (postconfluent cultures grown in standard conditions) for the expression of osteoblast-related transcripts: ALP, Col I, OP, and OC, with GAPDH as control. Both clones expressed osteoblast-related genes, although hTBC-2 had a higher ALP mRNA level, consistent with its high ALP staining (see also FIG. 2).

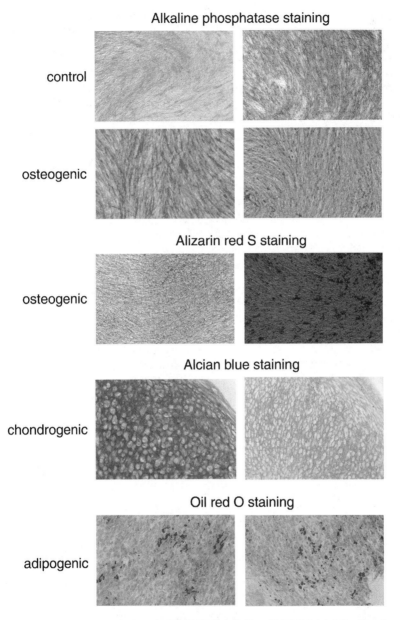

FIGURE 2. Histochemical analysis of hTBC-1 (*left*) and hTBC-2 (*right*) cells cultured in control (postconfluent cultures grown in standard conditions) and osteogenic, chondrogenic, and adipogenic conditions: hTBC-1 cells did not mineralize in osteogenic conditions (alizarin red S staining), but displayed chondrogenic potential in high-density aggregate cultures (Alcian blue staining); hTBC-2 cells differentiated into mature osteoblasts and mineralized, but did not undergo chondrogenic differentiation. Both clonal populations were able to undergo adipogenic differentiation (oil red O staining).

expression of ALP, Col I, OP, and OC. Although in control conditions these clonal populations differed in ALP staining (FIG. 2, alkaline phosphatase staining), in osteogenic conditions they both stained strongly for ALP. However, mineralized matrix (alizarin red S staining) was observed only in cultures of hTBC-2, whereas hTBC-1 did not mineralize. In contrast, in chondrogenic conditions, TGF-β1-treated hTBC-1 cell aggregates elaborated significant amounts of cartilaginous matrix (Alcian blue staining), while hTBC-2 did not. Nevertheless, in adipogenic conditions, both clonal populations were able to differentiate into adipocytes containing intracellular lipid droplets (oil red O staining). Overall, the characteristics of the hTBC-1 and hTBC-2 clonal populations suggest a heterogeneous nature of hTBC. Specifically, while hTBC-1 cells display chondrogenic potential, hTBC-2 cells do not undergo chondrogenic differentiation; instead, they differentiate into mature osteoblasts and mineralize. Our study is the first to characterize adult human trabecular bone–derived osteoblastic cell lines for their chondrogenic and/or adipogenic differentiation.[8,12,18] We conclude that the E6/E7 transduced hTBC populations consist of both osteo- and chondroprogenitor cells that are equally capable of adipogenic differentiation, but differ in their osteo- and chondrogenic pathways.

ACKNOWLEDGMENTS

We would like to thank K. Yoon for assistance in viral transduction of primary cell cultures. This study was supported in part by NIAMS RO3 AR 47396 and NIH Grant Nos. CA 71602, AR 44501, DE 12864, AR 39740, DE 11327, and AR 45181. U. Nöth was supported by a postdoctoral fellowship from the Deutsche Forschungsgemeinschaft (DFG), Germany (Grant No. NO-37111).

REFERENCES

1. ROBEY, P.G. & J.D. TERMINE. 1985. Human bone cells *in vitro*. Calcif. Tissue Int. **37:** 453–460.
2. NUTTALL, M.E., A.J. PATTON, D.L. OLIVERA *et al.* 1998. Human trabecular bone cells are able to express both osteoblastic and adipocytic phenotype: implications for osteopenic disorders. J. Bone Miner. Res. **13:** 371–382.
3. GUNDLE, R., C.J. JOYNER & J.T. TRIFFIT. 1995. Human bone tissue formation in diffusion culture *in vivo* by bone-derived cells and marrow stromal fibroblastic cells. Bone **16:** 597–601.
4. NÖTH, U., A.M. OSYCZKA, K.G. DANIELSON & R.S. TUAN. 2000. Multipotential mesenchymal cells derived from human trabecular bone explants [abstract]. J. Bone Miner. Res. **15(S1):** S384.
5. SINHA, R.K., F. MORRIS, S.A. SHAH & R.S. TUAN. 1994. Surface composition of orthopaedic metals regulates cell attachment, spreading, and cytoskeletal organization of primary human osteoblasts *in vitro*. Clin. Orthop. **305:** 258–272.
6. HALBERT, C.L., G.W. DEMERS & D.A. GALLOWAY. 1991. The E7 gene of human papilloma virus type 16 is sufficient for immortalization of human epithelial cells. J. Virol. **65:** 473–478.
7. PITTENGER, M.F., A.M. MACKAY, S.C. BECK *et al.* 1999. Multilineage potential of adult human mesenchymal stem cells. Science **284:** 143–147.
8. LOMRI, A., O. FROMIGUE, M. HOTT & P.J. MARIE. 1999. Genomic insertion of the SV-40 large T oncogene in normal adult human trabecular osteoblastic cells induces cell growth without loss of the differentiated phenotype. Calcif. Tissue Int. **64:** 394–401.

9. JAISWAL, N., S.E. HAYNESWORTH, A.I. CAPLAN & S.P. BRUDER. 1997. Osteogenic differentiation of purified, culture-expanded human mesenchymal stem cells *in vitro*. J. Cell. Biochem. **64:** 295–312.
10. GUNDLE, R., K. STEWART, J. SCREEN & J.N. BERESFORD. 1998. Isolation and culture of human bone-derived cells. *In* Marrow Stromal Cell Culture, pp. 43–66. Cambridge University Press. London/New York.
11. AUBIN, J.E. & A. HERBERTSON. 1998. Osteoblast lineage in experimental animals. *In* Marrow Stromal Cell Culture, pp. 88–110. Cambridge University Press. London/New York.
12. BODINE, P.V., M. TRAILSMITH & B.S. KOMM. 1996. Development and characterization of a conditionally transformed adult human osteoblastic cell line. J. Bone Miner. Res. **11:** 806–819.
13. JOHNSTONE, B., M. HERING, A.I. CAPLAN *et al.* 1998. *In vitro* chondrogenesis of bone marrow–derived mesenchymal progenitor cells. Exp. Cell Res. **238:** 265–272.
14. MACKAY, A.M., S.C. BECK, J.M. MURPHY *et al.* 1998. Chondrogenic differentiation of cultured human mesenchymal stem cells from marrow. Tissue Eng. **4:** 415–428.
15. YOO, J.U., T.S. BARTHEL, K. NISHIMURA *et al.* 1998. The chondrogenic potential of human bone marrow–derived mesenchymal progenitor cells. J. Bone Jt. Surg. **80A:** 1745–1757.
16. DENKER, A.E., A.R. HAAS, S.B. NICOLL & R.S. TUAN. 1999. Chondrogenic differentiation of murine C3H10T1/2 multipotential mesenchymal cells: I. Stimulation by bone morphogenetic protein-2 in high density micromass cultures. Differentiation **64:** 67–76.
17. HAAS, A.R. & R.S. TUAN. 1999. Chondrogenic differentiation of murine C3H10T1/2 multipotential mesenchymal cells. II. Stimulation by bone morphogenetic protein requires modulation of *N*-cadherin expression and function. Differentiation **64:** 77–89.
18. YUDOH, K., H. MATSUNO, F. NAKAZAWA *et al.* 2001. Reconstituting telomerase activity using the telomerase catalytic subunit prevents the telomere shorting and replicative senescence in human osteoblasts. J. Bone Miner. Res. **16:** 1453–1464.

Therapeutic Potential of Implanted Tissue-Engineered Bioartificial Muscles Delivering Recombinant Proteins to the Sheep Heart

Y. LU,[a] J. SHANSKY,[b] M. DEL TATTO,[a,b] P. FERLAND,[b] S. McGUIRE,[b]
J. MARSZALKOWSKI,[b] M. MAISH,[c] R. HOPKINS,[c] X. WANG,[a,b] P. KOSNIK,[b]
M. NACKMAN,[b] A. LEE,[b] B. CRESWICK,[a,b] AND H. VANDENBURGH[a,b]

[a]Department of Pathology, Brown University School of Medicine and Miriam Hospital, Providence, Rhode Island 02906, USA

[b]Cell Based Delivery Incorporated, Providence, Rhode Island 02906, USA

[c]Department of Surgery, Brown University School of Medicine and Miriam Hospital, Providence, Rhode Island 02906, USA

ABSTRACT: Tissue-engineered primary adult sheep muscle cells genetically engineered to express either rhVEGF or rhIGF-1 secreted the bioactive proteins locally in the sheep heart for at least 30 days.

KEYWORDS: gene therapy; tissue engineering; angiogenesis; ischemic heart disease

INTRODUCTION

Delivery of therapeutic proteins or their genes into the myocardium or surrounding pericardial space is a new strategy for the treatment of heart disease.[1] Recombinant vascular endothelial growth factor (rVEGF) and insulin-like growth factor-1 (rIGF-1) have been delivered as proteins, plasmid DNA, and genetically engineered cells with different pharmacokinetic and physiological responses. For therapeutic angiogenesis, it is best to deliver angiogenic factors in highly localized regions of ischemic heart tissue. For cardioprotective proteins such as IGF-1,[2] chronic delivery to the pericardial fluid would be most effective. In a mouse ischemic hind limb model, implanted tissue-engineered bioartificial muscles (BAMs) genetically engineered to express rVEGF rapidly increased local capillary density without increasing systemic VEGF levels.[3] In the current study, we examined delivery of rhVEGF and rhIGF-1 to the sheep heart using implantable BAM tissue made from genetically modified primary ovine skeletal myoblasts. Our goal was to assess the possibility of implantation of BAMs on the heart in a large animal model and to determine cell survival and protein delivery for up to 30 days.

Address for correspondence: H. Vandenburgh, c/o Cell Based Delivery Inc., 4 Richmond Square, Providence, RI 02906. Voice: 401-454-3540; fax: 401-454-3157.
herman_vandenburgh@brown.edu

Ann. N.Y. Acad. Sci. 961: 78–82 (2002). © 2002 New York Academy of Sciences.

PRIMARY SHEEP MYOBLAST CULTURES AND
RETROVIRAL TRANSDUCTIONS

Skeletal myoblasts were isolated from muscle samples of the vastus lateralis and stably transduced with replication-deficient retroviral vectors containing the cDNA for hVEGF or hIGF-1. Transduced cells were assayed for hVEGF or hIGF-1 secretion and expanded for BAM formation. All experimental animal procedures were approved by the Institutional Animal Care and Use Committee and conformed to the guiding principles of the American Physiological Society.

TISSUE-ENGINEERING OF OVINE
BIOARTIFICIAL MUSCLES (oBAMs)

Ovine bioartificial muscles (oBAMs) were made by suspending hVEGF- or hIGF-1-secreting or control sheep muscle cells in an extracellular matrix solution (1×10^6 cells/mL) and casting the suspension into silicone rubber molds (1 mL/ mold) as previously described.[3] oBAMs were implanted on the heart directly or first fibrin-glued to 50×50 mm patches of bovine pericardium (CardioFix pericardium, Sulzermedica).

SURGICAL PROTOCOL

Sheep were intubated, anesthetized, and mechanically ventilated, and a left thoracotomy was performed in the fourth intercostal space. Four techniques for implanting oBAMs were tested: (1) oBAMs were fibrin-glued (Tisseel VH) into the atrioventricular groove (4 oBAMs/AV groove, $n = 5$ animals), along the side of the pulmonary artery (2 oBAMs/animal) or the aorta sulcus (2 oBAMs/PA sulcus, $n = 2$ animals); (2) oBAM patches were sutured onto the anterior wall of the left ventricle (2–3 oBAMs/patch, 2–3 patches/animal, $n = 5$ animals); (3) oBAMs were fibrin-glued directly on the epicardial surface of the left ventricle (4 oBAMs/heart, $n = 3$ animals); (4) oBAMs were inserted directly into the myocardium of the left ventricle with a 12-gauge angiocatheter (2 oBAMs, $n = 1$ animal). Each animal was implanted with oBAMs in multiple sites, and both control and hVEGF-secreting oBAMs were implanted into the same animal. Blood and pericardial fluid were collected on the day of surgery and the day of explantation. Four to 30 days after surgery, oBAMs and local myocardium were removed and either fixed for histology or placed in growth medium to monitor secretion of growth factors.

GROWTH FACTOR ANALYSES AND TISSUE HISTOCHEMISTRY

hVEGF and hIGF-1 protein levels in culture medium, sheep serum, and pericardial fluid were measured using ELISA kits. Cryostat sections from BAMs and myocardium were stained for capillary cells using an indoxyl-tetrazolium method and analyzed using a Zeiss KS300 Image Analysis System. Other sections were immunocytochemically stained for skeletal muscle cells or hVEGF.

RESULTS

Tissue-Engineered oBAMs Express hVEGF and hIGF-1 in Vitro

Transduced sheep skeletal myoblasts were tissue-engineered into oBAMs approximately 1–2 mm in diameter by 20 mm in length and containing parallel arrays of multinucleated postmitotic myofibers similar to those in rodent BAMs.[3] The oBAMs *in vitro* secreted consistent levels of hVEGF (30–150 ng/BAM/day) or hIGF-1 (184–596 ng/BAM/day).

oBAMs Can Be Maintained When Implanted into the Sheep Heart

All of the oBAM patches were recoverable at 4–30 days after surgery. Half of the oBAMs in the AV groove and one-third of the oBAMs on the left ventricle surface were recoverable at 4–30 days *in vivo*. One oBAM was harvested at 30 days after surgery from the intramuscular site. No oBAMs were located in the PA sulcus or pulmonary artery after 4–30 days.

oBAMs Survive on the Heart and Secrete IGF-1 for up to 30 Days

Four days after implantation of hIGF-1 oBAMs, the IGF-1 concentration increased 6-fold over preimplant levels in the pericardial fluid (92 to 505 ng/mL). After 30 days *in vivo*, explanted hIGF-1 oBAMs still secreted 35% of their preimplant hIGF-1 levels and were therefore still viable.

Angiogenesis Is Accelerated by rhVEGF-oBAMs

After 7–19 days *in vivo*, hVEGF was present in the myofibers of hVEGF oBAMs and the surrounding space of the neighboring host myocardium as determined immunocytochemically. Capillary ingrowth in the myocardium was significantly increased as early as 1 week in sheep receiving rhVEGF-secreting implants (FIG. 1B) compared to control BAMs (FIG. 1A), and this was maintained for up to 19 days

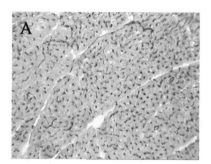

FIGURE 1. Capillary density was increased in the neighboring myocardium of sheep implanted with rhVEGF oBAMs (**B**) compared to control BAMs (**A**), as shown by immunocytochemical staining for endothelial cells.

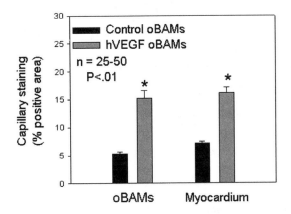

FIGURE 2. Quantification of the area staining positive for capillaries in explanted oBAMs and adjacent myocardium showed a significant increase in animals implanted with oBAMs compared to control BAMs. This increase was maintained for 19 days.

(FIG. 2). Vascularization within implanted hVEGF oBAMs was also elevated, showing a significantly higher capillary density than in control oBAMs (FIG. 2). There was no difference in the capillary density of the underlying myocardium between hVEGF oBAM patch implants and fibrin-glued hVEGF oBAMs (15.4 ± 1.1% vs. 16.9 ± 0.9% positive staining area, $P > 0.05$). There were no detectable levels of hVEGF in serum or pericardial fluid in any of the animals either pre- or post-hVEGF oBAM implantation.

SUMMARY

Preliminary clinical data suggest that therapeutic angiogenesis can improve blood flow to ischemic myocardium,[1] indicating a potential new strategy for no-option patients with ischemic heart disease. In this study, we have shown that primary adult sheep myoblasts can be genetically engineered to secrete rhVEGF or rhIGF-1 and tissue-engineered into BAM constructs that can survive and continue to secrete recombinant proteins for up to 30 days. This improves on the limited protein expression of up to several weeks seen with other gene therapy approaches.[1] Implantation of rhVEGF-secreting BAMs into the sheep heart stimulated localized angiogenesis with undetectable levels of hVEGF in serum or pericardial fluid, indicating a local rather than systemic effect. The techniques described here may eventually play a role in treating cardiac ischemic disease.

ACKNOWLEDGMENTS

This work was supported by NIH HL60502, AG15415, NASA NAG2-1205, and NIST 70NANB9H3011.

REFERENCES

1. LAHAM, R.J., M. SIMONS & F. SELLKE. 2001. Gene transfer for angiogenesis in coronary artery disease. Annu. Rev. Med. **52:** 485–502.
2. GIUSTINA, A., M. VOLTERRANI, F. MANELLI et al. 1999. Endocrine predictors of acute hemodynamic effects of growth hormone in congestive heart failure. Am. Heart J. **137:** 1035–1043.
3. LU, Y., J. SHANSKY, M. DEL TATTO et al. 2001. Recombinant vascular endothelial growth factor secreted from tissue-engineered bioartificial muscles promotes localized angiogenesis. Circulation **104:** 594–599.

Emerging Design Principles in Biomaterials and Scaffolds for Tissue Engineering

LINDA G. GRIFFITH

Biological Engineering Division and Department of Chemical Engineering, Massachusetts Institute of Technology, Cambridge, Massachusetts 02139, USA

ABSTRACT: Biomaterials and scaffolds play an essential role in tissue engineering by guiding new tissue growth *in vivo* and *in vitro*. While adaptation of existing surgical materials has fulfilled some needs in the field, new applications demand better control of bulk properties such as degradation and of surface properties that control cell interactions. Advances in molecular cell biology are driving the incorporation of new biological moieties into materials, and a set of design principles based on quantitative analysis of key cellular processes involved in regeneration is emerging. At the same time, new materials-processing methodologies are emerging to allow fabrication of these fragile materials into devices appropriate for delivery.

KEYWORDS: polymers; degradation; tissue engineering; biomaterials

INTRODUCTION

Broadly defined, tissue engineering is the process of creating living, physiological, three-dimensional tissues and organs utilizing specific combinations of cells, cell scaffolds, and cell signals, both chemical and mechanical. The process starts with a source of cells derived from the patient or from a donor. The cells may be immature cells, in the stem cell stage, or cells that are already capable of carrying out tissue functions; often, a mixture of cell types (e.g., liver cells and blood vessel cells) and cell maturity levels are needed. Coaxing cells to form tissue is inherently an engineering process as they need physical support (typically in the form of some sort of 3-D scaffold) as well as chemical and mechanical signals provided at appropriate times and places to form the intricate hierarchical structures that characterize native tissue. Tissue engineering applications can be broadly classified into *therapeutic applications*, where the tissue is either grown in a patient or grown outside the patient and transplanted, and *diagnostic applications*, where the tissue is made *in vitro* and used for testing drug metabolism and uptake, toxicity, pathogenicity, etc. The material needs for each of these broad categories are distinct, but overlapping.

The process of forming tissues from cells is a highly orchestrated set of events that occur over timescales ranging from seconds to weeks and with dimensions

Address for correspondence: Linda G. Griffith, Biological Engineering Division and Department of Chemical Engineering, Massachusetts Institute of Technology, Room 66-466, 77 Massachusetts Avenue, Cambridge, MA 02139. Voice: 617-253-0013; fax: 617-258-5042.

griff@mit.edu

Ann. N.Y. Acad. Sci. 961: 83–95 (2002). © 2002 New York Academy of Sciences.

ranging from 0.0001 cm to 10 cm. Current research and development in biomaterials and scaffolds address problems across this spectrum. At one end, studies of basic biological and biophysical processes at the molecular and cellular level are required so that we understand what processes the cells need help with and what events they can accomplish themselves. Studies at this end of the spectrum have led to the development of new tools for biologists to use in fundamental studies of cell behavior, which in turn lead to better bioactive biomaterials. At the other end of the spectrum, scaffolds are needed to direct the macroscopic process of tissue formation. Scaffolds used in therapeutic applications are made from polymers that are resorbed or degraded in the body. Two key challenges exist here. First, the first generation of degradable polymers widely used in tissue engineering was adapted from other surgical uses and has some deficiencies in terms of mechanical and degradation properties. New classes of degradable materials are being developed. The second challenge is how to fabricate these relatively delicate polymers into scaffolds that have defined shapes and a complex, porous internal architecture that can direct tissue growth. A variety of new approaches are being developed under the classical engineering constraints of cost, reliability, government regulation, and societal acceptance.

Finally, tissue engineering applications have begun to extend beyond therapeutic implants to creation of physiological models of human tissue *in vitro*. Greater latitude is possible in the choice of materials for such applications as the structural scaffold does not need to be degradable or even have mechanical compliance with human tissue. Scaffold fabrication remains highly challenging, and perhaps even more so than with therapeutic structures, as the natural *in vivo* remodeling that is a part of the tissue formation process does not take place fully *in vitro*.

Biomaterials used in tissue engineering can be broadly divided into categories of *synthetic* or *naturally derived*, with a middle ground of *semisynthetic* materials rapidly emerging. Most materials commonly in use in tissue engineering today—in clinically approved products or in applications at an initial research stage—are adapted from other surgical uses, such as sutures, hemostatic agents, and wound dressings. These include synthetic materials such as polylactide-co-glycolide polymers (component of Dermagraft™) and naturally derived materials such as collagen (component of Appligraf™). Adaptation of materials that have already been used in prior applications in humans can have some advantages from the regulatory perspective as the safety and toxicity profiles of the materials in humans are already defined. Thus, there can be confidence that new devices will be safe from the perspective of the composition of materials. Other performance aspects (cell-material interactions, degradation properties, etc.), however, are not assured and the need for substantially higher performance characteristics is driving research and development toward the design of new materials that meet specific performance criteria in tissue engineering. The focus in this presentation is on synthetic and semisynthetic materials, recognizing that naturally derived materials also play an important role in tissue engineering, but the scope of what can be covered here must be limited. The structure, strength, and processability of biomaterials that are used in the engineering of cell scaffolds will be considered, as well as the bioactive properties of scaffolds in terms of architecture and spatial organization of adhesion, growth, and differentiation signals. Design principles can now be formulated and, as they are refined through future study, will greatly aid achievement of the goals of reparative medicine.

BIOMATERIALS

Degradable Bulk Materials

For therapeutic applications, one of the most desirable material properties is degradation or resorption.[1] Although some tissues, particularly bone, can tolerate very slowly degrading or permanent materials of specific compositions, permanent implants almost always elicit a chronic inflammation called a foreign body response.[2–4] This response is characterized by formation of a poorly vascularized fibrous layer analogous to a scar at the material-tissue interface and will almost always compromise tissue function. Even in bone, where bone tissue formation may be observed adjacent to a nondegradable implant, the mechanical properties of the final hybrid tissue may be poor compared to native tissue. Thus, degradable materials are strongly preferred. Materials and their degradation products must also be nontoxic and nonimmunogenic upon implantation. Further aspects of basic biocompatibility are context-dependent and must be determined by experimentation.

A wide variety of both solid and hydrogel-type polymers have been developed.[1,5] "Degradable" polymers undergo extensive chain scission to form small soluble oligomers or monomers in the presence of body fluids. This degradation may proceed by a biologically active process (e.g., enzymes present in body fluids participate) or by passive hydrolytic cleavage. The term "biodegradable" typically refers to materials in which active biological processes are involved, and new classes of materials are being developed that respond to local enzyme release by cells involved in tissue engineering. "Resorbable" polymers gradually dissolve and are eliminated through the kidneys or other means.

In the early days of tissue engineering, degradable polyesters based on lactide and glycolide monomers were the workhorse synthetic polymers for scaffold construction. These materials were utilized in surgery 30 years ago as materials for sutures and bone fixation devices[6] and were available commercially in several porous 3D forms (mainly fabrics) that could readily be adapted for feasibility studies for a specific tissue. Polymers made from lactide and glycolide monomers are characterized by degradation times ranging from days to years, depending on specific formulation and initial M_w. Lactic acid is a chiral molecule, existing in L and D isomers (the L isomer is the biological metabolite), and thus "polylactic acid" actually refers to a family of polymers: pure poly-L-lactic acid (L-PLA), pure poly-D-lactic acid (D-PLA), and poly-D,L-lactic acid (DL-PLA). Polymers synthesized from pure L-lactide (the monomer based on the physiologically occurring version of lactic acid) or pure D-lactide are crystalline, while copolymers synthesized from a mixture of monomers are mostly amorphous. While there are no adverse physiological effects reported from the D isomer, the L isomer is used preferentially when a crystalline polymer is desired. The various lactide monomers can also be copolymerized with glycolide monomers to create a family of materials that are completely amorphous and relatively quickly degrading to materials that are highly crystalline and very slowly degrading. All polymers in this family are insoluble in water, but are degraded upon hydrolytic attack of the ester bond. The mechanical and degradation properties are affected by the combined effects of the crystallinity, the molecular weight (M_w), the glass transition temperature (Tg), and the monomer hydrophobicity.[7–12]

Degradable polyesters are still widely used as research materials in tissue engineering and are used in the scaffold of at least one successful clinical product: Dermagraft™. However, their limitations have also become apparent. The mechanical properties of the classical degradable polyesters are not always suitable for tissue engineering due to their relative inflexibility and tendency to crumble upon degradation, sometimes creating particles that remain in the site for years, or they break in ways that mechanically damage nascent tissue. Lactic and glycolic acid breakdown products produced by degradation of commonly used degradable polyesters have also been associated with adverse tissue reactions when used in fixation devices in bony sites,[13,14] presumably due to the rapid release of degradation from relatively large devices.

These limitations are being addressed by synthesis of polymers that yield less acidic degradation products and yet still have suitable strength and degradation properties. Notable among these are pseudo-poly(amino acids), a class of polymers based on natural amino acids linked by nonamide bonds including ester, iminocarbonate, urethane, and carbonate.[15–18] They offer improved mechanical properties, processing, stability, and ease of synthesis over poly(amino acids) joined by traditional peptide bonds. Kohn and colleagues have used a range of monomers and bonds to obtain polymers with desired properties such as Tg, crystallinity, and hydrophobicity.[19] Among the many polymers in this family, poly(tyrosine carbonates) show promise in orthopedic applications due to improved behavior of degradation products in bony sites.[15] These promising materials have been licensed and are being developed commercially. Other new formulations of bulk polymers are also being targeted primarily to bony applications.[20]

The limited range of physical properties of classic degradable polyesters has also led to synthesis of new block copolymers with polyurethane linkages, allowing a far greater range of mechanical properties to be achieved.[21,22] These multiblock copolymers have crystallizable hard segments and noncrystallizing oligoesters (adipic acid, ethylene glycol, 1,4-butanediol, and diol-terminated PCL) as soft segments. Other degradable polyesters are also being developed for clinical applications, notably poly(hydroxybutyrate) (PHB) and copolymers of hydroxybutyrate with hydroxyvalerate.[7] These polymers are naturally produced by biological processes (typically in microorganisms) rather than classical synthesis techniques and they exhibit complex degradation patterns. PHB and copolymers with hydroxyvalerate are produced in microorganisms and processed postpurification; they are included in this category due to their chemical simplicity and similarity to classical degradable polymers. Finally, a host of new polymers developed for drug delivery, including polyanhydrides[23] and polyorthoesters,[24] have interesting degradation properties, but their mechanical suitability for tissue engineering applications is less clear.

The use of synthetic degradable gels is emerging primarily as a way to deliver cells or scaffolds *in situ*. A predominant approach, pioneered by Hubbell and coworkers, is formation of photopolymerizable gels using polyethylene oxide (PEO)–based macromer substrates.[25] This approach is particularly amenable to inclusion of biologically active ligands.[26,27] Langer and coworkers have pioneered a process of forming a gel by shining light through the skin on injected monomers to form a gel, providing a means for improved minimally invasive delivery.[28] This approach may be particularly useful for applications such as "injectable cartilage."[29]

Bioactive Biomaterials

Synthetic materials used in tissue engineering start out with minimal or no intrinsic biological activity. They generally must acquire bioactivity either through passive means when placed in biological fluids or by active means through design and control of specific molecular interactions between cells and the material. Bioactive materials for tissue engineering fall roughly into two categories—(i) those used to surface-modify classical bulk materials such as the degradable polyesters and (ii) gels that may form in the presence of cells or tissues either *in vitro* or *in vivo*.

Most degradable polyesters are at least moderately cell-adhesive, with adhesion resulting from adsorption of proteins from serum in culture medium or body fluids.[30-33] This intrinsic adhesiveness generated by adventitious protein adsorption is adequate for some applications. One of the more visible early applications of tissue engineering was generation of cartilage in the shape of a human ear.[34] Isolated cartilage cells (chondrocytes) appear to have little requirement for specific surface properties in order to reform a tissue structure, other than that the surface be at most moderately adhesive, an observation demonstrated in many labs.[35-37] Likewise, a skin product based on human fibroblasts cultured on a degradable polyester matrix *in vitro* was recently given FDA approval.[31]

Cell adhesion to degradable polyesters and other water-insoluble scaffold materials can be increased or altered by modifying the surfaces to present ligands for cell surface adhesion receptors.[38-43] The ligands are typically small peptide adhesion domains that bind to integrin receptors; tens of such domains have been identified by mapping extracellular matrix molecules. A simple approach to accomplishing surface modification is attachment of a long hydrophobic peptide tail and spacer to the adhesion peptide to allow modification of generic hydrophobic substrates by adsorption of the peptide from solution.[41] This approach, as well as some involving covalent linkage of peptides to the polymer backbone,[38] can introduce additional sites for adhesion without necessarily blocking nonspecific protein adsorption and adhesion or allowing precise control over the surface density of ligand. Because some cell behaviors, such as migration and differentiation, can be inhibited by strong cell-substrate adhesion induced by high ligand surface densities,[44,45] interest is growing in implantable polymers that can surface-modify bulk materials to present ligand at controlled densities against a relatively inert background. Enhanced stability and ability to control ligand density and spacing may come from use of branched or comb PEO-containing polymers that can be blended with the bulk phase and induced to surface-segregate.[43]

Although many applications in tissue engineering may ultimately require a porous scaffold that can be handled during surgical procedures and retain a specific shape, approaches to nerve, blood vessels, bone, and many other tissues may best be addressed by using a gel scaffold formed *in situ*, either to deliver cells or to induce migration of cells into a wound site. Synthetic and semisynthetic gels that not only stimulate cells through adhesion and growth factor moieties, but also respond to cells by degrading in the presence of specific cell cues are being developed in response to these needs.[27] In one particularly well developed family of gels, the basic macromer unit is a linear or branched PEO end-capped with chemically reactive groups. Gelation may be accomplished by free radical polymerization of the monomers using nontoxic visible light in the presence of an initiator[25] or by Michael-type addition.[46] This system is particularly flexible for tissue engineering as it is intrinsically non-adhesive for cells and the gel properties can be tailored: the permeability of the gel can be controlled by the size of the monomers and the gel thickness; controlled degradation has been demonstrated by including hydrolyzable polyester segments or enzyme-cleavable peptides at the chain ends; and adhesion peptides can be included in the gel at a range of concentrations to control cell interactions.[1,27,47] This flexi-bility in both biological and physical properties is being used to address at least one complex and long-standing problem in tissue engineering—blood vessel healing following angioplasty, which denudes the vessel wall of the endothelial lining. Healing is often accompanied by growth factor–mediated hyperproliferation of the smooth muscle underlying the endothelium, leading to ultimate reocclusion of the vessel, a process partly mediated by proteins in the blood diffusing into the damaged vessel. Photopolymerizable gels offer the means to address several facets of this problem: formation of a gel on the luminal surface of the vessel can provide a non-clotting barrier to blood-borne stimulators of hyperproliferation;[48,49] proteins can be delivered from the gel to further control the behavior of the underlying tissue;[50] and endothelial cell–specific adhesion peptides can be incorporated into the gel to induce a normal repopulation of the vessel surface with endothelium.[51]

Scaffolds

Scaffolds for tissue engineering must direct the 3-D organization of cells *in vitro* or *in vivo*. Many of the initial scaffolds used in tissue engineering were, like bio-materials, adapted from other uses in surgery. Surgical fabrics, including woven and nonwoven meshes, have been widely adapted for tissues ranging from skin and cartilage to liver to bladder.[34,52–55] Approaches based on particulate-leaching and foaming techniques have expanded the range of porous synthetic degradable scaffolds to include sponges with a reasonable degree of systematic variation in pore sizes. Many tissues have a hierarchical structure that varies over length scales of 0.1– 1 mm and require regeneration in specific macroscopic shapes. Thus, new polymer processing approaches are needed to create degradable porous scaffolds that meet these needs. Ideally, the scaffold would offer three levels of control of both archi-tecture and composition. At the highest level, macroscopic shape and composition are important on a scale of millimeters to centimeters. At an intermediate level, the size, orientation, and surface chemistry of pores and channels for tissue ingrowth are crucial, on a scale of hundreds of microns. Finally, locally surface texture and porosity are important, on a scale of tens of microns. While processing techniques have been developed that address individual scales, no single technique can yet encompass all scales, particularly with control over both architecture and surface chemistry. For example, it may be desirable in a bone tissue engineering device to create a scaffold that has relatively large channels (0.5–1 mm) for rapid tissue penetration, but has local porosity for tissue development, and to create this structure in a specific shape based on a magnetic resonance imaging (MRI) or computerized tomography (CT) scan of the patient's defect. One approach to this is use of solid free-form fabrication (SFF) methods.

In SFF techniques, devices are built as a series of thin sequential layers, with a different physical structure. One SFF technique, the 3-Dimensional Printing (3DP™)

process, offers unique flexibility in creating structures that meet these demands. This fabrication method was originally developed for fabrication of ceramics and metals and has been adapted to a range of polymeric and composite organic/inorganic materials important in tissue engineering. In this process, a thin layer of polymer powder is spread on a piston, and a solvent that locally dissolves the polymer particles is printed onto the powder bed in specific locations where it is desired to join the particles into a structure.[56,57] Once a layer is printed, the solvent evaporates, leaving physically joined structures in the desired shape. Another layer of powder is then spread and the process repeated. Very complex shapes can be created in this fashion, and addition of polymers to the printed binder or variation of the composition of the polymer in the powder bed allows variation in scaffold degradation rates and composition.

Silicon microfabrication is also being used to make molds for 3D polymer scaffolds and for tissue organization. Deep reactive ion etching (DRIE) has a resolution on the scale of a single cell and can be used to create microvascular structures.[58]

Finally, many therapeutic applications of tissue engineering involve disease processes that might be prevented or treated if better drugs were available or if the processes could be better understood. For example, hepatitis C is currently the leading cause of liver transplantation. Development of new drugs to treat hepatitis C is severely impaired by the inability to infect liver cells with the virus in culture—cells taken from the body lose many of their liver-specific functions, including susceptibility to viral infection and ability to metabolize drugs. We postulate that hepatocytes, the cells associated with most liver-specific functions, lose function because they are deprived of the complex set of signals that exist *in vivo*, including interactions with several other cell types in liver as well as fluid shear forces. Scaffolds that allow the development of key tissue features *in vitro*—such as the capillary bed of a tissue—may become extremely useful as physiological models. Here, DRIE is also being used to create scaffolds for 3-D self-organization of tissues for use in *ex vivo* microfabricated bioreactors.[59,60]

Design Principles for New Biomaterials

Design principles are beginning to emerge for the use of ligands to control cell behavior on biomaterials for tissue engineering based on a physical and quantitative understanding of how cells respond to molecular signals and integrate multiple inputs to generate a given response. One example is the way in which the physical aspects of ligand presentation control cell motility behaviors such as migration and neurite extension. Cell motility across a substratum is strongly affected by the biophysical nature of adhesive interactions. It has been shown both theoretically and experimentally that the average speed of cell locomotion typically exhibits a biphasic dependence on the strength of cell substratum adhesion for a number of cell types, with a maximal value for cell migration speed occurring at an intermediate value of cell-substrate adhesive strength.[45,61,62] Systematic variation in cell-substratum adhesion strength has most commonly been achieved by variation of the surface density of ECM proteins, thus changing the number or strength of bonds between integrins and the extracellular environment. Comparable effects have been achieved in a 3D system by adding systematically varied amounts of fibronectin to collagen gels.[63] More generally, quantitative modulation of adhesion strength may be achieved via multiple approaches that include a change in the occupancy of integrins on the cell surface (either by a change in the number of integrins expressed on the cell surface or by a change in the surface density of adhesion ligands that interact with the integrins), a change in the affinity of the integrin for the adhesion ligand, or a change in the avidity of interaction between the integrin and the adhesion ligand. One could speculate that comparable changes in adhesion strength accomplished by any of these means might result in comparable changes in cell motility if adhesion strength is a primary variable. Supporting this concept, Palecek et al.[62] showed that when integrin-mediated cell motility was modulated by altering adhesion ligand surface density, cell integrin expression level, or integrin/ligand-binding affinity, cell speed ultimately depended on a single parameter—the cell-substrate adhesion strength. In each case, cell migration speed showed a maximum at intermediate values of adhesion strength.

These observations provide a design basis for modifying cell adhesive behavior to control cell motility in tissue engineering. Hubbell and coworkers have shown that the extent of neurite extension in fibrin gels modified with Arg-Gly-Asp (RGD) peptides shows biphasic behavior in vitro, with an intermediate concentration of peptide leading to maximal values of extension.[64,65] A biphasic response to adhesion peptides coated on degradable polymer scaffolds has also been observed for bone ingrowth in vivo, with maximal values of ingrowth at an intermediate peptide density.[40] Because bone ingrowth is a metric that lumps together many individual phenomena, including cell growth and differentiation, it is not yet possible to attribute the in vivo results solely to migration effects. Many biological phenomena are biphasic and thus additional studies are needed to resolve the components of the phenomena contributing to the observed result.

Design principles are also emerging from biophysical analysis of extracellular matrix properties. The multimeric structure of ECM molecules such as fibronectin (a dimer with dual adhesion sites) and tenascin-C (a hexabrachion that presents 6 identical cell adhesion domains within ~100 nm) suggests that receptor clustering may be influenced by the physical layout of the extracellular matrix components. This inference is supported by in vitro studies using synthetic RGD adhesion peptides.[66,67] Peptides grafted to albumin at 1–20 peptides/albumin molecule[66] or to synthetic star-configured PEO tethers at 1–9 peptides/polymer molecule[67] are more effective in promoting cell adhesion when the valency is high: peptides presented singly (1 peptide/molecule) are poor substrates for adhesion, whereas peptides presented at cluster sizes of 9 peptides/molecule or higher induce comparable adhesion to matrix proteins. Furthermore, when RGD adhesion ligands are presented in a non-clustered fashion, fibroblast migration is significantly impaired, even at identical average ligand densities.[67] In addition to responding to specific molecular cues organized at a nanoscale level, cells are also sensitive to the supramolecular organization of matrix as evidenced by the angiogenic response of ligand-modified fibrin matrices.[68]

In addition to sensing ECM spatial organization, cells exert forces on matrix and respond to mechanical properties of their surroundings by regulating adhesive interactions, and these behaviors are also important to consider in the design of biomaterials. Compared to compliant matrices of identical composition, rigid matrices have been shown to enhance cell-surface assembly of fibronectin[69] and to provide a preferential substrate for directional cell migration,[70] and they are associated with increased levels of protein phosphorylation at sites of cell-matrix contact.[71] Compliant matrices, on the other hand, promote cell motility.[71] Cell speeds are also greater on substrates where fibronectin is adsorbed and thus compliant than on substrates where fibronectin is covalently immobilized and thus inflexible.[72] Localized stress on integrin-ligand bonds leads to reinforcement in adhesion to ECM, suggesting a molecular role for substrate compliance as a parameter in cell migration.[73]

Analysis of ECM also leads to design principles for modulating growth factor interactions with cells. Many growth factors, including the angiogenic factors, FGF and VEGF, are strongly bound to ECM. At least one ligand for the epidermal growth factor receptor (EGFR) is embedded in the large extracellular matrix molecule, tenascin, and may thus be considered a potential immobilized ligand that signals through the cell surface.[74] From a biomaterials perspective, it is thus perhaps appropriate to present some growth factors as components of matrix rather than in soluble form. This approach has proven effective for influencing the function of smooth muscle cells in arterial tissue engineering[75] and nerve growth.[76] The building blocks are now in place to address adhesion, growth, and degradation, with synthesis of gels that incorporate a complete compendium of active sites.[77,78] Additional issues, such as mechanical compliance, are also being addressed.

Finally, the cells themselves may be viewed as a "macromaterials" component of forming a complete tissue structure via self-assembly.[79] Pioneering work by Steinberg suggests that the sorting out of intermixed embryonic cells, the spreading of one tissue over another, and the specific inside/outside tissue stratification that arises by any of these processes are guided in large part by the differential intercellular adhesivities of the cell types involved.[80,81] Therefore, barring kinetic barriers, the number and strength of cell-cell and cell-substrate bonds formed—parameters that can be systematically modulated by solid or soluble biomaterials—will dictate the final arrangement of a cell mixture into a tissue-like structure.[79] This perspective is being used to overcome kinetic barriers to forming neural cell aggregates (e.g., by the design of bifunctional PEO macromers bearing cell adhesion ligands)[82,83] and to control formation of liver cell structures via modulation of cell-substrate adhesion strength.[79–84]

CONCLUSIONS AND FUTURE DIRECTIONS

Traditional applications in polymeric biomaterials continue to be robust and derive new directions from advances in materials science. A growing number of applications require intimate integration of polymer science with molecular cell biology, both for developing new approaches to human therapies and as basic tools in biological research. The age of using polymeric materials to controllably manipulate cell function is still in its infancy and expansion of this field will likely enable development of new therapies and technologies for biomedical research.

REFERENCES

1. GRIFFITH, L.G. 2000. Polymeric biomaterials. Acta Mater. **48:** 263–277.

2. ANDERSON, J.M. & J.J. LANGONE. 1999. Issues and perspectives on the biocompatibility and immunotoxicity evaluation of implanted controlled release systems. J. Controlled Release **57:** 107–113.
3. ANDERSON, J.M. 1988. Inflammatory response to implants. Trans. Am. Soc. Artif. Intern. Organs **34:** 101–107.
4. BABENSEE, J.E., *et al.* 1998. Host response to tissue engineered devices. Adv. Drug Delivery Rev. **33:** 111–139.
5. SALTZMAN, W.M. 1996. Growth-factor delivery in tissue engineering. MRS Bull. **21:** 62–65.
6. KULKARNI, R.K., *et al.* 1971. Biodegradable poly(lactic acid) polymers. J. Biomed. Mater. Res. **5:** 169–181.
7. AMASS, W., A. AMASS & B. TIGHE. 1998. A review of biodegradable polymers: uses, current developments in the synthesis and characterization of biodegradable polyesters, blends of biodegradable polyesters, blends of biodegradable polymers, and recent advances in biodegradation studies. Polym. Int. **47:** 89–144.
8. ENGELBERG, I. & J. KOHN. 1991. Physicomechanical properties of degradable polymers used in medical applications: a comparative study. Biomaterials **12:** 292–304.
9. GOPFERICH, A. 1996. Mechanisms of polymer degradation and erosion. Biomaterials **17:** 103–114.
10. LI, S. 1999. Hydrolytic degradation characteristics of aliphatic polyesters derived from lactic and glycolic acid. J. Biomed. Mater. Res. (Appl. Biomater.) **48:** 342–353.
11. SCHMITT, E.A., D.R. FLANAGAN & R.J. LINHARDT. 1994. Importance of distinct water environments in the hydrolysis of poly (DL-lactide-co-glycolide). Macromolecules **27:** 743–748.
12. VERT, M., J. MAUDUIT & S. LI. 1994. Biodegradation of PLA/GA polymers: increasing complexity. Biomaterials **15:** 1209–1213.
13. BÖSTMAN, O., *et al.* 1990. Foreign-body reactions to fracture fixation implants of biodegradable synthetic polymers. J. Bone Jt. Surg. **72B:** 592–596.
14. SUGANAMA, J., *et al.* 1992. Biologic response of intramedullary bone to poly-L-lactic acid. *In* Tissue-Inducing Biomaterials. Vol. 252, pp. 339–343. Materials Research Society. Pittsburgh.
15. ERTEL, S.I., *et al.* 1995. Evaluation of poly(D-carbonate), a tyrosine-derived degradable polymer, for orthopedic applications. J. Biomed. Mater. Res. **29:** 1337–1348.
16. JAMES, K. & J. KOHN. 1996. New biomaterials for tissue engineering. MRS Bull. **21:** 22–26.
17. KOHN, J. & R. LANGER. 1984. A new approach to the development of bioerodible polymers for controlled release applications employing naturally-occurring amino acids. Polym. Mater. Sci. Eng. **51:** 119–121.
18. PACHENCE, J.M. & J. KOHN. 1997. Biodegradable polymers for tissue engineering. *In* Principles of Tissue Engineering, pp. 273–294. R. G. Landes. Austin.
19. BROCCHINI, S., *et al.* 1998. Structure-property correlations in a combinatorial library of degradable biomaterials. J. Biomed. Mater. Res. **42:** 66–75.
20. ANSETH, K.S., V.R. SHASTRI & R. LANGER. 1999. Photopolymerizable degradable polyanhydrides with osteocompatibility. Nat. Biotechnol. **17:** 156–159.
21. HIRT, T.D., P. NEUENSCHWANDER & U.W. SUTER. 1996. Synthesis of degradable, biocompatible, tough block co-polyesterurethanes. Macromol. Chem. Phys. **197:** 4253–4268.
22. SAAD, B., *et al.* 1997. Development of degradable polyesterurethanes for medical applications: *in vitro* and *in vivo* evaluations. J. Biomed. Mater. Res. **36:** 65–74.
23. TAMADA, J. & R. LANGER. 1992. The development of polyanhydrides for drug delivery applications. J. Biomater. Sci. Polym. Ed. **3:** 315–353.
24. MERKLI, A., *et al.* 1996. Purity and stability assessment of a semi-solid poly(ortho ester) used in drug delivery systems. Biomaterials **17:** 897–902.
25. HAN, D.K. & J.A. HUBBELL. 1996. Lactide-based poly(ethylene glycol) polymer networks for scaffolds in tissue engineering. Macromolecules **29:** 5233–5235.
26. HERN, D.L. & J.A. HUBBELL. 1998. Incorporation of adhesion peptides into nonadhesive hydrogels useful for tissue resurfacing. J. Biomed. Mater. Res. **39:** 266–276.
27. HUBBELL, J.A. 1999. Bioactive biomaterials. Curr. Opin. Biotechnol. **10:** 123–129.

28. ELISSEEFF, J., *et al.* 1999. Transdermal photopolymerization for minimally invasive implantation. Proc. Natl. Acad. Sci. U.S.A. **96:** 3104–3107.
29. PAIGE, K.T., *et al.* 1995. Injectable cartilage. Plast. Reconstr. Surg. **96:** 1390–1398.
30. BURG, K.J.L., *et al.* 1999. Parameters affecting cellular adhesion to polylactide films. J. Biomater. Sci. Polym. Ed. **10:** 147–161.
31. COOPER, M.L., *et al.* 1991. *In vivo* optimization of a living dermal substitute employing cultured human fibroblasts on a biodegradable polyglycolic acid or polygalactin mesh. Biomaterials **12:** 243–248.
32. PARK, A. & L.G. GRIFFITH-CIMA. 1996. *In vitro* cell response to differences in poly-L-lactide crystallinity. J. Biomed. Mater. Res. **31:** 117–130.
33. VAN SLIEDREDGT, A., *et al.* 1992. *In vitro* biocompatibility testing of polylactides: Part 1. Proliferation of different cell types. J. Mater. Sci. Mater. Med. **3:** 365–370.
34. VACANTI, C.A., *et al.* 1992. Tissue engineering of new cartilage in the shape of a human ear using specially configured polymers seeded with chondrocytes. *In* Tissue-Inducing Biomaterials. Vol. 252, pp. 367–374. Materials Research Society. Pittsburgh.
35. FREED, L.E., *et al.* 1997. Tissue engineering of cartilage in space. Proc. Natl. Acad. Sci. U.S.A. **94:** 13885–13890.
36. MA, P.X. & R. LANGER. 1999. Morphology and mechanical function of long-term *in vitro* engineered cartilage. J. Biomed. Mater. Res. **44:** 217–221.
37. PUELACHER, W.C., *et al.* 1994. Tissue-engineered growth of cartilage: the effect of varying the concentration of chondrocytes seeded onto the synthetic polymer matrices. Int. J. Oral Maxillofac. Surg. **23:** 49–53.
38. BARRERA, D.A., *et al.* 1993. Synthesis and RGD peptide modification of a new biodegradable copolymer: poly(lactic acid-co-lysine). JACS **115:** 11010–11015.
39. COOK, A.D., *et al.* 1997. Characterization and development of RGD-peptide-modified poly(lactic acid-co-lysine) as an interactive, resorbable biomaterial. J. Biomed. Mater. Res. **35:** 513–523.
40. EID, K., *et al.* 2001. Effect of RGD-coating on osteocompatibility of PLGA-polymer disks in a rat tibial wound. J. Biomed. Mater. Res. **57:** 224–231.
41. GLASS, J., *et al.* 1994. Cell attachment and motility on materials modified by surface-active RGD-containing peptides. Ann. N.Y. Acad. Sci. **745:** 177.
42. TWEDEN, K.S., *et al.* 1995. Accelerated healing of cardiovascular textiles promoted by an RGD peptide. J. Heart Valve Dis. **4**(suppl. I)**:** S90–S97.
43. IRVINE, D.J., *et al.* 2001. Nanoscale clustering of RGD peptides at surfaces using comb polymers: 2. Surface segregation of comb polymers in polylactide. Biomacro-molecules. In press.
44. INGBER, D.E. & J. FOLKMAN. 1989. Mechanicochemical switching between growth and differentiation during fibroblast growth factor–stimulated angiogenesis *in vitro*: role of extracellular matrix. J. Cell Biol. **109:** 317–330.
45. LAUFFENBURGER, D.A. & A.F. HORWITZ. 1996. Cell migration: a physically integrated molecular process. Cell **84:** 359–369.
46. LUTOLF, M.P., *et al.* 2001. Systematic modulation of Michael-type reactivity of thiols through the use of charged amino acids. Bioconjugate Chem. **12:** 1051–1056.
47. HUBBELL, J.A. 1996. *In situ* material transformations in tissue engineering. MRS Bull. **21:** 33–35.
48. HUBBELL, J.A. & J.L. WEST. 1995. Blood contact is necessary for intimal thickening, medial repopulation, and liberation of medial basic fibroblast growth-factor in the rat—investigation with hydrogel barriers. Thromb. Haemostasis **73:** 1350.
49. WEST, J.L. & J.A. HUBBELL. 1996. Separation of the arterial wall from blood contact using hydrogel barriers reduces intimal thickening after balloon injury in the rat: the roles of medial and luminal factors in arterial healing. Proc. Natl. Acad. Sci. U.S.A. **93:** 13188–13193.
50. HUBBELL, J.A. 1996. Hydrogel systems for barriers and local drug delivery in the control of wound healing. J. Controlled Release **39:** 305–313.
51. SLEPIAN, M.J. & J.A. HUBBELL. 1997. Polymeric endoluminal gel paving: hydrogel systems for local barrier creation and site-specific drug delivery. Adv. Drug Delivery Rev. **24:** 11–30.

52. OBERPENNING, F., *et al.* 1999. *De novo* reconstitution of a functional mammalian urinary bladder by tissue engineering. Nat. Biotechnol. **17:** 149–155.
53. VACANTI, J.P., *et al.* 1988. Selective cell transplantation using bioabsorbable artificial polymers as matrices. J. Pediatr. Surg. **23:** 3–9.
54. HALBERSTADT, C., *et al.* 1992. Physiological cultured skin substitutes for wound healing. *In* Tissue-Inducing Biomaterials. Vol. 252, pp. 323–330. Materials Research Society. Pittsburgh.
55. SODIAN, R., *et al.* 2000. Early *in vivo* experience with tissue-engineered trileaflet heart valves. Circulation **102:** 11122–11129.
56. GRIFFITH, L., *et al.* 1997. *In vitro* organogenesis of liver tissue. Ann. N.Y. Acad. Sci. **831:** 382–397.
57. ZELTINGER, J., *et al.* 2001. Effect of pore size and void fraction on cellular adhesion, proliferation, and matrix deposition. Tissue Eng. **7:** 557–572.
58. KAIHARA, S., *et al.* 2000. Silicon micromachining to tissue engineer branched vascular channels for liver fabrication. Tissue Eng. **6:** 105–117.
59. POWERS, M., *et al.* 2001. Functional behavior of primary rat liver cells in a 3D perfused microarray bioreactor. Tissue Eng. In press.
60. POWERS, M.J., *et al.* 2001. A microarray perfusion bioreactor for 3D liver culture. Biotechnol. Bioeng. In press.
61. DIMILLA, P.A., K. BARBEE & D.A. LAUFFENBURGER. 1991. Mathematical model for the effects of adhesion and mechanics on cell migration speed. Biophys. J. **60:** 15–37.
62. PALECEK, S.P., *et al.* 1997. Integrin ligand-binding properties govern cell migration speed through cell-substratum adhesiveness. Nature **385:** 537–540.
63. KUNTZ, R.M. & W.M. SALTZMAN. 1997. Neutrophil motility in extracellular matrix gels: mesh size and adhesion affect migration speed. Biophys. J. **72:** 1472–1480.
64. SCHENSE, J.C., *et al.* 2000. Enzymatic incorporation of bioactive peptides into fibrin matrices enhances neurite extension. Nat. Biotechnol. **18:** 415–419.
65. SCHENSE, J.C. & J.A. HUBBELL. 2000. Three dimensional migration of neurites is mediated by adhesion site density and affinity. J. Biol. Chem. **275:** 6813–6818.
66. DANILOV, Y.N. & R.L. JULIANO. 1989. (Arg-Gly-Asp)N-albumin conjugates as model substratum for integrin-mediated cell-adhesion. Exp. Cell Res. **182:** 186–196.
67. MAHESHWARI, G., *et al.* 2000. Cell adhesion and motility depend on nanoscale RGD clustering. J. Cell Sci. **113:** 1677–1686.
68. HALL, H., T. BAECHI & J.A. HUBBELL. 2001. Molecular properties of fibrin-based matrices for promotion of angiogenesis *in vitro*. Microvasc. Res. **62:** 315–326.
69. HALLIDAY, N.L. & J.J. TOMASEK. 1995. Mechanical properties of the extracellular matrix influence fibronectin fibril assembly *in vitro*. Exp. Cell Res. **217:** 109–117.
70. LO, C-M., *et al.* 2000. Cell movement is guided by the rigidity of the substrate. Biophys. J. **79:** 144–152.
71. PELHAM, R.J. & Y-L. WANG. 1997. Cell locomotion and focal adhesions are regulated by substrate flexibility. Proc. Natl. Acad. Sci. U.S.A. **94:** 13661–13665.
72. KATZ, B.Z., *et al.* 2000. Physical state of the extracellular matrix regulates the structure and molecular composition of cell-matrix adhesions. Mol. Biol. Cell **11:** 1047–1060.
73. CHOQUET, D., D.P. FELSENFELD & M.P. SHEETZ. 1997. Extracellular matrix rigidity causes strengthening of integrin-cytoskeleton linkages. Cell **88:** 39–48.
74. SWINDLE, C.S., *et al.* 2001. EGF-like repeats of human tenascin-C as ligands for epidermal growth factor receptor (EGFR). J. Cell Biol. **154:** 459–468.
75. MANN, B.K., R.H. SCHMEDLEN & J.L. WEST. 2001. Tethered-TGF-beta increases extracellular matrix production of vascular smooth muscle cells. Biomaterials **22:** 439–444.
76. ZISCH, A.H., *et al.* 2001. Covalently conjugated VEGF-fibrin matrices for endothelialization. J. Controlled Release **72:** in press.
77. MANN, B., *et al.* 2001. Smooth muscle cell growth in photopolymerized hydrogels with cell adhesive and proteolytically degradable domains: synthetic ECM analogs for tissue engineering. Biomaterials **22:** 3045–3051.
78. WEST, J.L. & J.A. HUBBELL. 1999. Polymeric biomaterials with degradation sites for proteases involved in cell migration. Macromolecules **32:** 241–244.
79. POWERS, M.J. & L.G. GRIFFITH. 1998. Adhesion-guided *in vitro* morphogenesis in pure and mixed cell cultures. Micros. Res. Tech. **43:** 379–384.

80. STEINBERG, M.S. 1963. Reconstruction of tissues by dissociated cells. Science **141:** 401–408.
81. STEINBERG, M.S. & M. TAKEICHI. 1994. Experimental specification of cell sorting, tissue spreading, and specific spatial patterning by quantitative differences in cadherin expression. Proc. Natl. Acad. Sci. U.S.A. **91:** 206–209.
82. BELCHEVA, N., S.P. BALDWIN & W.M. SALTZMAN. 1998. Synthesis and characterization of polymer-(multi)-peptide conjugates for control of specific cell aggregation. J. Biomater. Sci. Polym. Ed. **9:** 207–226.
83. MAHONEY, M.J. & W.M. SALTZMAN. 1999. Cultures of cells from fetal rat brain: methods to control composition, morphology, and biochemical activity. Biotechnol. Bioeng. **62:** 461–467.
84. POWERS, M.J., R.E. RODRIGUEZ & L.G. GRIFFITH. 1997. Cell-substratum adhesion strength as a determinant of hepatocyte aggregate morphology. Biotechnol. Bioeng. **53:** 415–426.

Biomaterials and Scaffolds in Reparative Medicine

ELLIOT L. CHAIKOF,[a] HOWARD MATTHEW,[b] JOACHIM KOHN,[c] ANTONIOS G. MIKOS,[d] GLENN D. PRESTWICH,[e] AND CHRISTOPHER M. YIP[f]

[a]*Department of Surgery, Emory University, Atlanta, Georgia, USA*

[b]*Chemical Engineering and Materials Science, Wayne State University, Detroit, Michigan, USA*

[c]*Department of Chemistry, Rutgers University, New Jersey, USA*

[d]*Department of Bioengineering, Rice University, Houston, Texas, USA*

[e]*Department of Chemistry, University of Utah, Salt Lake City, Utah*

[f]*Department of Chemical Engineering and Applied Chemistry, University of Toronto, Toronto, Canada*

ABSTRACT: Most approaches currently pursued or contemplated within the framework of reparative medicine, including cell-based therapies, artificial organs, and engineered living tissues, are dependent on our ability to synthesize or otherwise generate novel materials, fabricate or assemble materials into appropriate 2-D and 3-D forms, and precisely tailor material-related physical and biological properties so as to achieve a desired clinical response. This paper summarizes the scientific and technological opportunities within the fields of biomaterials science and molecular engineering that will likely establish new enabling technologies for cellular and molecular therapies directed at the repair, replacement, or reconstruction of diseased or damaged organs and tissues.

KEYWORDS: bioengineered scaffolds; tissue repair; bioartificial materials

The versatility of many of the approaches currently pursued or contemplated within the framework of this symposium on reparative medicine, including cell-based therapies, artificial organs, and engineered living tissues, are dependent on our ability to: (1) synthesize or otherwise generate novel materials; (2) fabricate or assemble materials into appropriate two- and three-dimensional forms; and (3) precisely tailor material-related physical and biological properties so as to achieve a desired clinical response. In this regard, there is little doubt that the development of innovative materials and scaffolds that are capable of modulating those cellular responses required for tissue repair and regeneration will only be achieved through multidisciplinary, collaborative interactions between investigators in diverse disciplines throughout the

Address for correspondence: Elliot L. Chaikof, M.D., Ph.D., Department of Surgery, Emory University School of Medicine, 1639 Pierce Drive, Room 5105, Atlanta, GA 30322. Voice: 404-727-8413; fax: 404-727-3660.
echaiko@emory.edu

Ann. N.Y. Acad. Sci. 961: 96–105 (2002). © 2002 New York Academy of Sciences.

physical and biological sciences. To further catalyze these interactions, a forum for scientists and engineers working in the fields of biomaterials science and molecular engineering was held at the June 2001 *NIH/BECON Symposium on Reparative Medicine*. The following report summarizes this session with a particular emphasis on both the scientific and technological opportunities, as well as those challenges whose solution would establish new enabling technologies for cellular and molecular therapies directed at the repair, replacement, or reconstruction of diseased or damaged organs and tissues.

NEW METHODOLOGIES FOR MACROMOLECULAR DESIGN AND SYNTHESIS

Molecular biologists, polymer chemists, and bio-organic chemists are designing new materials with controlled properties, including chemical diversity, macromolecular architecture, and biostability through a variety of complementary approaches. The application of *genetic engineering strategies* have already demonstrated that the yield of a *naturally occurring* proteins can be significantly increased over that which can be achieved by its extraction from animal tissues. In the process, human amino acid sequences can be used so as to avoid adverse immunological responses. However, the most profound impact of biosynthetic methodologies with regards to biomaterials development lies in the potential to introduce precise changes in the amino acid sequence and/or to construct new proteins based upon the assembly of *de novo* peptide sequences. A case in point is the generation of structural proteins, referred to as *protein polymers* that consist of sequentially repeated amino acid blocks.[1–3]

Typically, the incorporation of repetitive oligopeptide sequences, derived from a consideration of the primary amino acid structure of a native protein, imparts critical structural properties from the parent protein to a recombinant polypeptide. However, biological, thermodynamic, and mechanical properties of protein polymers may be further modulated by engineering alterations of peptide chain length, consensus repeat sequence, and/or through the introduction of additional functional groups or oligopeptide units. For example, substituting different amino acids for those ordinarily occurring in the sequence can profoundly affect the susceptibility of the protein to proteolytic degradation or facilitate the placement of crosslinks at well-defined intervals along the polypeptide chain. It is also significant that the uniformity of macromolecular structure achieved by recombinant strategies provides exquisite control over macroscopic polymer properties, including material processability. Nevertheless, challenges remain in the design of protein polymers with properties that exceed those associated with native proteins. This will likely require special provisions to overcome the usual restriction to 20 natural amino acids. Examples of efforts in this regard include investigations to expand the genetic code to include non-natural amino acids or the incorporation of non-natural side chains, such as fluorinated, electroactive, or allyl functional groups.[4,5]

In contrast to molecular bioengineering strategies, the power of *polymer chemistry* lies in the potential to generate polysaccharide- and protein-based materials with a high level of chemical diversity. For example, a range of synthetic polymers has been produced with peptide and saccharide groups, as both main chain and pendant group components.[6,7] Functional properties of macromolecules, however, are not only influenced by the diversity of backbone and main chain chemistries, but also by their size, architecture, conformation, and their capacity to form multicomponent assemblies of discrete shape, size, and activity through self-organization. Thus, a significant challenge remains a requirement to develop polymers with a high degree of control over architecture and conformation. In this respect, recent opportunities include the development of "living" and other related "controlled" polymerization techniques that reduce chain transfer and termination reactions, the investigation of strategies to incorporate protein-like secondary structure, such as helical conformation into generated biomimetic polymers, and approaches that promote multicomponent self-assembly.[8]

In between the fields of polymer chemistry and molecular bioengineering lies *bioorganic chemistry*, in which iterative solution or solid-phase methods are applied to generate complex natural products, as well as non-natural protein or polysaccharide mimics. Currently, the synthesis of peptides comprising 75–100 residues can be routinely achieved. However, useful properties generally require protein chain lengths in excess of 100 monomer units. Thus, opportunities exist in the application of ligation methods to generate larger macromolecules from generated fragments. As an example, peptoids (*N*-acylated glycine polymers), a promising class of bioinspired, poorly degradable material with engineered properties, have been developed for intracellular delivery and as collagen-mimetics. In the area of carbohydrate-based materials, recent developments in enzymatic and chemoenzymatic methods, as well as automated solid-phase and one-pot synthetic approaches have facilitated rapid synthesis of complex oligosaccharides.[9,10] Other challenges include the design of new protecting groups, simplified protecting group manipulations, and the rapid assembly of oligosaccharides.

Novel polymer architectures, which are not necessarily biologically inspired, continue to provide opportunities for enhancing bioactivity, while tailoring desired physical properties. For example, polymers with repeating pendant functional groups are ideal for preparing combs and brushes in which biologically active or physiochemically responsive groups decorate a core.[11] The development of a new generation of dendrimers constructed from glycerol and lactate building blocks promises biodegradable polyether-polyester nanoparticles. Likewise, new hyper-branched polymeric micelles and polymerized liposomes offer possibilities in drug delivery that may mesh with synthetic extracellular matrices and tissue-targeted polymers. Finally, the generation and application of *combinatorial approaches* for material discovery, as well as the utilization of *computational modeling*, including the development of quantitative structure–activity relationships (QSAR) have already demonstrated the potential for generating libraries of materials with focused properties.[12]

COMPOSITE MATERIALS VIA SELF-ASSEMBLY AND BIOMIMETIC PROCESSES

Complex Bioorganic Composites

The necessity for unique material property combinations with respect to both biological and physiochemical functionality often demands material solutions that rely on the generation of multicomponent hybrid, composite, and otherwise complex bio-materials. For example, most natural "biomaterials" possess a composite microstructure in the sense that they contain two or more chemically and structurally defined components, which play specific roles and combine to generate the physical and biological properties characteristic of the particular matrix or tissue. Thus, the creative use and combination of natural and synthetic biopolymers, ceramics, and inorganic materials offers a convenient bridge between chemical and biosynthetic approaches, often within the context of thermodynamically driven assembly processes.

Since tissue repair naturally occurs in an extracellular matrix (ECM) environment rich in glycosaminoglycans (GAGs) and glycoproteins it should not be surprising that modifications of the GAGs and matrix proteins continue to provide useful scaffolds and polymers for targeted delivery and wound healing. For example, both collagen and hyaluronan–derived materials are in current use in clinical settings.[13,14] Similarly, chitosan and alginate–based systems hold promise as alternatives to synthetic polylactate and polyglycolate materials. In many instances, however, it remains difficult to achieve all targeted properties from a material composed of a single macromolecular species. Consequently, opportunities exist in the development of materials, which include more than one class of macromolecule, often with the incorporation of a synthetic polymer component.

In the case of hybrid or composite materials, the explicit goal remains synergy between the desirable mechanical properties of one component with the biological compatibility and physiological relevance of the other component. For example, novel materials based on human elastin, spider silk, the adhesive mussel byssus proteins, the muscle protein titin, and collagen have been developed. Importantly, producing hybrids, blends, or interpenetrating networks of protein elastomers with synthetic hydrophilic polymers or with natural polymers offers new strategies for re-engineering mechanically improved replacements for the ECM. Enzymatic methods are also being explored for the incorporation of bioactive peptides into fibrin matrices to improved cell growth and proliferation.[15] Of interest, the growth of sheets of cells on surfaces that contain a temperature-sensitive polymer has led to the production of cellular sheets that can be removed and assembled to produce a layered tissue.[16]

Notably, materials science creates an intimate interface that links drug delivery to reparative medicine. Antigen–antibody hydrogels, temperature- and pH-sensitive gels for drug release, polymer–protein conjugates with temperature-switchable ligand binding or mechanical properties, and cell-adhesive delivery systems are among many examples.[17] Current challenges include the design of tissue-targeted polymers with control over subcellular trafficking that may be necessary under certain conditions for optimized drug responses. It bears emphasis that the pharmacokinetics and pharmacodynamics of macromolecular therapeutics clearly differ from those of the small-molecule drugs, each with its own set of unique problems.

A necessary requirement for tissue repair includes the ability of repopulating cells to adhere to the surface of a scaffold. Furthermore, depending upon the setting, progenitor cells may need to mature into a tissue-specific phenotype, and fully differentiated cells will need to operate with appropriate functional responses. In this regard, the importance of preventing unwanted microbial biofilms, while achieving desired cell-substrate interactions is evident. While this problem remains unsolved, plastic and metal surfaces have been made non-adherent with hyaluronan (HA) and poly(ethylene oxide) (Pluronics), antibacterial with N-alkylated poly(vinylpyridine) coatings, or bioadhesive and bioconductive with RGD peptide attachment. Plasma etching and other chemical modification techniques continue to provide interesting options for activating "inert" surfaces so as to introduce growth factors or other bioactive groups with designed "patchiness" providing some control over the spatial organization of cells in two and three dimensions. Finally, self-assembly of rod-like protein polymers, uses of polymerized liposomes, and production of self-assembled monolayers and other nanostructures on surfaces will also assist in recapitulation of an appropriate environment for tissue repair.

Ceramics and Complex Inorganic Composites

While it may not be possible to completely predict future needs in the area of scaffolds for mineralized tissues, some generally desirable features can be established. Thus, while recent advances in processing have generated protein-hydroxyapatite composites with considerable strength, these constructs still fall far short of the properties of bone.[18] Moreover, there remains a requirement for bone regeneration scaffolds that are either injectable or can be delivered through minimally invasive procedures, conform to the contours of the defect, and effect cure *in situ*. Developing mineralized tissue scaffolds that recapitulate the microstructure of bone and associated biomechanical properties will require a significant improvement in our understanding of the molecular-level interactions controlling biomineralization processes. In particular, further insight into the role of mineral-associated proteins such as osteopontin and bone sialoproteins is needed.[19,20] For example, the time sequence and kinetics of protein involvement in mineral deposition remains incompletely defined. In addition, while collagen alone can induce nucleation of oriented apatite crystals, it is not clear how the intimate organization of mineral and organic phases occurs and accessory proteins or other molecules may provide the key. In addition, the role of post-translational modifications of these proteins (e.g., phosphorylation, sialylation and sulfation) in modulating nucleation, and the significance of properties such as charge density and spatial arrangement have not been elucidated. Finally, the possible involvement of free amino acids or small peptides in nucleation processes is an intriguing area of investigation, given recent observations that such molecules can alter the morphology of growing calcite crystals in a chirality-dependent manner.[21] The potential to apply natural or synthetic small molecules to achieve entirely new biomineral architectures with properties that approach or exceed those of native bone is an important goal of the field.

There is a need to improve understanding of interactions between proteins and mineral phases at the molecular and atomic level. For example, why does ceramic grain size influence the protein-binding selectivity and osteocyte functions?[22,23] What is the role of protein denaturation in this process and does binding of small molecule modulators contribute? Finally, how can we best use this knowledge to design molecular structures for generating composite bioceramics *in vitro* or stimulating controlled mineralization of scaffolds *in vivo*? Our understanding of these phenomena and biomimetic materials design methods in general will benefit from the broader use of computational studies aimed at elucidating molecular structure–property relationships. Beyond the issues of conceptualization and design, the testing and evaluation of composite structures presents many challenges, which are not faced by single-component materials.[24] In addition, tissue responses to composite materials are more complex than the response to individual component materials. As a result, the prediction of their long-term performance and remodeling characteristics is an area of research that presents broad opportunities for further study.

FABRICATION AND PROCESSING FROM THE NANOSCALE TO THE MACROSCALE

Many strategies in regenerative medicine have focused on the use of biodegradable polymers as temporary scaffolds for cell transplantation or tissue induction. The success of a scaffold-based strategy is highly dependent on the properties of the material, requiring at a minimum that it be biocompatible, easy to sterilize, and degradable over an appropriate time scale into products that can be metabolized or excreted. Furthermore, a polymer scaffold should be easily and reproducibly processed into a desired shape and structure, which can be maintained after implantation, thus defining the ultimate shape of the regenerated tissue. Mechanical properties are also of crucial importance in polymer scaffold design for the regeneration of load-bearing tissues such as bone, cartilage, or blood vessels. In this regard, the scaffold material nanostructure and microstructure are both important determinants of these properties. Thus, developing new chemical and physical techniques for inducing well-defined biomimetic and biocompatible 2-D and 3-D structures within synthetic materials remains a significant challenge in the field.

For optimal performance, scaffold degradation rates must be tuned to match the rate of tissue regeneration in order to maintain the mechanical strength of the transplant so as to avoid collapse or stress shielding. Since the desired rate is application-specific, an important goal entails developing "families" of scaffolding materials possessing similar biological properties but varying in their rates of degradation. Ideally, degradable scaffolding polymers should yield soluble, resorbable products that do not induce an adverse inflammatory response.

Since cell morphology correlates closely with function, it is now possible to *control cell shape* and hence the expression of differentiated cell phenotypes by creating specific patterns of surface chemistry and/or texture. For example, rounded cell morphology can be maintained by confining cells with physical or chemical microstructures, and recent investigations have demonstrated that biomaterial surface topography and patterning can be used to control cellular activity. Most of these principles have been demonstrated in planar systems. Extending these principles to three-dimensionally patterned or textured structures represents a major hurdle and will require the extension of fabrication efforts derived from the fields of computer-aided design/computer-aided manufacturing (CAD/CAM), solid free form fabrication processes, and microelectromechanical systems (MEMs).

In addition to insoluble signals mimicking ECM molecules or domains, cells can also respond to soluble bioactive molecules such as cytokines, growth factors, and angiogenic factors. Although these molecules alone can be used for tissue induction, it is possible that combining cell transplantation with drug and/or growth factor delivery may provide an opportunity for accelerating the tissue regeneration process. For example, tissue-inductive factors can be incorporated into biodegradable polymers during scaffold processing.[25,26] Alternatively, biodegradable microparticles or nanoparticles loaded with these molecules can be embedded into the substrates. The release of bioactive molecules *in vivo* is then governed by both diffusion and scaffold degradation. Retention of bioactivity has been a major concern in delivering large molecules such as proteins, and improvements in this area are clearly needed.

Porosity, pore size, and pore structure are important factors since they influence nutrient supply to transplanted cells. To regenerate highly vascularized organs such as liver, porous scaffolds with large void volumes and large surface-area-to-volume ratios are desirable for maximal cell seeding, growth, and vascularization. Small-diameter pores are preferred to yield high surface area per unit volume so long as the pore size is greater than the diameter of a cell in suspension (typically 10–15 μm). However, topological constraints may require larger pores for optimizing cell growth. For example an optimal pore size of 200 to 400 μm was found to be optimal for maximizing bone ingrowth. Scaffold architecture also affects the extent and *dynamics* of tissue induction; for example, the rate of fibrovascular tissue invasion in porous biodegradable polymers depends on pore size. Issues of porosity and tissue colonization have been examined in detail for a number of years, and it is now possible to induce capillary ingrowth to implants.[27] Nonetheless, the ability to generate robust and durable arterial and venous blood supplies needed to sustain an entire engineered organ remains unsolved.

BIORELEVANT STRUCTURE–PROPERTY ANALYSIS

The clinical performance characteristics of artificial organs, engineered living tissues, and cell-based therapies are inherently tied to the structural and functional analysis of material components at molecular- and cellular-length scales. This mandates a consideration of static and dynamic materials analysis in both two and three dimensions, as well as under both model and physiologically relevant systems. In order to effect these measurements, appropriate instrumentation and techniques are needed to accurately measure these properties over a wide range of dimensional-length scales. Ideally, measurements will be performed in real-time, *in situ*, and under physiologically realistic conditions. However, beyond simply identifying arbitrary physical or chemical parameters for correlative analyses, it will be necessary to address several issues of particular importance.

First, are the properties of interest static or dynamic in nature? If the latter, what time scale should be considered, and how do we best reconcile the need to evaluate biomaterial performance characteristics with regard to both acute and chronic tissue responses? The desired tissue and biomaterial performance parameters may dictate forces of a defined magnitude and rate that must be accommodated or transmitted. Such forces may induce specific cellular responses, which may in turn feed back and modulated the force. In turn, there is a need to consider how physical property development occurs during the tissue development process and how the biomechanical microenvironment influences not only the function of differentiated cells, but also the generation of mature cells from progenitor cell populations.

Fundamentally, an argument can be made that all bulk properties are derived from the atomic-level interactions between molecules and that there is a structure–function hierarchy. However, the link between molecular scale, mesoscale, and macro-

TABLE 1. Challenges in biorelevant structure–property analysis

Category	Challenges
Instrumentation	Implementing analytical capability over a broad range of scales (force, length, time) *in vivo*
	Conducting property and response measurements under physiologically relevant conditions
Imaging	Functional mapping capability for mechanical and chemical responses
	Combining functional imaging with physical property evaluation
	Integration of characterization tools for tracking multiple cellular responses with time
Cell/tissue properties	Evaluating functional characteristics at the cellular level
	Use of internal markers/probes to evaluate system mechanical and chemical responses
The uncertainty principle	Reconciling the fact that measurement techniques inherently perturb the system

scopic properties is not always obvious. Furthermore, there is abundant evidence that altering the size scale of a system's components can produce vastly different performance characteristics. With this in mind, investigations are required to determine how the knowledge of molecular-scale measurements of cell adhesion or protein–protein interactions can be optimally harnessed in the design of new biomaterials. For example, is it always mandatory to measure properties at the molecular scale or, in the design of "larger" structures, is it more appropriate to map these interactions at the "in-use" length scale?

While not intended to be all-inclusive, several key challenges in the area of structure–property analysis are summarized in TABLE 1. The development of new instruments, as well as the integration of existing tools, will be essential for establishing new analytical paradigms. In particular, the ability to simultaneously correlate temporal variations of several functional characteristics within specific cellular or subcellular locations and to conduct these analyses under conditions relevant to the *in vivo* situation, will be a task of particular importance. As an example, imaging tools are required for quantifying cell function and cell–cell adhesion forces as a function of applied mechanical load (tensile, flexural, shear). The utilization of systems integrating total internal reflection fluorescence (TIRF) or surface plasmon resonance (SPR) microscopy with spectroscopy and local tensile loading may provide opportunities for defining cross-talk between intracellular processes. Moreover, the of use internal markers or probes to assist in the noninvasive evaluation of local cellular mechanical and/or chemical responses to implanted biomaterials will be an important component of these efforts.

SUMMARY

There is little doubt that the development, fabrication, and analysis of novel biomaterials and scaffolds will constitute a centerpiece of the investigational efforts that will define the field of reparative medicine. This brief review has sought to identify a number of challenges that lie along the path ahead. It is anticipated that solutions will be derived from principles and methodologies that either currently exist or will evolve at the interface between the biological and physical sciences. While some of these investigations might be considered of a more "basic" nature and other efforts fundamentally "applied," all of these endeavors remain motivated by the unique needs of patients that are defined within a specific clinical context.

REFERENCES

1. MCMILLAN, R.A., T.A.T. LEE & V.P. CONTICELLO. 1999. Rapid assembly of synthetic genes encoding protein polymers. Macromolecules **32:** 3643–3648.
2. FERRARI, F.A. & J. CAPPELLO. Biosynthesis of protein polymers. *In* Protein-Based Materials. K. McGrath & D. Kaplan, Eds. :36–60. 1997. Birkhäuser. Boston.
3. PANITCH, A., T. YAMAOKA, M.J. FOURNIER, *et al.* 1999. Design and biosynthesis of elastin-like artificial extracellular matrix proteins containing periodically spaced fibronectin CS5 domains. Macromolecules **32:** 1701–1703.
4. VAN HEST, J.C.M. & D.A. TIRRELL. 1998. Efficient introduction of alkene functionality into proteins in vivo. FEBS Lett. **1428:** 68–70.
5. LIU, D.R. & P.G. SCHULTZ. 1999. Progress towards the evolution of an organism with an expanded genetic code. Proc. Natl. Acad. Sci. USA **96:** 4780–4785.
6. GRANDE, D., S. BASKARAN, C. BASKARAN & E.L. CHAIKOF. 2000. Glycosaminoglycan-mimetic biomaterials. 1. Non-sulfated and sulfated glycopolymers by cyanoxyl-mediated free-radical polymerization. Macromolecules **33:** 1123–1125.
7. DEMING, T.J. 1997. Facile synthesis of block copolypeptides of defined architecture. Nature **390:** 386–389.
8. KIRSHENBAUM, K., R. ZIMMERMAN & K. DILL. 1999. Designing polymers that mimic biomolecules. Curr. Opin. Struct. Biol. **9:** 530–535.
9. SEARS, P. & C.H. WONG. 2001. Toward automated synthesis of oligosaccharides and glycoproteins. Science **291:** 2344–2350.
10. PLANTE, O.J., E.R. PALMACCI & P.H. SEEBERGER. 2001. Automated solid-phase synthesis of oligosaccharides. Science **291:** 523–527.
11. BANERJEE, P., D.J. IRVINE, A.M. MAYES & L.G. GRIFFITH. 2000. Polymer latexes for cell-resistant and cell-interactive surfaces. J. Biomed. Mat. Res. **50:** 331–339.
12. BROCCHINI, S., K. JAMES, V. TANGPASUTHADOL & J. KOHN. 1997. A combinatorial approach for polymer design. J. Am. Chem. Soc. **119:** 4553–4554.
13. LUO, Y., K.R. KIRKER & G.D. PRESTWICH. 2000. Cross-linked hyaluronic acid hydrogel films: new biomaterials for drug delivery. J. Control. Rel. **69:** 169–184.
14. SILVER, F.H. & A.T. GARG. 1997. Collagen: characterization, processing, and medical applications. *In* Handbook of Biodegradable Polymers. :319–346. Harwood. Amsterdam.
15. SAKIYAMA, S.E., J.C. SCHENSE & J.A. HUBBELL. 1999. Incorporation of heparin-binding peptides into fibrin gels enhances neurite extension: an example of designer matrices in tissue engineering. FASEB J. **13:** 2214–2224.
16. KWON, O.H., A. KIKUCHI, M. YAMATO, *et al.* 2000. Rapid cell sheet detachment from poly(N-isopropylacrylamide)-grafted porous cell culture membranes. J. Biomed. Mat. Res. **50:** 82–89.
17. WANG, C., R.J. STEWART & J. KOPECEK. 1999. Hybrid hydrogels assembled from synthetic polymers and coiled-coil protein domains. Nature **397:** 417–420.
18. KIKUCHI, M., S. ITOH & S. ICHINOSE. 2001. Self-organization mechanism in a bone-like hydroxyapatite/collagen nanocomposite synthesized *in vitro* and its biological reaction *in vivo*. Biomaterials **22:** 1705–1711.

19. GIACHELLI, C.M. & S. STEITZ. 2000. Osteopontin: a versatile regulator of inflammation and biomineralization. Matrix Biol. **19:** 615–622.
20. GORSKI, J.P. 1992. Acidic phosphoproteins from bone matrix: a structural rationalization of their role in biomineralization. Calcif. Tissue Int. **50:** 391–396.
21. ORME, C.A., A. NOY, A. WIERZBICKI, *et al.* 2001. Formation of chiral morphologies through selective binding of amino acids to calcite surface steps. Nature **411:** 775–779.
22. WEBSTER, T.J., C. ERGUN, R.H. DOREMUS, *et al.* 2000. Enhanced functions of osteoblasts on nanophase ceramics. Biomaterials **21:** 1803–1810.
23. WEBSTER, T.J., C. ERGUN, R.H. DOREMUS, *et al.* 2001. Enhanced osteoclast-like cell functions on nanophase ceramics. Biomaterials **22:** 1327–1333.
24. EVANS, S.L. & P.J. GREGSON. 1998. Composite technology in load-bearing orthopaedic implants. Biomaterials **19:** 1329–1342.
25. BABENSEE, J.E., L.V. MCINTIRE & A.G. MIKOS. 2000. Growth factor delivery for tissue engineering. Pharm. Res. **17:** 497–504.
26. BANCROFT, G.N. & A.G. MIKOS. 2001. Bone tissue engineering by cell transplantation. *In* Tissue Engineering for Therapeutic Use. Y. Ikada & N. Ohshima, Eds.: 151–163. Elsevier. New York.
27. MOONEY, D.J. & A.G. MIKOS. 1999. Growing new organs. Scientific American **280:** 60–65.

Hybrid, Composite, and Complex Biomaterials

GLENN D. PRESTWICH[a] AND HOWARD MATTHEW[b]

[a]Department of Medicinal Chemistry, University of Utah,
Salt Lake City, Utah 84112-5820, USA

[b]Departments of Chemical Engineering and Materials Science, Wayne State University,
Detroit, Michigan 48202, USA

KEYWORDS: biomaterials; polymers; tissue engineering

This presentation is concerned with chemical modifications of natural materials and their uses in composites and for surface modification. The ability to re-engineer naturally occurring polymers and to create biologically inspired polymers is redefining and diversifying the kinds of biocompatible materials that can be developed for reparative medicine and other medical applications. Ultimately, these materials must possess physicochemical, mechanical, and biological properties that facilitate the recapitulation of embryonic cell differentiation, growth, and proliferation in the adult organism. Moreover, surface modification and surface self-assembly methodologies can now improve the compatibility of implanted nonbiological medical devices. The discussion is focused primarily on the new materials themselves—what they are, what their purpose is, and what key research areas remain.

MODIFIED NATURAL POLYMERS

Tissue repair naturally occurs in an extracellular matrix (ECM) environment rich in glycosaminoglycans (GAGs), particularly hyaluronan (HA) and ECM proteins such as collagen, fibronectin, and elastin. Ongoing work involves preparing new materials by chemical or genetic modifications of polypeptides, oligonucleotides, oligosaccharides, GAGs, and combinations thereof.

Bacterial expression of novel proteins with unnatural amino acids will provide new polypeptide building blocks with polymerizable and cross-linkable groups, and artificial proteins can be the basis for creating ordered nanostructures. Production of intractable polymers such as spider silk can be accomplished in plants, an important clue for other problematic polypeptides of interest. Oligonucleotides have been used to make knots, boxes, and repeating structures; the value of such constructs in repar-

Address for correspondence: Glenn D. Prestwich, Department of Medicinal Chemistry, University of Utah, 419 Wakara Way, Suite 205, Salt Lake City, UT 84108. Voice: 801-585-9051; fax: 801-585-9053.
gprestwich@pharm.utah.edu

Ann. N.Y. Acad. Sci. 961: 106–108 (2002). © 2002 New York Academy of Sciences.

ative medicine remains unexplored. Modifications of the GAGs have provided many useful scaffolds and polymers for targeted delivery and wound healing; however, improvements in mechanical properties are urgently needed. Reconstructive surgery and engineered tissues already employ HA–derived materials. Similarly, chitosan- and alginate-based systems can provide alternatives to synthetic polylactate and polyglycolate materials. Cellular engineering by modification of surface oligosaccharides turns the tables, that is, modifying cells for the matrix rather than the matrix for the cells. Thus, while the modification of single natural polymers has created new horizons for biomaterials, hybrid materials which better mimic Nature represent the important direction for the future.

BIOINSPIRED AND BIODERIVED POLYMERS

Many new materials are hybrids, incorporating more than one class of macromolecule, and often including a synthetic polymer component. In these cases, the goal is to achieve synergy between the desirable mechanical properties of one component with the biological compatibility and physiological relevance of the other component. These bioinspired and bioderived materials could be classified into a variety of categories, including brush, comb, and dendrimeric polymers, peptidomimetic systems, tissue-targeted polymers, smart (stimulus-responsive) polymers, bioconductive tissue-engineering scaffolds, enzymatically processed hybrid materials, and elastic materials.

Polymers with repeating pendant functional groups are ideal for preparing combs and brushes, in which biological activity or physicochemically responsive groups decorate a core. This classical approach to graft co-polymers will now incorporate many new biological building blocks and produce ordered, patterned structures such as those possible with polystyrene brushes on silicate surfaces. A new generation of biodegradable polyether-polyester dendrimers constructed from glycerol and lactate building blocks promises nanoparticles in place of traditionally processed PLLG materials. New hyperbranched polymeric micelles and polymerized liposomes offer possibilities in drug delivery that may mesh with synthetic ECMs and tissue-targeted polymers. Peptoids (N-acylated glycine polymers) have been developed for intracellular delivery and as collagen mimetics, the beginning of a promising class of bioinspired, poorly degradable material with engineered properties.

Tissue-targeted polymers should have a specific disease focus and employ cell-uptake and subcellular trafficking control for effective drug delivery. This interface intimately links reparative biomaterials with drug delivery. The pharmacokinetics and pharmacodynamics of macromolecular therapeutics will differ from those of the small molecule drugs, and will require support along with the development of the new biomaterials themselves. In general, reparative medicine needs a strong commitment to advances in drug formulation and delivery.

Stimulus-responsive ("smart") polymers have proliferated in the laboratory with rising expectations for practical utility. Antigen–antibody hydrogels, temperature- and pH-sensitive gels for drug release, smart polymer–protein conjugates with temperature-switchable ligand binding or mechanical properties, and cell-adhesive delivery systems are among many examples. An important direction for this field is the realization of clinical applications that are unique to the smart polymer hybrids now in laboratory development. Beyond drug delivery, can smart polymer hybrids be developed to provide control of cell adhesion, staged cell differentiation, and biomechanical properties *in vivo*? An elastin-like polypeptide has been targeted to tumors with hyperthermia. Enzymatic methods may be used to incorporate bioactive peptides into fibrin matrices to improved cell growth and proliferation.

Exciting and novel elastic materials based on human elastin, spider silk, the adhesive mussel byssus proteins, the muscle protein titin, and collagen have been developed and in some cases commercialized. One current direction is production of elastomeric polypeptide repeats with cross-linkable groups for incorporation into hybrid materials. Design, structural, and functional studies must all be supported for this promising field to flourish. Importantly, producing hybrids of protein elastomers with synthetic hydrophilic polymers, for example, poly(HPMA), or with natural polymers such as GAGs, offers a novel strategy for re-engineering a mechanically improved ECM replacement.

SURFACE MODIFICATION

Most cells and tissues must adhere to surfaces to achieve biological function. In this regard, we emphasize the importance of studying hybrid and complex biomaterials at solid interfaces, and the prevention of unwanted microbial biofilms on implanted devices. Plastic and metal surfaces may be made non-adherent with HA and Pluronics, antibacterial with new *N*-alkylated poly(vinylpyridine) coatings, or bioadhesive and bioconductive with RGD peptide attachment. The patchiness of adhesive peptides is receiving fresh new attention for tissue engineering in two- and three-dimensional scaffolds. Plasma etching and other chemical modification techniques activate "inert" surfaces for the introduction of growth-compatible materials. A novel tissue-engineering strategy involves growth of sheets of cells on a surface with a temperature-sensitive polymer; sheets can be removed and assembled to produce a layered tissue. Self-assembly of rod-like protein polymers, uses of polymerized liposomes, and production of self-assembled monolayers and other nanostructures on surfaces will assist in recapitulating biological tissues.

SELECTED READING

1. CARNAHAN, M.A. & M.W. GRINSTAFF. 2001. Synthesis and characterization of polyether-ester dendrimers composed of glycerol and lactic acid. J. Am. Chem. Soc. **123:** 2905–2906.
2. LUO, Y. & G.D. PRESTWICH. 2001. Novel biomaterials for drug delivery. Exp. Opin. Ther. Pat. **11:** 1395–1410.
3. SUH, J.K.F. & H.W.T. MATTHEW. 2000. Application of chitosan-based polysaccharide biomaterials in cartilage tissue engineering: a review. Biomaterials **21:** 2589–2598.

Biomaterials in Reparative Medicine

Biorelevant Structure–Property Analysis

CHRISTOPHER YIP

Department of Chemical Engineering and Applied Chemistry, Institute of Biomaterials and Biomedical Engineering, and Department of Biochemistry, University of Toronto, Toronto, Ontario M5S 3G9 Canada

KEYWORDS: biomaterials; tissue repair

The clinical performance characteristics of artificial organs, engineered living tissues, and cell-based therapies are inherently tied to the structural and functional analysis of material components at molecular, cellular, and tissue-length scales. This mandates a consideration of static and dynamic materials analysis in both two- and three-dimensions, as well as under both model and physiologically relevant systems.

If we are to develop innovative biomaterials for use in reparative medicine, we must identify the key structure–property parameters that describe the *in situ* physical situation. In order to effect these measurements, we need to develop the appropriate instrumentation and techniques that can measure these properties or attributes over a wide range of dimensional length scales. Ideally these instruments will adopt an integrated strategy such that the measurements are performed in real-time, *in situ*, and under nominally physiologically realistic conditions.

Designing new biomaterials requires careful consideration of the necessary biorelevant structure–function–property relationships. However, more than simply identifying the physical or chemical parameters—strength, frictional characteristics, chemical inertness / reactivity—we may need to also consider the following:

- Are the properties of interest static (i.e., time-invariant)? Or are they dynamic in nature? And, if so, what time scale are we interested in? How do we reconcile the need to measure the *long-term* durability of the biomaterial (retention of bulk properties) and the *short-term* response?

- What are the forces—magnitude and rate—that need to be accommodated or transmitted?

- Do the forces induce a specific cellular response and are these forces then summarily modulated by this response?

Address for correspondence: Christopher Yip, University of Toronto, 407 Rosebrugh Building, 4 Taddle Creek Road, Toronto, Ontario M5S 3G9 Canada. Voice: 416-976-7853; fax: 416-978-4317.
christopher.yip@utoronto.ca

Ann. N.Y. Acad. Sci. 961: 109–111 (2002). © 2002 New York Academy of Sciences.

- If one were constructing a tissue-engineering construct or scaffold for use as a support for a cell, what mechanical characteristics are needed to emulate a *cellular* support matrix. Do the cells expect a specific type of *dynamic* response from their supporting substrate?

- Should we consider how physical property development occurs during the developmental cycle of a cell? One might argue that requirements for a successful biomaterial are drawn from mature tissue models. What about the situation where the biomaterial is serving as a scaffold for a stem cell–based construct? What are the necessary properties required for such a situation?

- What can we learn from molecular-scale measurements of cell adhesion or protein–protein interactions in our design of new biomaterials?[1]

(1) Fundamentally, an argument can be made that all bulk properties are derived from the atomic-level interactions within a given molecule and that there is a structure–function hierarchy.

(1a) What is the link between molecular scale $\leftrightarrow$ mesoscale $\leftrightarrow$ macroscopic properties?

(1b) Is this a valid argument to apply for biomaterials design in general? Do we need to really measure properties at the molecular scale or, if we are designing "larger" structures, is it more appropriate to map these interactions at the "in-use" length scale?

(2) Single molecule measurements may track different relaxation or stiffening pathways. Is this a key consideration?

- How about fatigue? What is the expected duty cycle of a specific structural element on the body? Do such properties negatively impact the end user in that the design of materials for longevity may be counterproductive if it accelerates the rate of wear or negatively impacts other areas?

Some key challenges in the development of approaches for characterizing the appropriate necessary structure–property relationships are shown in the following:

- Do instruments exist that can perform the necessary *in vivo* measurements across the domains of force, length, and time?

- Do we have instruments that can map both functional mechanical and chemical responses across these same time scales for our biomaterials? Can these tests be performed in a physiologically relevant environment?

- There is a need to couple functional imaging with physical property determination.

(3) As an example—can we develop imaging tools for looking at cell function/cell–cell adhesion forces as a function of applied (tensile/flexural/shear) load?

(3a) total internal reflectance fluorescence/surface plasmon resonance microscopy/spectroscopy coupled with local loading (point source);

(3b) confocal microscopy coupled with surface characterization tools;

(3c) localized three-dimensional (spatial) spectroscopy for looking at force transmission.

- Systems integration is a logical step forward. Many characterization tools tend to operate in isolation. Should we now be looking at coupled or integrated approaches? Are such tools amenable for tracking cellular responses with time? Can we implement these as new instrument design parameters?

- Examine from the other perspective: how can we accurately characterize the structure–function characteristics of living tissues and structures *at the cellular level* so that we can best mimic those with engineered biomaterials? How can we use *in situ* internal markers to measure the mechanical/chemical responses of these biomaterials?

- How do we reconcile that our measurement techniques perturb the system? Is this a key concern? And if so, can we rely on inherent markers in our studies?

REFERENCE

1. YIP, C.M. 2001. Atomic force microscopy of macromolecular interactions. Curr. Opin. Struct. Biol. **11:** 567–572.

.

Bioscaffolds for Tissue Repair

Breakout Session Summary

Moderators

ELLIOT CHAIKOF, *Emory University*

HOWARD MATTHEW, *Wayne State University*

Panelists

JOACHIM KOHN, *Rutgers University*

GLENN D. PRESTWICH, *University of Utah*

ANTONIO MIKOS, *Rice University*

CHRISTOPHER YIP, *University of Toronto*

Rapporteur

ELLIOT CHAIKOF, *Emory University*

BROAD STATEMENT

The versatility of many of the approaches currently pursued or contemplated within the framework of the field of reparative medicine is dependent on our ability to: (1) synthesize or otherwise generate novel materials; (2) fabricate or assemble materials into appropriate two- and three-dimensional forms; and (3) precisely tailor material-related physical and biological properties so as to achieve a desired clinical response.

VISION

The development of *innovative materials and scaffolds* that are capable of modulating those cellular responses required for tissue repair and regeneration will be achieved through multidisciplinary, collaborative interactions between investigators working in the physical and biological sciences.

OBJECTIVES

To facilitate this vision, future research should address four different areas. The first is methodologies for macromolecular design and synthesis. There is the need to

Address for correspondence: Elliot L. Chaikof, M.D., Ph.D., Department of Surgery, Emory University School of Medicine, 639 Pierce Drive, Room 5105, Atlanta, GA 30322.

Ann. N.Y. Acad. Sci. 961: 112–113 (2002). © 2002 New York Academy of Sciences.

(1) design recombinant matrix proteins with properties that exceed those associated with native proteins; (2) develop strategies for the synthesis of polymers with controlled molecular weight, architecture, and conformation; and (3) optimize solution and solid-phase methods to generate complex natural products, as well as non-natural biomacromolecule mimics. The second critical area is the development of hybrid, composite, and complex biomaterial/scaffolds. There is the need to (1) extend techniques for the utilization of naturally occurring biopolymers either alone or in combination with other natural or synthetic materials and (2) to characterize the effect of material degradation and related byproducts on reparative processes. The third critical area is the fabrication and processing from the nanoscale to the macroscale. There is the need to (1) optimize fabrication methods that operate on nano-, micro- and macroscale levels via two- and three-dimensional assembly processes and (2) to elucidate the effects of processing conditions on material properties and function. The fourth critical area is biorelevant structure–property analysis. There is the need (1) to develop appropriate instrumentation and techniques that can measure physical or biological properties or other material attributes over a wide range of dimensional length scales and (2) to design integrated strategies that evaluate material properties in real-time, in tissue, and under physiologically relevant conditions.

CHALLENGES

Rational approaches for material design and fabrication will require collaborative interactions between biologists, engineers, and physical scientists. Elucidation of structure/function relationships of tissues and natural matrices is required to serve as a starting point and reference base for material design. Novel synthetic approaches will be needed for enhancement of the materials design process. Identification of promising biomaterial candidates will require the development of analytical tools and physiologically relevant testing methods.

ACKNOWLEDGMENTS

The Biomaterials and Bioscaffolds for Tissue Repair Breakout Session was coordinated by Dr. Dharam Dhindsa, CSR, NIH and Dr. Teresa Nesbitt, CSR, NIH.

Manufacturing and Characterization of 3-D Hydroxyapatite Bone Tissue Engineering Scaffolds

T.-M.G. CHU,[a,b] S.J. HOLLISTER,[a,d] J.W. HALLORAN,[c] S.E. FEINBERG,[b,d] AND D.G. ORTON[e]

[a]Department of Biomedical Engineering, [b]School of Dentistry, [c]Department of Materials Science and Engineering, and [d]Department of Surgery, University of Michigan, Ann Arbor, Michigan 48109, USA

[e]Terumo Cardiovascular Systems, Ann Arbor, Michigan, USA

ABSTRACT: Internal architecture has a direct impact on the mechanical and biological behaviors of porous hydroxyapatite (HA) implants. However, traditional processing methods provide very minimal control in this regard. This paper reviews a novel processing technique developed in our laboratory for fabricating scaffolds with controlled internal architectures. The preliminary mechanical property and *in vivo* evaluation of these scaffolds are also presented.

KEYWORDS: bioscaffolds; hydroxyapetite; bone tissue engineering

INTRODUCTION

Bone replacement has been an important subject in the the field of reparative medicine. One of the most-studied materials in bone replacement is porous hydroxyapatite (HA). The internal architectures, including pore size, pore shape, and pore connection pattern, play several important roles in the performance of the porous HA scaffolds. They control the degree of bone regeneration,[1] influence the path of bone regeneration,[2] and determine the mechanical properties of the scaffolds.[3] However, traditional processing methods, mainly hydrothermal exchange technique and organic particle embedding technique, provide very minimal control over the internal architecture of the scaffolds. To address this issue, we have developed a technique to implement designed internal architectures inside the HA scaffolds.[4] In this paper, we will briefly summarize the manufacturing and characterization of our 3-D HA bone tissue engineering scaffolds.

Address for correspondence: Dr. T.-M.G. Chu, Department of Biomedical Engineering, University of Michigan, Ann Arbor, MI 48109. Voice: 734-764-3377; fax: 734-763-4788.
ctmin@engin.umich.edu

Ann. N.Y. Acad. Sci. 961: 114–117 (2002). © 2002 New York Academy of Sciences.

MANUFACTURING

The manufacturing technique we have developed is a lost mold technique combining the use of (*a*) scaffold designs generated from computer-aided design (CAD) software or other imaging techniques, (*b*) thermal-curable HA/acrylate suspensions, and (*c*) epoxy molds made by stereolithography (SL). The application of SL in making epoxy models from computer-generated images have been depicted in detail in the literature[5] and will not be discussed here. To manufacture a 3-D HA scaffold the structure is first designed using CAD software or computer tomography (CT) data.[6] The negative image of the design is then used to build an epoxy mold on the stereolithography apparatus (SLA). A thermal-curable HA/acrylate suspension is subsequently cast into the epoxy mold and cured at 85°C. The cured part is placed in a furnace at high temperature to simultaneously burn out the mold and the acrylate binder. Left from the burn-out process is the HA green body in the designed 3-D structure. The HA green body is then fully sintered at 1350°C into a 3-D HA scaffold.

MECHANICAL PROPERTY EVALUATION

We first evaluated the mechanical property of our scaffolds. A cube with straight channels penetrating the cube orthogonally in X, Y, and Z directions was designed to mimic a simple three-dimensional interconnected structure. The cube was 7.4 mm in its width, length, and height. HA specimens were fabricated as described previously. The averaged channel size and total porosity of these specimens are listed in TABLE 1. The compressive strength of the specimens was investigated with an Instron machine (Instron 8521, Instron Corp. Canton, MA.). A total of twelve specimens were tested. The averaged compressive strength of the specimens was 30 ± 8 MPa, comparable to that of the coralline HA of 25–35 MPa.[7] The averaged compressive modulus of the specimens from the stress–strain curve was 1.4 ± 0.4 GPa.

IN VIVO EVALUATION

The *in vivo* performance of these 3-D HA scaffolds was also evaluated in miniature pigs. Two scaffold architecture designs were used in the *in vivo* study. The external geometry of both designs was identical: a cylinder with a diameter of 8 mm and a height of 6 mm. The orthogonal design contained channels of 450 μm penetrating in all X, Y, and Z directions. The radial design contained a central column of about 3 mm in diameter running in Z direction with channels of 380 μm extending from the center toward outside surface in radial directions. The channel size, wall size, and the final total porosity of these *in vivo* specimens are listed in TABLE 1.

Four Yucatan miniature pigs (6 months old) were used. Four defect sites were created in each hemi-mandible in the retromandibular area. Three defects on each side were filled with HA scaffolds, with the fourth site being left as either an empty control defect, or filled immediately with the cored bone to serve as a positive control. Two animals were sacrificed at 5 weeks and two at 9 weeks. The resected pieces were then embedded, sectioned, and stained with toludine blue. Quantita-

TABLE 1. The channel size, wall size, and total porosity of the mechanical and *in vivo* specimens

	Mechanical specimens	Orthogonal design	Radial design
Channel size (μm) 334 (width) 469 (height)		444	366
Wall size (μm)	500	748	
Porosity (%)	39	44	38

TABLE 2. Amount of bone regeneration divided by penetration zone and central zone

	Orthogonal design	Radial design
Penetration zone	55 ± 14%	41 ± 23%
Central zone	2 ± 2%	2 ± 3%

tive histomorphometry on the stained sections were performed using a square grid on a microscope.

In the retrieved specimens, normal regenerated bone tissue was found at 5 weeks and 9 weeks in both designs. Macroscopically, the geometry of the regenerated tissue was significantly influenced by the scaffold design. The regenerated bone tissue formed a single large piece at the center in the radial design, while the regenerated bone in the orthogonal design constituted an interpenetrating matrix with the HA scaffold. At 5 weeks, the average bone ingrowth was 19±9% in the orthogonal design and 14±3% in the radial design. At 9 weeks, the average bone ingrowth increased to 45±21% in the orthogonal design and 23±28% in the radial design. A further study[8] revealed that the distribution of the regenerated bone was not homogeneous among the tissue sections. A penetration zone of approximately 1.4 mm was found underneath the surface of both orthogonal and radial design scaffolds. The amount of bone regeneration corrected for the zone difference showed that the amount of bone regeneration was significantly different ($P < 0.05$) between the two zones (TABLE 2). Average bone regeneration in the orthogonal design was higher than the radial design, although not statistically significant, probably due to the relatively small number of animals (4) used in this pilot study. Larger-scale study is currently under way to investigate the effect of architectural design on the amount of bone regeneration.

CONCLUSIONS

HA scaffolds demonstrated biocompatibility and osteoconductivity in the *in vivo* animal model. The result indicates that the manufacturing process did not impose adverse effects on the biological property of HA.

The size and shape of the regenerated bone pieces were significantly different in orthogonal and radial designs. The result shows that it is possible to control the macroscopic morphology of the regenerated bone tissue through scaffold architecture design.

A 1.4-mm penetration zone was found underneath the surface in both orthogonal and radial design scaffolds at 9 weeks. Amount of bone regeneration between the penetration zone and the central zone was significantly different in both architecture designs. The result suggests that spatial distribution should be considered in evaluating the amount of bone regeneration inside the scaffolds.

No significant difference was found in the amount of bone regeneration between the two architectural designs, probably due to the small sample size.

ACKNOWLEDGMENT

This work was supported by the Rackham Faculty Grant from the Rackham School of Graduate Study, University of Michigan, and by NIH Grant # RO1 DE 13416-01A1.

REFERENCES

1. RIPAMONTI, U. *et al.* 1992. The critical role of geometry of porous hydroxyapetite delivery system of bone by osteogenin, a bone morphogenic protein. Matrix **12:** 202–212.
2. CHANG, B.-S. *et al.* 2000. Osteoconduction at porous hydroxyapetite with various pore configurations. Biomaterials **21:** 1291–1298.
3. LIU, D. 1996. Control of pore geometry on influencing the mechanical property of porous hydroxyapetite bioceramic. J. Mater. Sci. Lett. **15:** 419–421.
4. CHU, T.-M.G. *et al.* 2001. Hydroxyapatite implants with designed internal architecture. J. Mater. Sci.: Mater. Med. **12:** 471–478.
5. BINDER, T. *et al.* 2000. Stereolithographic biomodeling to create tangible hard copies of cardiac structures from echocardiographic data: *in vitro* and *in vivo* validation. J. Am. Coll. Cardiol. **35:** 230–237.
6. HOLLISTER, S.J. *et al.* 1998. Image based design and manufacture of scaffolds for bone reconstruction. *In* IUTAM Synthesis in Biosolid Mechanics. P. Pedersen & M. Bendsoe, Eds.: 163–174. Kluwer. Amsterdam.
7. SHORS, E. & R. HOLMES. 1993. Porous hydroxyapatite. *In* An Introduction to Bioceramics. L. Hench & J. Wilson, Eds.: 181–198. World Scientific. Singapore.
8. CHU, T.-M.G. *et al.* 2002. Mechanical and *in vivo* performance of hydroxyapatite implants with controlled architectures. Biomaterials **23:** 1283–1293.

Biological Response of Chondrocytes to Hydrogels

J. H. ELISSEEFF,[a] A. LEE,[b] H. K. KLEINMAN,[b] AND Y. YAMADA[b]

[a]Whitaker Institute of Biomedical Engineering, Department of Biomedical Engineering, Johns Hopkins University, Baltimore, Maryland 21218, USA

[b]Craniofacial Development and Regeneration Branch, National Institute of Dental and Craniofacial Research, Bethesda, Maryland 20892, USA

ABSTRACT: Primary bovine chondrocytes were encapsulated in alginate and alginate combined with cartilage matrix extract, Cartrigel, for the purpose of cartilage tissue engineering. The cell constructs were incubated *in vitro* and gene expression of cartilage-specific extracellular matrix molecules was quantitated and localized with *in situ* hybridization with a decrease in expression observed in the alginate–Cartrigel constructs. Further understanding of cell response to scaffolds will allow rational design and development of hydrogels for cartilage tissue engineering.

KEYWORDS: cartilage replacement; hydrogels; chondrocytes; bioscaffolds

INTRODUCTION

Cartilage tissue lost by trauma, disease, or congenital abnormalities does not have the ability to regenerate or heal.[1] More than one million surgical operations are performed in the United States each year to treat knee, hip, and shoulder dysfunction related to cartilage loss. Tissue engineering has emerged as a potential method to replace cartilage.[2] One general strategy of tissue engineering is to seed cells on a biomaterial scaffold. The cell–scaffold construct is incubated and the resulting tissue composition and functionality are analyzed. Previous research has shown the utility of hydrogels as a scaffold material for cell encapsulation and tissue engineering.[3,4] Understanding fundamental mechanisms of tissue engineering, including how cells respond to a scaffold, is necessary for intelligent design of scaffold materials. The purpose of this research was to understand chondrocyte response and tissue development in hydrogels, and specifically to quantitate and localize gene expression of cartilage-specific markers.

Address for correspondence: Jennifer Elisseeff, Ph.D., 3400 N. Charles Street, Clark Hall, Department of Biomedical Engineering, Johns Hopkins University, Baltimore, MD 21218. Voice: 410-516-4915; fax: 410-516-8152.
jhe@bme.jhu.edu

Ann. N.Y. Acad. Sci. 961: 118–122 (2002). © 2002 New York Academy of Sciences.

EXPERIMENTAL METHODS

Hydrogel Formation

Chondrocytes were isolated from the femoropatellar groove of calves (6–8 weeks old, Research 87, Marlboro, MA) as previously described.[5] Polymer solutions were composed of alginate alone (1.2% alginate, Kelco, Chicago, IL, in 0.15 M NaCl, 20 mM HEPES) or 50/50 v/v alginate/cartrigel. Cartrigel is a urea extract of bovine femoropatellar cartilage highly enriched in cartilage extracellular matrix components prepared as previously described.[6] Polymer solutions were added to a cell pellet to make a final cell concentration of 20×10^6 cells/cc. One hundred fifty–microliter aliquots of cell–polymer suspension were placed in tissue culture inserts with a semi-permeable membrane. The inserts were placed in 1 ml of $CaCl_2$ (102 mM, 20mM HEPES) for 20 minutes to cause gelation. The gels were removed from the inserts and incubated in complete DMEM media at 37°C, 5% CO_2 for one week.

Hydrogel Analysis

Gels for RNA extraction were dissolved in EDTA for 10 min (50 mM, 10 mM HEPES), centrifuged ($1000 \times g$, 10 min), washed in PBS, and placed in 1 ml Trizol (Sigma, St. Louis, MO). Manufacturer's instructions for RNA precipitation were followed with modification for tissues with high polysaccharide content. cDNAs were generated from 1 μg of RNA using random hexamers (first-strand synthesis system for RT-PCR, Life Technologies). Relative quantitation of aggrecan and types I and II collagen was performed using real-time PCR (Sequence Detector 7700, PE Biosystems) and SYBR Green PCR kit (Applied Biosystems, Foster City, CA). Primers for aggrecan, type I and II collagen and GAPDH were used as previously described.[7,8] The housekeeping gene GAPDH was monitored as an internal standard. Bovine cartilage and freshly isolated chondrocytes were used as controls and for normalization.

Hydrogels for immunohistochemistry and *in situ* hybridization were preserved in 10% formalin and 4% paraformaldehyde, respectively. Samples were embedded in paraffin following standard histological technique, with RNAse precautions used for *in situ* samples. For immunostaining, endogenous peroxidase activity was inactivated with 10% H_2O_2 in methanol. Sections were digested with 2.5% hyaluronidase or 0.005U/μl chondroitinase for 30 minutes. Immunohistochemistry was performed using Histostain (Zymed) and antibodies for aggrecan (mouse monoclonal, University of Iowa Hybridoma Bank) and collagen (rabbit polyclonal antibodies for type I and II collagen, Research Diagnostics).[7] Probes for *in situ* hybridization were prepared and used as previously described.[9]

RESULTS AND DISCUSSION

Response of bovine chondrocytes to hydrogels used for tissue engineering was studied using quantitative real-time PCR and *in situ* hybridization. Aggrecan and type II collagen, two markers of differentiated chondrocytes, were examined in alginate and in alginate enriched with extracellular matrix components extracted from cartilage (Cartrigel). Alginate is a polysaccharide that forms an ionic gel in the pres-

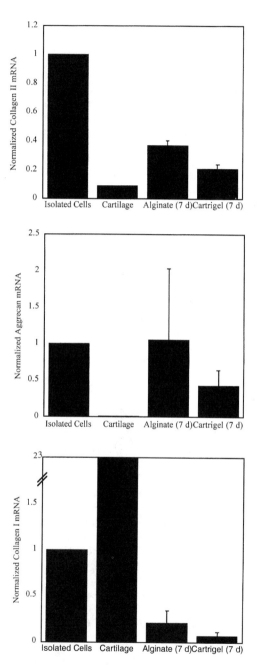

FIGURE 1. Quantitative levels of mRNA determined using type II collagen (*top*), aggrecan (*middle*), and type I collagen (*bottom*) for enzymatically isolated bovine chondrocytes, bulk bovine cartilage, and cells encapsulated in alginate and alginate–Cartrigel composite gels. Values are normalized to GAPDH and a standard curve created from isolated chondrocytes.

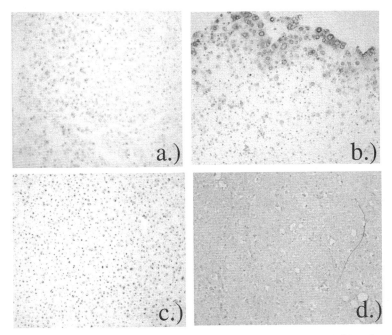

FIGURE 2. *In situ* hybridization and immunostaining for aggrecan (**a,b**) and type II collagen (**c,d**). Aggrecan RNA and protein are found preferentially in the periphery of the gel, while collagen RNA and protein is homogenous throughout the gel.

ence of a divalent cation such as calcium chloride. Chondrocytes encapsulated in alginate remain differentiated compared to cells cultured in monolayer, which become fibroblastic.[10,11] This has provided a useful tool to expand and study chondrocyte and matrix biology *in vitro*. Biological signals, in the form of growth factors or proteins, are often incorporated into tissue engineering scaffolds to promote cell–scaffold interactions.[12,13] Alginate was enriched with a urea extract of cartilage extracellular matrix as a biological signal to direct chondroycte function.

Type II collagen and aggrecan mRNA levels increased over 5- and 10-fold, respectively, after isolation from cartilage (FIG. 1, *top* and *middle*). Type I collagen mRNA decreased significantly after isolation (FIG. 1, *bottom*). After one week of incubation in alginate type II collagen mRNA decreased to less than 40% of the freshly isolated chondrocytes, but still remained 3-fold higher than native cartilage. Aggrecan mRNA levels remained similar to the isolated chondrocytes after one week's incubation, but demonstrated a large standard deviation. It is hypothesized that this large standard deviation is caused by chondrocytes having variable aggrecan expression throughout the gel. Incorporation of cartrigel decreased mRNA levels for aggrecan and type I and II collagen. While aggrecan mRNA levels decreased in cartrigel, the variability in expression throughout the gel, or standard deviation, also decreased. The cartrigel may be serving the purpose of promoting the chondrocytes to a similar phenotype with more uniform aggrecan expression. Aggrecan *in situ* and immunostaining (FIG. 2a and b) reveal increased RNA localization and protein deposition, respectively, in the periphery of the hydrogel. Type II collagen demonstrated a homogenous expression and protein deposition throughout the hydrogel. Chondrocytes regulate cartilage extracellular matrix through cell–matrix interactions using integrins and other adhesion receptors and subsequent production or degradation of matrix. It is hypothesized that the chondrocytes in the alginate–Cartrigel constructs decreased expression of extracellular matrix molecules in response to their presence in the Cartrigel. Further understanding of cell response in hydrogel scaffolds will provide useful information for the rational design of biomaterials for cartilage tissue engineering.

ACKNOWLEDGMENT

We gratefully acknowledge Eva Roque for her technical assistance.

REFERENCES

1. BUCKWALTER, J. & H. MANKIN. 1997. Articular cartilage. Part II: Degeneration and osteoarthrosis, repair, regeneration,and transplantation. J. Bone Joint Surg. **79-A:** 612–632.
2. LANGER, R. & J. VACANTI. 1993. Tissue engineering. Science **260:** 920–926.
3. ELISSEEFF, J. *et al.* 2000. Photoencapsulation of chondrocytes in poly(ethylene oxide)-based semi- interpenetrating networks. J. Biomed. Mater. Res. **51(2):** 164–171.
4. ROWLEY, J.A., G. MADLAMBAYAN & D.J. MOONEY. 1999. Alginate hydrogels as synthetic extracellular matrix materials. Biomaterials **20(1):** 45–53.
5. FREED, L. & G. VUNJAK-NOVAKOVIC. 1995. tissue engineering of cartilage. *In* The Biomedical Engineering Handbook. J. Bronzind, Ed. :1778–1796. CRC Press. Boca Raton, FL.
6. KLEINMAN, H.K. *et al.* 1986. Basement membrane complexes with biological activity. Biochemistry **24:** 312–318.
7. WATANABE, H. *et al.* 1997. Dwarfism and age-associated spinal degeneration of heterozygote cmd mice defective in aggrecan. Proc. Natl. Acad. Sci. USA **94(13):** 6943–6947.
8. KALUZ, S. *et al.* 1996. Structure of an ovine interferon receptor and its expression in endometrium. J. Mol. Endocrinol. **17(3):** 207–215.
9. ARIKAWA-HIRASAWA, E. *et al.* 1999. Perlecan is essential for cartilage and cephalic development. Nat. Genet. **23(3):** 354–358.
10. HAUSELMANN, H.J. *et al.* 1994. Phenotypic stability of bovine articular chondrocytes after long-term culture in alginate beads. J. Cell Sci. **107**(Pt. 1): 17–27.
11. PAIGE, K. *et al.* 1996. De novo cartilage generation using calcium alginate-chondrocyte constructs. Plast. Reconstruct. Surg. **97:** 168–178.
12. SALTZMAN, W.M. 1996. Growth factor delivery in tissue engineering. MRS Bull. **21:** 62.
13. PARK, A., B. WU & L.G. GRIFFITH. 1998. Integration of surface modification and 3D fabrication techniques to prepare patterned poly(L-lactide) substrates allowing regionally selective cell adhesion. J. Biomater. Sci. Polym. Ed. 9: 89–110.

Engineering a Biological Joint

J. GLOWACKI, K.E. YATES, S. WARDEN, F. ALLEMANN, G. PERETTI,
D. STRONGIN, R. MacLEAN, AND D. ZALESKE

*MGH Skeletal Biology Research Center,Massachusetts General Hospital,
Boston, Massachusetts, USA*

*Department of Orthopedic Surgery, Brigham and Women's Hospital,
Harvard Medical School, Boston, Massachusetts 02115, USA*

KEYWORDS: bioengineering; joints

Repair of congenital or acquired joint deformities requires re-creation of compound structures with complex anatomical and functional features. Innovative solutions are needed for many such clinical problems, in early as well as in late stages of degeneration. Progress in musculoskeletal cell, developmental, and molecular biology, advances in *in vitro* histogenesis, and innovations in materials and manufacturing processes have not yet resulted in widespread clinical applications. An engineered biological joint would enable reconstruction of articular loss resulting from congenital, traumatic, neoplastic, and degenerative conditions. The overall goal of this program is to engineer a biological joint using a devitalized joint from one species as a morphogenic scaffold for chondrocytes from another species and to apply it in a small-animal model.

We report the use of a devitalized biological knee as a scaffold for repopulation with chondrocytes. Bovine articular chondrocytes (bACs) were delivered in a porous collagen sponge, previously shown to support chondrogenesis,[1] chondroinduction,[2] osteogenesis,[3] and *in vivo* compatibility.[1,4] Constructs were engineered *in vitro* and their ectopic *in vivo* fate was examined. Chimeric joints were made by affixing porous collagen sponges that contained bACs to opposing shaved articular femoral and tibial surfaces of devitalized embryonic (19 day) chick knees. Membranes of expanded polytetrafluoroethylene (ePTFE) were positioned between the femoral and tibial sponges in some constructs (FIG. 1). The constructs were cultured and subsequently transplanted into severe-combined-immunodeficiency-disease (SCID) mice. The constructs were analyzed histologically at intervals *in vitro* and *in vivo* and by gene expression analyses for bovine markers.

After 1 week *in vitro,* collagen sponges with bACs were adherent to the shaved articular surfaces of the devitalized chick joints. Initial penetration of neocartilage into the chick scaffold occurred in preexistent canals. Subsequently, metachromatic

Address for correspondence: Julie Glowacki, Ph.D., Orthopedic Research, Brigham and Women's Hospital, 75 Francis Street, Boston, MA 02115. Voice: 617-732-5397; fax: 617-732-6937.
jglowacki@rics.bwh.harvard.edu

Ann. N.Y. Acad. Sci. 961: 123–125 (2002). © 2002 New York Academy of Sciences.

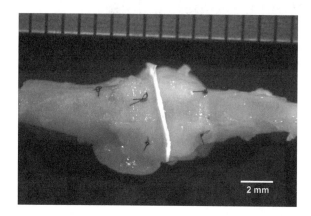

FIGURE 1. A chimeric joint (anterior view). Knees were harvested from embryonic chicks (19 days post fertilization, Charles River SPAFAS, Inc., North Franklin, CT) and were cleaned while maintaining the collateral and cruciate ligaments. The articular surfaces were exposed and shaved. The chick knees were devitalized by lyophilization for 18 hours. Chimeric joints were created by application of a "sandwich" of two collagen sponges seeded with bovine chondrocytes and sutured to the rehydrated chick knees. ePTFE membranes (GoreTex™, generously provided by W.L. Gore Associates, Flagstaff, AZ) were cut to match the size of the joint and were inserted into prepared joints. Scale is in millimeters.

Weeks in culture

	1	2	3
Type II collagen			
Aggrecan core			
Aggrecan link			
G3PDH			

FIGURE 2. Expression of bovine cartilage genes in joints from chimeric knees cultured for 1, 2, and 3 weeks. At each time point, two chimeric knees from the experimental groups without ePTFE membranes were removed from culture. The joints were excised and were pooled for RNA extraction by homogenization in 3 ml Trizol reagent (Life Technologies, Inc., Grand Island, NY). Total RNA was diluted to 100 ng/ml and treated with DNAse I (Roche Molecular Biochemicals, USA.) to eliminate any contaminating genomic DNA. One microgram of DNase-treated RNA was used in random hexamer-primed cDNA synthesis according to manufacturer's instructions (Superscript II; Life Technologies, Inc.). Two microliters of cDNA were used in each PCR with primers for bovine genes as described.[5, 6]

neocartilage accumulated in the sponges. The bovine chondrocytes were the source of the neocartilage as demonstrated by RT-PCR analysis of extracted RNA. Molecular analysis showed expression of bovine cartilage genes for the 3-week observation period (FIG. 2). The ePTFE membranes were successful in maintaining a joint space between opposing sponges.

Some constructs were cultured for 1 week before transplantation into dorsal subcutaneous pouches of 5-week-old immunodeficient mice (male, Fox Chase SCID strain). Chimeric joints exhibited dramatic changes *in vivo*. The bACs invaded the articular matrix of the devitalized knees. At 6 weeks, bovine neocartilage replaced much of the chick cartilage. At 8 weeks, bAC proliferation and neocartilage formation transgressed the scaffolds and created a synchondrosis around the joint. This duration is considered to be excessive in this unloaded site. Thus, the *in vivo* subcutaneous environment in SCID mice was conducive to cartilage formation in joint constructs. Seeded bACs repopulated the devitalized chick scaffold and ePTFE membranes maintained knee joint spaces.

ACKNOWLEDGMENT

This research was supported by NIH/NIAMS Grant No. RO1 AR45870.

REFERENCES

1. MIZUNO, S. & J. GLOWACKI. 1996. A three-dimensional composite of demineralized bone powder and collagen for *in vitro* analysis of chondroinduction of human dermal fibroblasts. Biomaterials **17:** 1819–1825.
2. MIZUNO, S. & J. GLOWACKI. 1996. Chondroinduction of human dermal fibroblasts by demineralized bone in three-dimensional culture. Exp. Cell Res. **227:** 89–97.
3. MUELLER, S.M., S. MIZUNO, L.C. GERSTENFELD & J. GLOWACKI. 1999. Medium perfusion enhances osteogenesis by murine osteosarcoma cells in three-dimensional collagen sponges. J. Bone Min. Res. **14:** 2118–2126.
4. GERSTENFELD, L.C., T. UPOROVA, J. SCHMIDT, *et al.* 1996. Osteogenic potential of murine osteosarcoma cells: comparison of bone-specific gene expression in *in vitro* and *in vivo* conditions. Lab. Invest. **74:** 895–906.
5. GLOWACKI, J., K. YATES, G. LITTLE & S. MIZUNO. 1998. Induced chondroblastic differentiation of human fibroblasts by three-dimensional culture with demineralized bone matrix. Mat. Sci. Eng. C **6:** 199–203.
6. ALLEMANN, F., S. MIZUNO, K. EID, *et al.* 2001. Effects of hyaluronan on engineered articular cartilage ECM gene expression in 3-dimensional collagen scaffolds. J. Biomed. Mat. Res. **55:** 13–19.

The Efficacy of Bone Marrow Stromal Cell-Seeded Knitted PLGA Fiber Scaffold for Achilles Tendon Repair

HONG WEI OUYANG,[a] JAMES C.H. GOH,[a] XIU MEI MO,[b] SWEE HIN TEOH,[b] AND ENG HIN LEE[a]

[a]Department of Orthopaedic Surgery, National University of Singapore, Singapore

[b]Laboratory of Biomedical Engineering, National University of Singapore, Singapore

INTRODUCTION

Tendons are frequently targets of injury in sports and aging trauma. Significant dysfunction and disability can result from suboptimal healing of tendon injuries. It is generally thought that tendons have some self-healing ability, but in cases where tendons are missing or the wound sites are too large to allow for reapposition of the ends, tendon replacement is necessary.[1] Various bioabsorbable materials have been used for Achilles regeneration;[2] however, no published data that we are aware of has addressed the use of D,L-lactide-co-glycolide (PLGA, 10:90) fibers or knitted structure to deliver bone marrow stromal cells for tendon repair. This study was conducted to evaluate the effect of marrow-stromal-cell-(bMSC)-seeded knitted PLGA scaffold for Achilles tendon repair in a rabbit model.

MATERIALS AND METHODS

In 32 legs of 16 adult female New Zealand White rabbits, a tendon defect was created according to the method used by Young et al.[3] In brief, from a lateral approach, the gastrocnemius tendon was separated from the plantaris and soleus tendons and a 1-cm-long gap defect was created in the midsubstance of the gastrocnemius tendon. In group I (12 legs), a knitted PLGA graft was sutured to the ends of tendon and seeded with 1×10^7 bone marrow stromal cells in 0.3 mL fibrin glue; group II (12 legs) was treated with a knitted PLGA scaffold infused with 0.3 mL fibrin glue; and group III (8 legs) was treated with a single suture as control. In all legs, the tension on the repaired tendon was returned to approximately normal. The rabbits were allowed to move freely in the cages postoperatively throughout the test period. Specimens were harvested at 2, 4, 8, and 12 weeks for macroscopic, histological, and immunohistochemical examination.

Author for correspondence: Associate Professor J.C.H. Goh, Orthopaedic Diagnostic Center, Level 3, Department of Orthopaedic Surgery, National University of Singapore, 5 Lower Kent Ridge Road, Singapore 119074. Voice: 0065-7724423; fax: 0065-7744082.

dosgohj@nus.edu.sg

Ann. N.Y. Acad. Sci. 961: 126–129 (2002). © 2002 New York Academy of Sciences.

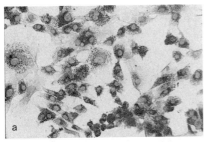 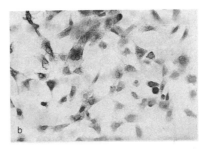

FIGURE 1. Immunohistochemical analysis of (**a**) collagen type I and (**b**) collagen type III in bone marrow stromal cells before transplantation. It illustrates the bMSC-synthesized collagen type I and III in culture. (Magnification 100×.)

RESULTS

Macroscopic Findings

All of groups I and II tendon repairs showed good attachment of the scaffold to the proximal and distal ends of tendon 2 weeks after surgery. The volume of space-filling fibrous tissue and enveloping soft tissue increased at 4 and 8 weeks, but the increase was not obvious at 12 weeks. It was observed that the volume of regenerated tissue was greater in group I compared with groups II and III.

Immunohistochemical Findings

It was found that the cells from the first passage were able to synthesize types I and III collagen. The mass of types I and III collagen fibers was found in the neotendon at 4 weeks postoperation (FIG. 1).

Histologic Findings

In the control group, the regenerated tissues were poorly organized and mixed with fat tissue. The bMSCs/PLGA-treated tendon repair showed what was apparently more eosinphilic tissue formation inside and around the scaffold as early as 2 weeks after implantation compared with group II (FIG. 2).

FIGURE 3 shows the neotendon regeneration and remodeling processes. Two weeks after the operation, a large number of cells (both round and spindle-shaped) were present in rich eosinphilic stained tissue, which filled and wrapped the scaffold. The vascular component was observed at this time. At 4 weeks postoperation, the number of cells was markedly reduced. The regenerated tissues were densely packed and exhibited organized bands of collagen with elongated cells. Crimped fibers were grouped in bounds along the axis of tensile load. The inflammatory reaction around the PLGA scaffold was light and the fibers were degraded into pieces. The vascular component was rarely present. The histological appearance of the 8- and 12-week

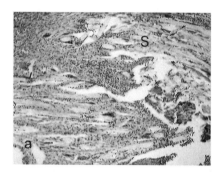
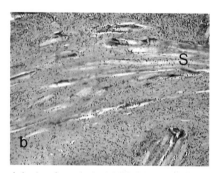

FIGURE 2. Histology of neotendon (HE staining) at 2 weeks in (**a**) PLGA-only-treated group, and (**b**) bMSC/PLGA-treated group. Knitted PLGA scaffold allowed a large number of cells to infiltrate and synthesize matrix, while bMSCs promoted more dense tissue formation. "S" indicates scaffold. (Magnification 100×.)

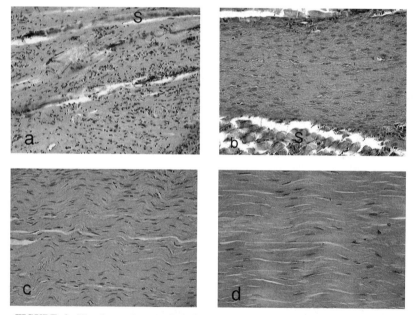

FIGURE 3. Histology of neotendon (HE staining) in bMSC/PLGA-treated group at (**a**) 2 weeks; (**b**) 4 weeks; (**c**) 12 weeks; and (**d**) normal tendon. An apparent crimp pattern was observed from 4 weeks. With increasing follow-up duration, the histology of neotendon became more similar to native tendon. "S" in (**a**) and (**b**) indicates PLGA scaffold. *Brackets* in (**c**) and (**d**) indicate length of crimp periods. (Magnification 200×.)

neotendons became more mature. All tissues showed organized bundles of highly crimped fibers. The volume of the PLGA scaffold was obviously reduced. Synovial cells were observed along the surface of the neotenodons. Compared to normal tendon, the length of the crimp patterns of 12-week neotendon were shorter and the cells presented in the neotendons were richer.

DISCUSSION

The results of this study showed that delivering bMSC-seeded PLGA scaffold was more effective than applying scaffold alone. The bMSC/PLGA-treated tendon repair showed more eosinphilic tissue formation inside and around the scaffold as early as 2 weeks after implantation; also, at 4 weeks the repaired tissues composed of bounds of collagen fibers with a crimp pattern were more mature in appearance compared to when PLGA was applied to the scaffold alone. This suggested that bMSCs have the potential for inducing early matrix production and remodeling, and to restore normal structure and function rapidly. Young *et al.* applied a bMSC-contracted collagen matrix for Achilles tendon repair and reported similar benefits of applying bMSCs, but they failed to use the collagen matrix as control because collagen gel without cells could not contract.[3]

On the other hand, the histology of groups I and II showed that the knitted PLGA scaffolds were able to allow cell infiltration, conduct tissue formation, and be absorbed gradually after the host neotissue formation. More important, the scaffold can resist the mechanical force *in vivo* shortly after implantation. It is well known that PLGA is commonly used as surgical suture material. In our other study, bMSCs were found to adhere and proliferate on PLGA film.[4] As for the scaffold structure, a knitting technique has been used for ligament reconstruction and showed promising results.[5]

Thus, the results indicated that bMSC-seeded knitted PLGA scaffold is a promising graft for Achilles tendon repair. Further study on the biomechanical and functional improvement of tendon repair by applying bMSC/PLGA composites will be carried out.

ACKNOWLEDGMENTS

The authors cordially thank Ms. S.W. Chong, Ms. J.W.K. Chan, and Dr. Z. Wang for their assistance in this study.

REFERENCES

1. LOUIE, L. 1998. Tissue engineered tendon. *In* Frontiers in Tissue Engineering. C.W. Patrick *et al.*, Eds.: 412–441. Pergamon. New York.
2. SATO, M. *et al.* 2000. Reconstruction of rabbit Achilles tendon with three bioabsorbable materials: histological and biomechanical studies. J. Orthop. Sci. **5:** 256–267.
3. YOUNG, R.G. *et al.* 1998. Use of mesenchymal stem cells in collagen matrix for Achilles tendon repair. J. Orthop. Res. **16:** 406–413.
4. OUYANG, H.W. *et al.* 2002. Characterization of anterior cruciate ligament cells and bone marrow stromal cells on various biodegradable polymeric films. Mater. Sci. Eng. C **763:** in press.
5. URBAN, J. *et al.* 2000. Experimental research on the usefulness of the Dallos polyster prostheses in reconstructive operations of knee joint ligaments. Polimery W Medycynie **30**(1–2)**:** 11–19.

Effects of Alginate Composition on the Growth and Overall Metabolic Activity of βTC3 Cells

CHERYL L. STABLER,[a,b] ATHANASSIOS SAMBANIS,[a,c] AND IOANNIS CONSTANTINIDIS[a,d]

[a]Georgia Tech/Emory Center for the Engineering of Living Tissues, Atlanta, Georgia 30332, USA

[b]Department of Biomedical Engineering, Georgia Institute of Technology, Atlanta, Georgia 30332, USA

[c]School of Chemical Engineering, Georgia Institute of Technology, Atlanta, Georgia 30332, USA

[d]Department of Radiology, Emory University, Atlanta, Georgia 30322, USA

The encapsulation of cells within semipermeable membranes for the purpose of immunoisolation from the host has many potential applications in tissue engineering, ranging from the treatment of Parkinson's disease to the encapsulation of insulin-secreting cells for the long-term treatment of diabetes.[1,2] Alginate is the biomaterial commonly used in the entrapment of cells. Additional layers of poly-L-lysine and alginate are typically added to coat the central alginate matrix to improve the stability of the gel[3] as well as to create an immunoisolation membrane.[4] The advantages of alginate/poly-L-lysine/alginate (APA) beads include structural integrity and at least partial immunoprotection to the entrapped cells, ease in manufacture, and manipulation of the molecular-weight cutoff of the membranes.

Alginate is a common term for a family of unbranched polymers composed of 1,4-linked β-D-mannuronic and α-L-guluronic acid residues in varying proportions, sequence, and molecular weight. The gelation of alginate takes place when multivalent cations (usually Ca^{2+}) interact with blocks of guluronic residues between two different chains, resulting in a three-dimensional network.[5] The strength of the network depends on the overall fraction of guluronic acid residues, the molecular weight of the polymer, and the Ca^{2+} ion concentration at the time of gelation.[6] The physical properties of alginate gels vary widely, depending on their chemical composition. In summary, alginates possessing a high guluronic acid content develop stiffer, more porous gels, which maintain their integrity for longer periods of time, whereas alginates rich in mannuronic acid have reciprocal properties. Therefore, high guluronic alginates have long been advocated for use in encapsulated cell systems.[7]

Address for correspondence: Ioannis Constantinidis, University of Florida, Department of Medicine, Division of Endocrinology and Metabolism, P.O. Box 100226, Gainesville, FL 32610-0226, USA. Voice: 352-846-2227; fax: 352-846-2231.

Ann. N.Y. Acad. Sci. 961: 130–133 (2002). © 2002 New York Academy of Sciences.

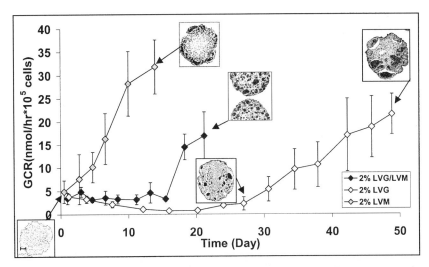

FIGURE 1. Temporal changes in the rates of metabolic activity (glucose consumption rate) for βTC3 cells encapsulated in 2% w/v LVG/LVM (*solid diamonds*), LVG (*open diamonds*), and 2% LVM (*gray diamonds*). Each point represents the average activity over a period of 24–48 h, depending on the feeding cycle. The *error bars* are the standard deviations for the measurements from three independent encapsulations performed with each alginate. Cross sections of the APA beads containing βTC3 cells and stained with H/E are placed at their respective sampling time on the graph.

Given the variety of physical properties associated with alginate composition, it is reasonable to hypothesize that encapsulated cells might be affected differently by changes in the alginate composition. In our laboratory, we have explored these effects by entrapping βTC3 cells within alginates of varying molecular weight, structure, and concentration.[8,9] Through these studies, we were able to determine that a high guluronic alginate at a 2% concentration, although advocated for use by other research groups due to its strength and stability, has detrimental effects on βTC3 cells by inhibiting their normal growth and overall metabolic activity for almost 60 days. A high mannuronic alginate permitted βTC3 cell growth at an exponential rate, until the microbeads broke due to the high cell density. Furthermore, the growth patterns of the βTC3 within the high guluronic alginates and the high mannuronic alginates, in general, differ significantly from an O-ring to a clustering pattern.

In the continuation of these studies, an encapsulation system that provided some control over excessive cell growth while maintaining structural integrity for *in vivo* implantation would prove to be advantageous. In an attempt to strike a balance between the strength of the high guluronic alginate and the cell-favorable environment of the high mannuronic alginate, a mixture of alginates was created using a 2% concentration of a high guluronic alginate (LVG) and a 2% concentration of a high mannuronic alginate (LVM) in a 50/50 ratio. The resulting alginate had 55% guluronic acid content and 45% mannuronic acid content. FIGURES 1 and 2 illustrate the temporal changes in the rates of glucose consumption (GCR) and insulin secretion (ISR) by βTC3 cells encapsulated in 2% w/vLVG/LVM (solid diamonds), compared with

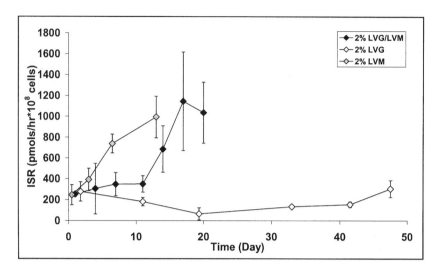

FIGURE 2. Temporal changes in the rates of secretion activity (insulin secretion rate) for βTC3 cells encapsulated in 2% w/v LVG/LVM (*solid diamonds*), LVG (*open diamonds*), and 2% LVM (*gray diamonds*). Each point represents the average activity over a period of 24–48 h, depending on the feeding cycle. The *error bars* are the standard deviations for the measurements from three independent encapsulations performed with each alginate.

2% w/v LVG alginates (open diamonds) and 2% w/v LVM (gray diamonds). Each point on the graphs represent the average overall metabolic and secretory rate from the three independent encapsulations, while the values determined for each encapsulation are the average rate exhibited by the culture over a period of two to three days, depending on the feeding cycle. The error bars represent the standard deviation from the mean. The graphs illustrate that βTC3 cells encapsulating the 50/50 mixture of LVG/LVM maintain their overall metabolic and secretory activity over approximately two weeks. This is compared to the 2% LVG alginate, which displayed a decrease in the overall metabolic and secretory activity over the course of almost one month, and the 2% LVM alginate, which showed an exponential increase in the overall metabolic and secretory rates following entrapment. H/E stained histology images also demonstrate the growth of the cells within the APA beads, where the βTC3 cells within the 2% LVG/LVM mixture exhibit a semiclustering growth pattern of cells throughout the bead. Therefore, through these experiments, an alginate was created that controls the proliferation rate of the βTC3 cells for two weeks and retains the mechanical properties necessary for safe implantation. Our data demonstrate that the appropriate selection of an extracellular matrix can provide at least partial control of cell growth.

ACKNOWLEDGMENTS

This work was supported by grants from the ERC Program of the National Science Foundation under Award Number EEC-9731643, and the NIH (DK47858 and

RR13003). C. Stabler was also supported by an NIH Training Grant (GM08433). This financial support is greatly appreciated.

REFERENCES

1. LANZA, R.P. & W.L. CHICK. 1997. Transplantation of encapsulated cells and tissues. Surgery **121:** 1–9.
2. HICKS, B.A. *et al.* 1991. Transplantation of beta cells from transgenic mice into nude athymic diabetic rats restores glucose regulation. Diabetes Res. Clin. Pract. **14:** 157–164.
3. BENSON, J.P., K.K. PAPAS, I. CONSTANTINIDIS & A. SAMBANIS. 1997. Towards the development of a bioartificial pancreas: effects of poly-L-lysine on alginate beads with βTC3 cells. Cell Transplant. **6:** 395–402.
4. LIM, F. & A.M. SUN. 1980. Microencapsulated islets as bioartificial endocrine pancreas. Science **210:** 908–910.
5. GRANT, G.T. *et al.* 1973. Biological interactions between polysaccharides and divalent cations: the egg-box model. FEBS Lett. **32:** 195–198.
6. MARTINSEN, A., G. SKJAK-BRAEK & O. SMIDSRED. 1989. Alginate as immobilization material: I. Correlation between chemical and physical properties of alginate gel beads. Biotechnol. Bioeng. **33:** 79–89.
7. COLTON, C.K. 1996. Engineering challenges in cell-encapsulation technology. Trends Biotechnol. **14:** 158–162.
8. STABLER, C., K. WILKS, A. SAMBANIS & I. CONSTANTINIDIS. 2001. The effects of alginate composition on encapsulated betaTC3 cells. Biomaterials **22:** 1301–1310.
9. CONSTANTINIDIS, I., I. RASK, R. C. LONG, JR. & A. SAMBANIS. 1999. Effects of alginate composition on the metabolic, secretory, and growth characteristics of entrapped beta TC3 mouse insulinoma cells. Biomaterials **20:** 2019–2027.

Polymer/Alginate Amalgam for Cartilage-Tissue Engineering

E.J. CATERSON, W.J. LI, L.J. NESTI, T. ALBERT, K. DANIELSON, AND R.S. TUAN

Department of Orthopaedic Surgery, Thomas Jefferson University, Philadelphia, Pennsylvania 19107, USA

ABSTRACT: Marrow stroma–derived cells (MSC) are highly proliferative, multipotential cells that have been considered as ideal candidate cells for autologous tissue engineering applications. In this study, we have characterized the chondrogenic potential of human MSCs in both a PLA/alginate amalgam and pure PLA macrostructure as model three-dimensional constructs to support both chondrogenic differentiation and proliferation following TGF-β treatment. MSCs were seeded in experimental groups that consisted of PLA-loaded constructs and PLA/alginate amalgams with and without recombinant human TGF-β1. Chondrogenesis of the PLA and the PLA/alginate amalgam cultures was assessed at weekly intervals by histology, immunohistochemistry, scanning electron microscopy, sulfate incorporation, and RT-PCR. Chondrogenic differentiation occurs within a polymeric macrostructure with TGF-β1 treatment as indicated by histological, immunohistochemical, sulfate incorporation, and gene expression profiles. This macrostructure can be further encased in an alginate gel/solution to optimize cell shape and to confine growth factors and cells within the polymer construct, while the polymeric scaffold provides appropriate mechanical/tissue support. The stable three-dimensional PLA/alginate amalgam represents a novel candidate system of mesenchymal chondrogenesis, which is amendable to investigation of mechanical and biological factors that normally modulate cartilage development and formation as well as a potential tissue engineering construct for cartilage repair.

KEYWORDS: marrow stromal cell; cartilage tissue engineering; bioresorbable polymers; alginate

Throughout life, continuous insults from trauma and/or disease serve to damage or destroy tissue. In response, nature has provided a complex interplay of matrix remodeling and matrix deposition in an effort to respond to and possibly repair these insults. However, some of these tissue responses are inappropriate and/or insufficient to respond to all the injuries; in the case of cartilage it is most likely due to the avascularity and low cellular nature of this tissue.[1] Tissue engineering, which has

Address for correspondence: Rocky S. Tuan, Ph.D., Department of Orthopaedic Surgery, Thomas Jefferson University, 1015 Walnut Street, Philadelphia, PA 19107. Voice: 215-955-5479; fax: 215-955-9159.

Rocky.S.Tuan@mail.tju.edu

Ann. N.Y. Acad. Sci. 961: 134–138 (2002). © 2002 New York Academy of Sciences.

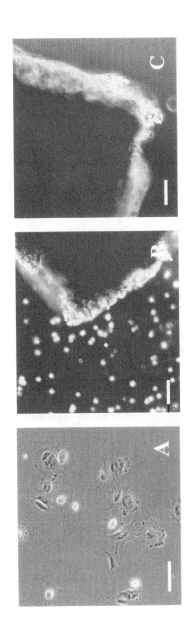

FIGURE 1. Phase-contrast microscopy of bone-marrow-derived cells in chondrogenic medium 2 h after cell seeding. (**A**) Bone-marrow-derived cells undergoing adhesion on tissue-culture plastic. *Bar* = 10 μm. (**B**) and (**C**) Lower magnification views of the surfaces of plain PLA (B) and PLA/ alginate (C) cell-laden constructs. Images were captured immediately after the cultures were disturbed by the addition of supplemental medium 2 h after cell seeding. In (B) not all of the cells had adhered to the PLA, and some cells foated away with the addition of the medium. (C) Shows that the alginate gel served to stably confine the cells within the construct even after the addition of the medium. *Bars* in (B) and (C) = 50 μm.

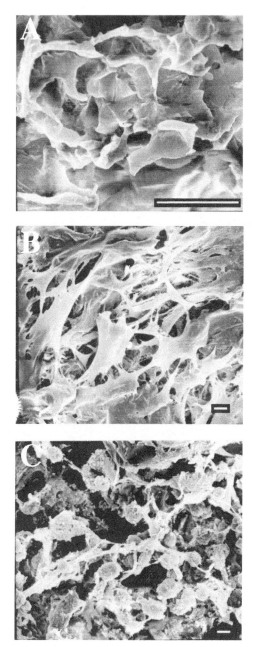

FIGURE 2. (A) SEM views of PLA scaffold alone, (B) PLA seeded with bone-marrow-derived cells, (C) and PLA/alginate amalgam seeded with bone-marrow-derived cells after 7 days in culture in chondrogenic medium. Bone-marrow-derived cells in the PLA/alginate amalgam construct (C) exhibited a round cell shape and produced extracellular matrix. Bone-marrow-derived cells on plain PLA (B) qualitatively appeared more extended and spindle-shaped. *Bar* (A) = 100 μm; *bars* (B,C) = 10 μm.

been an emerging field, is now becoming a mainstream methodology that has yet to be realized in clinical practice. It combines molecular and engineering principles to bolster, enhance, and harness the natural reparative tissue response that has failed to adequately respond to injury.[2]

Tissue engineering principles have been and are currently being employed in cell-based efforts to restore damaged or diseased cartilage. These principles encompass several of the components—cells capable of producing a functional matrix, an appropriate scaffold for transplantation and support of the cell-laden graft, and iatrogenic or autogenous bioactive molecules—that drive the processes of differentiation and maturation.[3] A number of previously defined systems utilize these basic tenets and use multipotential human-bone-marrow-derived cells for cartilage-tissue engineering.[1,3,4] These cartilaginous systems are models for and the building blocks of cartilage-specific matrix implants, which can direct cellular proliferation, promote differentiation, and prevent inappropriate matrix deposition in an attempt to restore function.

The system utilized in these studies consists of a three-dimensional amalgam scaffold composed of the biodegradable polymer, poly-L-lactic acid, and the macromolecular polysaccharide gel, alginate, seeded with human-bone-marrow-derived cells. The bone-marrow-derived cells are multipotential cells, nonimmunogenic, readily accessible, and highly responsive to environmental and external cues for differentiation and maturation. The biodegradable PLA/alginate amalgam scaffold construct is an appropriate carrier for *in vitro* differentiation and subsequent transplantation of the marrow-derived cells, because of alginate's ability to enhance cell shape and cell retention within the construct, providing for a high-density growth format of cells encased in a removable and temporary matrix. This amalgam scaffold can also direct chondrogenic differentiation when the appropriate bioactive molecules are supplied.[5,6] In addition, alginate prevents chondrocytes from undergoing dedifferentiation, and when implanted *in vivo,* induces only a minimal inflammatory reaction.[7,8] PLA is a rational selection for a biodegradable polymer scaffold to bolster the alginate amalgam, because polyester scaffolds of α-hydroxy acids, such as poly-L-lactic acid, have been shown to confer mechanical stability to regenerating tissue and simultaneously support cartilage and connective tissue ingrowths.[9]

Cells seeded in the plain PLA and the PLA/alginate amalgams were treated with recombinant human transforming growth factor-β1 (TGF-β1) (10 ng/mL) with every medium change to induce chondrogenesis. Cartilage formation was noted in the (TGF-β1)-treated groups, with chondrogenic-specific gene expression assessed by reverse transcription-polymerase chain reaction (RT-PCR). In addition, histological and immunohistochemical analysis revealed a cartilaginous phenotype in both the plain PLA and the PLA/alginate amalgam. Cellular proliferation and sulfate incorporation were measured by scintillation counting of whole constructs, as well as in extracted protein and proteoglycan fractions. The location of cellular proliferation was demonstrated with immunohistochemical staining for proliferating cell nuclear antigen (PCNA) and 5-bromo-2′deoxyuridine (BrdU). The location of cell proliferation was restricted to the perimeter of the tissue-engineered constructs. This region of cellular proliferation was also associated, by immunohistochemistry, with a fibrous shell of collagen type I surrounding the plain PLA constructs. This intervening fibrous tissue could be deleterious to tissue integration by creating a barrier between host and tissue-engineered neocartilage. The PLA/alginate amalgam showed no signs of fibrous-tissue deposition, and thus is potentially a more ideal combination of cells, scaffold, and bioactive molecules, which can direct appropriate cartilage-specific differentiation and inhibit inappropriate fibrous-tissue deposition for cartilage-tissue engineering.

REFERENCES

1. WAKITANI, S. et al. 1994. Mesenchymal cell-based repair of large, full-thickness defects of articular cartilage. J. Bone Jt. Surg. Am. **76:** 579–592.
2. BONASSAR, L.J. & C.A. VACANTI. 1998. Tissue engineering: the first decade and beyond. J. Cell. Biochem. Suppl. **31:** 297–303.
3. PERKA, C., O. SCHULTZ, R.S. SPITZER & K. LINDENHAYN. 2000. The influence of transforming growth factor beta1 on mesenchymal cell repair of full-thickness cartilage defects. J. Biomed. Mater. Res. **52:** 543–552.
4. PONTICIELLO, M.S., R.M. SCHINAGL, S. KADIYALA & F.P. BARRY. 2000. Gelatin-based resorbable sponge as a carrier matrix for human mesenchymal stem cells in cartilage regeneration therapy. J. Biomed. Mater. Res. **52:** 246–255.
5. PITTENGER, M.F. et al. 1999. Multilineage potential of adult human mesenchymal stem cells. Science **284:**143–147.
6. JOHNSTONE, B. et al. 1998. In vitro chondrogenesis of bone marrow-derived mesenchymal progenitor cells. Exp. Cell Res. **238:** 265–272.
7. MARIJNISSEN, W.J. et al. 2000. Tissue-engineered cartilage using serially passaged articular chondrocytes. Chondrocytes in alginate, combined in vivo with a synthetic (E210) or biologic biodegradable carrier (DBM). Biomaterials **21:** 571–580.
8. BONAVENTURE, J. et al. 1994. Reexpression of cartilage-specific genes by dedifferentiated human articular chondrocytes cultured in alginate beads. Exp. Cell Res. **212:** 97–104.
9. FREED, L.E. et al. 1994. Joint resurfacing using allograft chondrocytes and synthetic biodegradable polymer scaffolds. J. Biomed. Mater. Res. **28:** 891–899.

Accelerated and Improved Osteointegration of Implants Biocoated with Bone Morphogenetic Protein 2 (BMP-2)

HERBERT P. JENNISSEN

Institute of Physiological Chemistry, University of Essen, D-45122 Essen, Germany

ABSTRACT: A concept and methodology are presented for the direct biocoating of implantable metals like titanium and stainless steel with bone morphogenetic protein 2 (BMP-2) for future applications as cementless bone or dental prostheses. Such bioactive surfaces can influence cells and tissues by chemotactic as well as juxtacrine mechanisms. Reference is made to first experiments in sheep and rabbits in which BMP-2 coatings impressively increased the osteoinductive potential of titanium implants.

KEYWORDS: osteointegration; bone implant healing; bone morphogenetic protein 2 (BMP-2); biocoating of implantable metals

INTRODUCTION

A biocompatible implant is expected to display a structural and a surface compatibility. Of these two the surface compatibility is of decisive importance, since the first reaction between the organism and the implant takes place on this level, primarily by involving proteins. Recently a concept and methodology were developed for the direct biocoating of implantable metals like titanium and 316L stainless steel with bioactive factors such as bone morphogenetic proteins (BMP)[1–3] for future application as cementless bone or dental prostheses. Such bioactive surfaces can influence cells and tissues by chemotactic as well as juxtacrine mechanisms.[1,2] BMP-2, which has been shown to be an anthelix protein,[5] is the most important representative of the bone morphogenetic proteins and leads to a proliferation and differentiation of bone precursor cells. The direct immobilization of BMP-2[4,5] on the implant surfaces appears to lead to a significant improvement in the *in vivo* application[2,6] of BMP-2 in comparison to present-day BMP-carrier systems based, for example, on bovine collagens, which are generally admixed or combined with the implant.

Address for correspondence: Prof. Dr. H.P. Jennissen, Institut für Physiologische Chemie, Universität-GHS-Essen, Hufelandstr. 55, D-45122 Essen, Germany. Voice: 49-201-723-4125; fax: 49-201-723-4694.

hp.jennissen@uni-essen.de

Ann. N.Y. Acad. Sci. 961: 139–142 (2002). © 2002 New York Academy of Sciences.

MATERIALS AND METHODS

Recombinant human bone morphogenetic protein 2 (rhBMP-2) was prepared as previously described[1] or was purchased from Biochrom (Berlin). The biological activity of rhBMP-2 was tested with MC3T3-E1 cells by the activation of the *de novo* synthesis of alkaline phosphatase (AP).[4] The half-activation constants ($K_{0.5}$) were in the range of 20–75 nM.[2] Flat electropolished titanium (cp Ti, grade 2) mini-plates (size: $0.8 \times 10 \times 15$ mm[1,2] or $0.8 \times 6 \times 10$ mm[6]) and porous titanium-alloy plasma spray-coated titanium-alloy cylinders (5 mm stem diameter, 10 mm in length)[2] were surface-enhanced by a novel procedure with chromosulfuric acid (CSA).[1] The treatment of metals with chromosulfuric acid (CSA-Ti-alloy)[1] leads to ultra-hydrophilic (contact angles 0–10°, no hysteresis)[3] bioadhesive surfaces. rhBMP-2 was immobilized by covalent and non-covalent methods[1,2,6] on these CSA-treated surfaces. The amount of rhBMP-2 immobilized has ranged from 10–300 ng/cm[2]. The animal experiments (sheep, rabbits) are described in Refs. 2 and 6. Microradiography (X-rays) is described in Ref. 2.

REVIEW AND DISCUSSION

For clinical applications in bone implant healing, rhBMP-2 is generally admixed with a carrier or delivery system for instillation with the implant (for review see Ref. 2). The methods are more or less empirical, with little information being available on the interaction between rhBMP-2 and the individual carrier or delivery system. Typical carrier/delivery systems range from collagen and its derivatives over osteoconductive materials, acidic gels, and bioactive glasses to polymers like poly(D,L-lactides) and osteopromotive membranes (see Ref. 2). The results with these delivery systems vary widely and the multiplicity as such including the high concentrations of rhBMP-2 instilled (in the mg/ml range) strongly indicate that an optimal method is still lacking. In fact we could show[4] that the dose-dependent effect of rhBMP-2 on AP-induction in MC3T3-E1 cells plateaus out into a maximal response at ca. 300–1000 nM BMP-2 (i.e., 8-25 µg/ml) obviating higher concentrations *in vivo*.

Therefore we have taken a new approach by directly immobilizing minute amounts of rhBMP-2 (local *in vivo* tissue engineering) on the metallic implant surface itself without a carrier intermediate.[1,2,6] Thus the danger of unwanted ectopic bone formation is minimized. We have two objectives with this approach: (1) The first objective is to obtain a limited and targeted release of rhBMP-2 from the surface to trigger chemotaxis of distant bone progenitor cells (*chemotactic stimulation*). Immobilization methods for such a targeted release are adsorption, coupling by hydrolyzable covalent bonds, or gene constructs leading to a controlled *in vivo* release. (2) The second objective is a stable covalent immobilization of rhBMP-2 for enhanced and prolonged stimulation of rhBMP-2 receptors on adhering bone cells by immobilized rhBMP-2 (*juxtacrine stimulation*). The aim is to promote specific cell adhesion and to trigger local osteogenesis with the induction of bone-cell-specific proteins such as collagen I, osteocalcin, BMP-4, and alkaline phosphatase (AP).

To prepare such novel surfaces on titanium and other metal implants we have developed a two-step procedure.[1-3] First we prepare an ultra-hydrophilic *priming coat* (showing an "inverse Lotus effect"[3]) by wet chemical means to endow the sur-

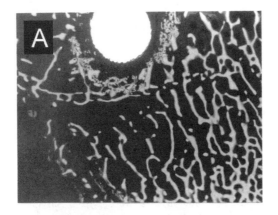

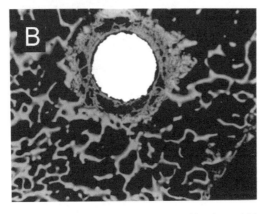

FIGURE 1. Osteointegration of a rhBMP-2-biocoated implant within 4 weeks in sheep. Cross-sections through implants in the distal femur condylus in a gap healing model (initial gap: 1000 μm). In the above microradiographs only the calcified areas of bone are visible. (**A**) Ti-alloy brushite control with a 300–500-μm wide gap surrounding the implant cylinder. (**B**) rhBMP-2 coated (ca. 213 ng/cm^2) CSA-Ti-alloy-APS (APS: aminopropyl triethoxy silane) implant with circumferential bone formation and integration with minimal residual gap. The titanium implant (bright object in center) has a diameter of 5 mm. (From Jennissen *et al.*[2] Reprinted by permission.)

face with bioadhesive properties (see Ref. 3). In the second step we prepare a *biocoat* consisting of immobilized rhBMP-2 to endow the surface with bioactive properties.[2,6]

FIGURE 1 shows the microradiographic results obtained with a rhBMP-2 coating of ca. 213 ng/cm^2 on a titanium cylinder in a gap-healing model in sheep.[2] In the brushite (i.e., calcium phosphate) control (FIG. 1A), it can be seen that after four weeks there remains a cicumferential residual gap of 300–500 μm. In the case of the BMP-2-coated implant (FIG. 1B) a circumferential osteointegration with minimal residual gap and no evidence of capsule formation is found. In other experiments neither the priming coat nor the chemical modifications involved in biocoating have led to any toxic side effects (inflammation) or to the origin of infections.[6]

ACKNOWLEDGMENTS

The verity of the article by Schiermeier and Weydt[7] on Deutsche Forschungsgemeinschaft (DFG) support is acknowledged.

REFERENCES

1. JENNISSEN, H.P., T. ZUMBRINK, M. CHATZINIKOLAIDOU & J. STEPPUHN. 1999. Biocoating of implants with mediator molecules: surface enhancement of metals by treatment with chromosulfuric acid. Materialwiss. Werkstofftech. **30:** 838–845 [online ISSN: 1521–4052).
2. JENNISSEN, H.P., M. CHATZINIKOLAIDOU, H.M. RUMPF, et al. 2000. Modification of metal surfaces and biocoating of implants with bone-morphogenetic protein 2 (BMP-2). In DVM Bericht. **313:** 127–140. DVM. Berlin. [ISSN 1615–2298].
3. JENNISSEN,H.P. (2001) Ultra-Hydrophile metallische Biomaterialien. Biomaterialien **2:** 45–53.
4. WIEMANN, M. H.M. RUMPF, D. BINGMANN & H.P. JENNISSEN. 2001. The binding of rhBMP-2 to the receptors of MC3T3-E1 cells and the question of cooperativity. Materialwiss. Werkstofftech. **32:** 931–936.
5. LAUB, M., T. SEUL, E. SCHMACHTENBERG & H.P. JENNISSEN. 2001. Molecular modelling of bone-morphogenetic protein 2 (BMP-2) by 3D-rapid prototyping. Materialwiss. Werkstofftech. **32:** 926–930.
6. VOGGENREITER, G., K. HARTL, M. CHATZINIKOLAIDOU, et al. 2001. Assessment of the biological activity of chemically immobilized rhBMP-2 on titanium surfaces in vivo. Materialwiss. Werkstofftech. **32:** 942–948.
7. SCHIERMEIER, Q. & P. WEYDT. 2000. German research agency stifles creativity. Nature **404:** 217.

Molecular Signaling in Bioengineered Tissue Microenvironments

DONALD P. BOTTARO,[a] ANDREA LIEBMANN-VINSON,[b] AND
MOHAMMAD A. HEIDARAN[b]

[a]Entremed, Inc., Rockville, Maryland 20850, USA

[b]BD Technologies, Inc., Research Triangle, North Carolina 27709, USA

ABSTRACT: Biological tissues and organs consist of specialized living cells arrayed within a complex structural and functional framework known generally as the extracellular matrix (ECM). The great diversity observed in the morphology and composition of the ECM contributes enormously to the properties and function of each organ and tissue. For example, the ECM contributes to the rigidity and tensile strength of bone, the resilience of cartilage, the flexibility and hydrostatic strength of blood vessels, and the elasticity of skin. The ECM is also important during growth, development, and wound repair: its own dynamic composition acts as a reservoir for soluble signaling molecules and mediates signals from other sources to migrating, proliferating, and differentiating cells. Artificial three-dimensional substitutes for ECM, called tissue scaffolds, may consist of natural or synthetic polymers or a combination of both. Scaffolds have been used successfully alone and in combination with cells and soluble factors to induce tissue formation or promote tissue repair. Appropriate numbers of properly functioning living cells are central to many tissue-engineering strategies, and significant efforts have been made to identify and propagate pluripotent stem cells and lineage-restricted progenitor cells. The study of these and other living cells in artificial microenvironments, in turn, has led to the identifcation of signaling events important for their controlled proliferation, proper differentiation, and optimal function.

KEYWORDS: molecular signaling; biomimetic microenvironments; tissue scaffolds

THE DEVELOPMENT OF BIOMIMETIC MICROENVIRONMENTS

Three-dimensional cell and tissue culture has a long history, starting with the culture of chick embryo heart tissue on silk veil, followed by the introduction of sponge matrices for the culture of tissue.[1] In 1969, it was observed that bone formed when pieces of a synthetic sponge, made out of polyhydroxyethylmethacrylate (poly-HEMA), were implanted into the skin of young pigs.[2] Since then, the idea of using bioengineered three-dimensional scaffolds for *in vitro* cell culture as well as for *in vivo* tissue replacement has received increasing attention and is today the most prom-

Address for correspondence: Donald P. Bottaro, EntreMed, Inc., 9640 Medical Center Drive, Rockville, MD 20850. Voice: 240-864-2777; fax: 240-864-2601.

donb@entremed.com

Ann. N.Y. Acad. Sci. 961: 143–153 (2002). © 2002 New York Academy of Sciences.

ising approach to mimic the complex three-dimensional cellular structure of living tissues. In contrast to conventional two-dimensional cell culture systems (e.g., culture dishes or multi-well tissue-culture plates), scaffolds not only provide an adhesive substrate, but they also act as a three-dimensional physical support for *in vitro* culture and, in most cases, for subsequent implantation.[3–25] In particular, the hypothesis that utilization of three-dimensional culture systems may improve the maintenance and manipulation of stem cells has drawn increasing attention to this area of research.[26–28]

For proper function, scaffolds for *in vitro* cell culture as well as for tissue engineering have to meet certain design criteria.[5,10,19,23,25,29,30] These include spatial and compositional properties that attract and guide the activity of reparative cells. The regeneration of lost or damaged tissue requires that reparative cells adhere, migrate, proliferate, and differentiate in a manner that results in the synthesis of proper new tissue. It has been well established that the specific interaction of cells with their surrounding extracellular matrix is primarily responsible for promoting and regulating these repair processes. For example, chondrocytes were found to maintain their chondrocytic phenotype when seeded into collagen type II, while they converted to a fibroblastic cell morphology when seeded into collagen type I.[22] The scaffold material not only defines the surface properties of the scaffold, important for interactions of this surface with proteins and cells, but also determines the mechanical properties of the three-dimensional structure and subsequently of the cell/scaffold construct. Current strategies for tissue regeneration focus on applying the basic principles of cell–ECM interaction to the development of implantable matrices that mimic natural tissues.

The most common biomaterials used in tissue grafting scaffolds fall into three major categories: natural polymers, synthetic polymers, and inorganic composites. Collagen- and glycosaminoglycan-based materials are the most widely used natural polymers in tissue engineering.[31,32] The prevalence of collagen in the majority of human tissues underlies its ability to support the growth, differentiation, and function of a wide variety of cell types. Similarly, glycosaminoglycans have physical and biological properties that make them attractive as tissue-grafting biomaterials. In particular, glycosaminoglycans have been shown to regulate cell functions and to play an important role in tissue development and repair.[15,33,34]

Hybrid materials, combining naturally derived and synthetic polymers, have been developed to combine the advantages and overcome the shortcomings of both organic material classes. Bovine articular chondrocytes were found to proliferate, regenerate a cartilaginous matrix, and maintain their phenotype for six weeks cultured in hybrid sponges composed of collagen microsponges in pores of sponges made from lactic acid polymers or lactic acid/glycolic acid copolymers.[18,35] Cultured neurons did not show nerve fiber growth unless fibronectin, collagen, or nerve growth factor was incorporated into the synthetic poly-HEMA hydrogel.[36,37] Poly-HEMA–gelatin composite hydrogels showed enhanced cellular interactions and tissue integration when implanted subcutaneously, in contrast to plain poly-HEMA gels that were encapsulated and showed no tissue ingrowth.[38]

Surface properties such as texture, roughness, hydrophobicity, charge, and chemical composition are known to affect cell adhesion and subsequent cell function on a polymer surface.[9,39–44] Surface properties and the control thereof also play a major role in creating scaffolds that interact biospecifically with cell type(s) of interest.

Ideally, the base material of a bioactive scaffold does not support cell adhesion. Cell adhesive properties are introduced to the scaffold by incorporation of bioactive molecules, such as peptides.[45] For example, Han *et al.* synthesized lactide-based poly(ethylene glycol) (PEG) networks, which show cell adhesion resistance due to the PEG and can be readily functionalized with biological ligands through the terminal hydroxyl moiety of the PEG chain.[45]

MOLECULAR SIGNALING IN BIOMIMETIC MICROENVIRONMENTS

The growth and differentiation of most cell types is regulated by the interplay of four major signaling sources: (1) soluble growth factors; (2) insoluble extracellular matrix and growth substrates; (3) environmental stress and physical cues; and (4) cell–cell interactions FIG. 1).

Soluble growth factors can have profound effects on cell growth and differentiation. They can feed into complex and overlapping signal transduction pathways, and may share common receptors, co-receptors, and/or receptor subunits. Moreover, it appears not only that the growth and differentiation of many cell types require the proper combination of soluble cytokines, but also that these factors must be present in appropriate amounts. Biphasic dose–response profiles are characteristic of several growth factors *in vitro*, suggesting that excessive amounts may have diminished or inappropriate biological impact. The interactions of growth factors and cytokines with shared receptors or extracellular binding partners such as glycosaminoglucans and proteoglycans can lead to a dynamic concentration of free molecules.

Adhesion and signaling between cells and supporting substrates, such as extracellular matrix molecules, are frequently mediated by cell surface receptors of the integrin family.[46] These adhesive interactions are particularly relevant to cell morphology, cell shape, trafficking of cells to different tissue compartments, and, ultimately, cell survival. In addition to providing adhesion, integrins transmit a variety of signals both into and out of the cell. Signals transmitted internally include activation of the mitogen-activated protein (MAP) kinase and phospholipase-C gamma (PLC-gamma) and phosphatidyl inositol 3-kinase (PI 3-kinase) pathways, which can lead to changes in cell motility, DNA transcription, and enhanced differentiation.[47] There can also be signal transduction from the interior of the cell to the cell surface to "activate" integrins so that they become capable of binding.[46]

An emerging concept in tissue biology is that there are signaling synergies between extracellular matrix molecules and soluble growth factors. In many, if not most, instances where the combined effects of soluble factors and integrins have been examined, synergistic intracellular pathway activation has been observed. Cell adhesion has been shown to greatly enhance the autophosphorylation of the epidermal growth factor (EGF) and platelet-derived growth factor (PDGF) receptors in response to their cognate ligands.[48] In other cells where ECM does not affect growth factor receptor function, the activation of protein kinase C (PKC) via hydrolysis of phosphoinositides depends on cell adhesion.[49] Cell adhesion regulates the transmission of signals to MAP kinase by altering the activation of the upstream protein kinases MEK-1 or Raf.[50] There is also evidence that growth factor–stimulated activation of PI 3-kinase and downstream effectors such as protein kinase B/AKT and

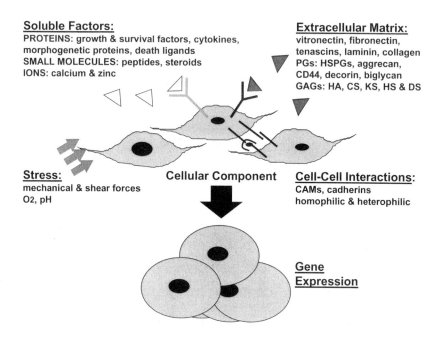

Soluble Factors:
PROTEINS: growth & survival factors, cytokines,
morphogenetic proteins, death ligands
SMALL MOLECULES: peptides, steroids
IONS: calcium & zinc

Extracellular Matrix:
vitronectin, fibronectin,
tenascins, laminin, collagen
PGs: HSPGs, aggrecan,
CD44, decorin, biglycan
GAGs: HA, CS, KS, HS & DS

Stress:
mechanical & shear forces
O2, pH

Cellular Component

Cell-Cell Interactions:
CAMs, cadherins
homophilic & heterophilic

Gene Expression

FIGURE 1. Modulation of cell fate by extrinsic factors. The growth and differentiation of many cell types is regulated by four important sources of external signaling: (1) soluble factors; (2) insoluble and soluble extracellular matrix components; (3) environmental stress; and (4) cell–cell interactions. Soluble factors include cytokines, growth factors, morphogenetic proteins, small-molecule agonists, steroid hormones, peptides, and ions. Insoluble and soluble extracellular matrix components include collagenous proteins (collagen types I, II, and XVIII) and a broad family of noncollagenous proteins that includes fibronectin, vitronectin, and tenascins. The ECM also contains proteoglycans and heparan sulfate proteoglycans (PGs and HSPGs; e.g., aggrecan, versican, neurocan, leucine-rich proteoglycans syndecan, and perlecan) as well as glycosaminoglycans (GAGs; e.g., hyaluronic acid [HA], chondroitin sulfate [CS], keratan sulfate [KS], heparan sulfate [HS], and dermatan sulfate [DS]). Environmental stress includes dynamic or static mechanical forces, shear forces, pH effects, and oxygen (O_2) tension. Cell–cell interactions are regulated by cellular adhesion molecules (CAMs) and cadherins, which mediate homophilic interactions between cells.

p70rsk depend on cell-substrate adhesion ligands.[48] Consistent with these findings, the adhesion of cells to fibronectin has also been shown to stimulate and enhance PDGF-induced inositol lipid breakdown.[51] Fibrillar collagen inhibits arterial smooth muscle proliferation through regulation of cdk2 inhibitors.[52] Thus, at least three important signaling pathways controlled by growth factors also require cell adhesion (FIG. 2).

Accumulating data suggest that the interplay of signals present in the cellular microenvironment prime cells for a rapid response to a limiting and often transient external cue. Thus the activation of growth factor and cytokine signaling pathways, occurring within minutes of ligand addition, leads to a variety of biological effects including migration, proliferation, differentiation, and morphogenesis. What is not

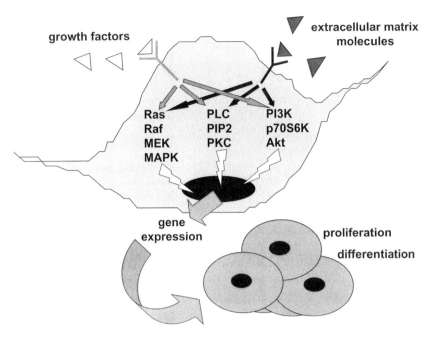

FIGURE 2. Integrin- and growth-factor-mediated biological responses are regulated by common intracellular signaling pathways. This includes activation of the Ras/Raf/MEK/MAP kinase (MAPK) cascade and phospholipase C gamma (PLC) and phosphatidylinositol 3-kinase (PI3K) pathways. PLC activation leads to hydrolysis of phosphatidylinositol 4,5-bisphosphate (PIP2) to produce diacyglycerol and inositol triphosphate leading to activation of protein kinase C (PKC). Ligand-induced activation of PI3K leads to p70s6 kinase (p70s6K) and subsequent Akt protein kinase activation. These pathways converge on the cell nucleus, affecting gene expression, and in turn, stimulating cell proliferation and/or differentiation.

yet clearly understood is the exact relationship between the activation of these pathways and the extent and direction of cellular growth and differentiation. FIGURE 3 shows a hypothetical timeline of events that can occur following growth factor/ECM addition, including modulation of cell adhesion, cell shape, migration, DNA synthesis, growth versus death, and differentiation.

Cell proliferation requires the activity of certain intracellular molecules that are collectively referred to as the cell-cycle machinery.[53] Within the past decade molecular analysis of growth factor and ECM regulation of mammalian cell proliferation has led to the identification of a novel class of molecules that are involved in controlling cell-cycle transition. Cyclin-dependent kinases (Cdks) are key regulators of cell-cycle progression[54] and their activities are positively regulated by their activating subunits, the cyclins. Cyclin molecules identified to date include cyclin A, B, C, D, E, F, G, and H. These molecules bind to Cdks with different affinities and contribute to their enzymatic activation. Accumulating evidence indicates that the G1/S transition is regulated by D-type cyclins (D1, D2, and D3) and cyclin E activating Cdk4/Cdk6 and Cdk2, respectively,[55] while cyclins A and B have been shown to

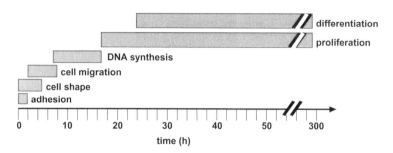

FIGURE 3. Hypothetical timeline of the consequences of growth factor and extracellular matrix signaling. *Horizontal bars* represent time windows commonly used for measuring cell adhesion, morphology, migration, DNA synthesis, proliferation, and differentiation. The impact of growth factors and extracellular matrix components on cell to matrix adhesion, cell shape, and cell migration may be quantitated within a few hours of their exposure to these molecules. In contrast, DNA synthesis is monitored 12–24 hours after initial experimental treatments. Cell proliferation is typically measured after several days. Differentiation may require the longest time period, typically 7–10 days after experimental challenge.

play a role in G2/M transition by activating Cdk1 (cdc2).[56] Cdk activity is negatively regulated by a family of proteins known as cyclin-dependent kinase inhibitors (CKIs).[57,58] On the basis of structural features, these inhibitors can be divided into two subfamilies which are made up of p16INK4a/p15INK4b/p18INK4c/p19INK4d and p21CIP/p27Kip1/p57Kip2. The main targets for p16 and related proteins are Cdk4 and Cdk6, and p16 family members act by inhibiting complex formation between these Cdks and D type cyclins.[59] In contrast, p21, p27 and p57 inhibit the function of multiple Cdk–cyclin complexes including Cdk2, Cdk3, Cdk4, Cdk6 and cdc2 without dissociating the Cdk–cyclin complex.[60]

Because cell cycle progression is checked by Cdks, whose activity is modulated by environmental stimuli, it is commonly assumed that cell division within any microenvironment is highly dependent on the right balance between the amount of Cdks, their modulators CKIs and/or cyclins. Indeed, in support of this notion it has been demonstrated that growth factors and extracellular matrices are able to affect cell-cycle progression by inducing such changes in the expression level of CDKs and their modulators. For example, PDGF is capable of increasing the level of cdc2 during one complete cell cycle. This increase in cdc2 level is accompanied by an increase in expression of cyclins A and B, which is followed by a decrease of p21 and p27, two inhibitors of cell-cycle progression.[61]

Another critical step in cell division is phosphorylation of the pRB/E2F1 complex, which leads to activation of E2F1, a transcription factor involved in cell-cycle progression through G2/M interphase. All of these findings suggest that the major target for the effect of many extrinsic factors appears to be the level and activity of Cdks that regulate the activity of pRB/E2F1 complexes. Certain major components of the cell-cycle machinery (e.g., cyclin-dependent kinase inhibitors and pRB/E2F1) have also been implicated in the process of cellular differentiation. For example, it has been demonstrated that myoblast differentiation is regulated by complex formation between pRB/E2F1 and MyoD, a transcription factor controlling muscle differ-

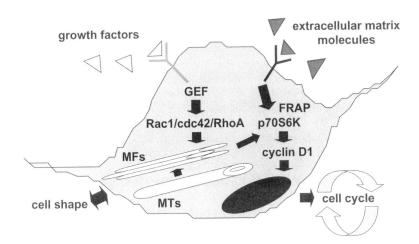

FIGURE 4. Proposed molecular mechanism by which cell shape regulates cell-cycle progression. Cell shape and morphology are regulated by extrinsic signaling from extracellular matrix components and growth factors. Activation of ECM and/or growth-factor-dependent pathways regulates cell-cycle progression through the activity of p70S6 kinase, which phosphorylates the ribosomal S6 submit and thereby regulates the synthesis of cyclin D1. Altering FRAP may also affect cyclin D1 expression, cyclin D1-dependent protein kinase activation, and cell cycle progression. Microtubules (MTs) and microfilaments (MFs) mediate extrinsic physical signals (e.g., tension) affecting cell shape, as well as internal signaling through small G-proteins such as Rac1, Cdc42, and RhoA, which affect MF polymerization and stress fiber formation. Refer to the text for discussion. (Adapted from Assoian[65]).

entiation.[62] Consistent with these findings, C/EBP complex formation with pRB/E2F1 has also been suspected of playing an important role in adipocyte differentiation.[63] Another link between the cell cycle and differentiation has been made in the case of p21, a known inhibitor of cyclin–Cdk activity. One recently published report[64] clearly demonstrated a decrease in the level of p21 protein, which was correlated with the end-stage differentiation of keratinocytes.

Cell shape and morphology, cell-cycle progression and cell differentiation are highly interdependent. FIGURE 4 shows a hypothetical mechanism by which cell shape regulates cell-cycle progression. According to this model, signals transmitted by extracellular matrix receptors affect cell-cycle progression by modulating the activity of the well-known signaling molecule p70s6 kinase. The activation of this enzyme by extracellular matrix receptors is likely to be mediated through FRAP and/or small GTP binding proteins such as Cdc42.[65] A substantial collection of experimental evidence supports the concept that ECM-mediated changes in cell shape lead, in turn, to alterations in the ability of cells to undergo ligand-dependent proliferation. The dynamic state of a cell culture, where adhesive connections between cells and between cells and the ECM are being made and broken, can lead to cellular responses not observed under static conditions.[66,67] The proper function of many cells grown *in vitro* is highly dependent on their state of differentiation. Differentiation is defined by changes in the gene expression profile, which can be monitored by several methods. Gene arrays perhaps provide the most convenient method of monitoring expression profiles of certain cell types grown in three-dimensional scaffolds. It is of importance to note that the gene expression patterns induced by three-dimensional cultures have been shown to be distinct from those found in two-dimensional monolayer cultures, suggesting that the three-dimensional architecture of the environment profoundly influences the state of differentiation of many cell types cultured *in vitro*.

Environmental factors such as shear forces and mechanical stress have also been shown to influence cell function *in vitro*. Equipment capable of applying a cyclic dynamic mechanical stimulation has been used to stimulate smooth muscle cells during culture in two scaffold materials. Tissues engineered in this way displayed enhanced tensile strength and Young's moduli.[8] Cyclic mechanical loading of fibroblasts cultured in three-dimensional collagen matrices was found to elicit complex and substantial changes in matrix modifying protease production, suggestive of a dynamic interplay between mechanical matrix modulation and matrix composition.

The complexity of the systems and principles discussed briefly in this review suggest that the development of biomimetic microenvironments for proper cell function will require the systematic evaluation of many distinct but highly interdependent extracellular signaling pathways. We have a deep understanding of the intracellular molecular mechanisms underlying several important cell functions. A better undertanding of the hierarchy of environmental cues, the integration of their respective cell receptor systems, and the connection of these pathways to vital multimolecular complexes and organelles is needed. The study of molecular signaling in three-dimensional multicellular model systems will move us toward that goal, and is now at the forefront of research in reparative and regenerative medicine.

REFERENCES

1. HOFFMAN, R.M. 1993. To do tissue culture in two or three dimensions? That is the question. Stem Cells **11:** 105–111.
2. WINTER, G.D. & B.J. SIMPSON. 1969. Heterotopic bone formed in a synthetic sponge in the skin of young pigs. Nature **223:** 88–89.
3. MOONEY, D.J. & A.G. MIKOS. 1999. Growing new organs. Scientific American **280:** 60–65.
4. THOMSON, R.C., M.C. WAKE, M.J. YASZEMSKI & A.G. MIKOS. 1995. Biodegradable polymer scaffolds to regenerate organs. Advan. Polymer Sci. **122:** 245–274.
5. MIKOS, A.G. & R.S. LANGER. 1996. Preparation of bonded fiber structures for cell implantation. U.S. Patent No. 5,512,600.
6. NIKLASON, L.E. & R.S. LANGER. 1997. Advances in tissue engineering of blood vessels and other tissue. Transplant. Immunol. **5:** 303–306.
7. VUNJAK-NOVAKOVIC, G. *et al.* 1998. Dynamic cell seeding of polymer scaffolds for cartilage tissue engineering. Biotechnol. Prog. **14:** 193–202.
8. KIM, B.S., J. NIKOLOVSKI, J. BONADIO & D.J. MOONEY. 1999. Cyclic mechanical strain regulates the development of engineered smooth muscle tissue. Nature Biotechnol. **17:** 979–983.
9. MAQUET, V. & R. JEROME. 1997. Design of macroporous biodegradable polymer scaffolds for cell transplantation. *In* Porous Materials for Tissue Engineering Materials Science Forum, 250. D.M. Liu & V. Dixit, Eds. :15–42. Trans Tech Publications. Enfield, UK.

10. PETERS, M.C. & D.J. MOONEY. 1997. synthetic extracellular matrices for cell transplantation. *In* Porous Materials for Tissue Engineering Materials Science Forum, 250. D.M. Liu & V. Dixit, Eds. :43–52. Trans Tech Publications. Enfield.

11. MUNIRUZZAMAN, M., Y. TABATA & Y. IKADA. 1997. Protein Interaction with gelatin hydrogels for tissue engineering. *In* Porous Materials for Tissue Engineering Materials Science Forum, 250. D.M. Liu & V. Dixit, Eds. :89–96. Trans Tech Publications. Enfield, UK.

12. AGRAWAL, C.M., K.A. ATHANASIOU & J.D. HECKMAN. 1997. Biodegradable PLA-PGA polymers for tissue engineering in orthopaedics. *In* Porous Materials for Tissue Engineering Materials Science Forum, 250. D.M. Liu & V. Dixit, Eds. :115–128. Trans Tech Publications. Enfield.

13. ASHIKU, S.K., M.A. RANDOLPH & C.A. VACANTI. 1997. Tissue engineered cartilage. *In* Porous Materials for Tissue Engineering Materials Science Forum, 250. D.M. Liu & V. Dixit, Eds. :129–150.. Trans Tech Publications. Enfield.

14. YASZEMSKI, M.J., R.G. PAYNE, W.C. HAYES, *et al.* 1996. Evolution of bone transplantation: molecular, cellular and tissue strategies to engineer human bone. Biomaterials **17:** 175–185.

15. PIEPER, J.S., A. OOSTERHOF, P.J. DIJKSTRA, *et al.* 1999. Preparation and characterization of porous crosslinked collagenous matrices containing bioavalable chondroitin sulphate. Biomaterials **20:** 847–858.

16. HEALY, K.E., A. REZANIA & R.A. STILE. 1999. Designing biomaterials to direct biological response. Ann. N.Y. Acad. Sci. **875:** 24–35.

17. HERN, D.L. & J.A. HUBBELL. 1998. Incorporation of adhesion peptides into nonadhesive hydrogels useful for tissue resurfacing. J. Biomed. Mater. Res. **39:** 266–276.

18. CHEN, G., T. USHIDA & T. TATEISHI. 2000. Hybrid biomaterials for tissue engineering: a preparative method for PLA or PLGA-collagen hybrid sponges. Advanced Materials **12:** 455–457.

19. EISELT, P. *et al.* 1998. Development of technologies aiding large-tissue engineering. Biotechnol. Prog. **14:** 134–140.

20. MADIHALLY, S.V. & H.W.T. MATTHEW. 1999. Porous chitosan scaffolds for tissue engineering. Biomaterials **20:** 1133–1142.

21. GLASS, J.R., K.T. DICKERSON, K. STECKER, & J.W. POLAREK. 1996. Characterization of a hyaluronic acid-Arg-Gly-Asp peptide cell attachment matrix. Biomaterials **17:** 1101–1108.

22. NEHRER, S. *et al.* 1997. Canine chondrocytes seeded in type i and type ii collagen implants investigated in vitro. J. Biomed. Mater. Res. (Appl. Biomater.) **38:** 95–104.

23. J.P. VACANTI & R.S. LANGER. 1998. Three-dimensional fibrous scaffold containing attached cells for producing vascularized tissue in vivo. U.S. Patent No. 5,759,830.

24. PUTNAM, A.J. & D.J. MOONEY. 1996. Tissue engineering using synthetic extracellular matrices. Nature Med. **2:** 824–826.

25. KIM, B.-S. & D.J. MOONEY. 1998. Development of biocompatible synthetic extracellular matrices for tissue engineering. Tibtech. **16:** 224–230.

26. ROSENZWEIG, M., M. PYKETT, D.F. MARKS & R.P. JOHNSON. 1997. Enhanced maintenance and retroviral transduction of primitive hematopoietic progenitor cells using a novel three-dimensional culture system. Gene Ther. **4:** 928–936.

27. MARTIN, I., A. MURAGLIA, G. CAMPANILE, *et al.* 1997. Fibroblast growth factor-2 supports ex vivo expansion and maintenance of osteogenic precursors from human bone marrow. Endocrinology **138:** 4456–4462.

28. TOMIMORI, Y., M. TAKAGI & T. YOSHIDA. 2000. The construction of an in vitro three-dimensional hematopoietic microenvironment for mouse bone marrow cells employing porous carriers. Cytotechnology **34:** 121–130.

29. THOMSON, R.C., M.C. WAKE, M.J. YASZEMSKI & A.G. MIKOS. 1995. Biodegradable polymer scaffolds to regenerate organs. Advan. Polymer Sci. **122:** 245–274.

30. THOMSON, R.C., M.J. YASZEMSKI & A.G. MIKOS. 1997. Polymer scaffold processing. *In* Principles of Tissue Engineering. R. Lanza, R. Langer & W. Chick, Eds. :263–272. Landes Publishing Co. Austin, TX.

31. RAMSHAW, J.A.M., J.A. WERKMEISTER & V. GLATTAUER. 1995. Collagen-based biomaterials. Biotechnol. Genet. Eng. Rev. **13:** 335–382.

32. FRIESS, W. 1998. Collagen: biomaterial for drug delivery. Eur. J. Pharmaceut. Biopharmaceut. **45:** 113–136.
33. LAROCHELLE, W.J., K. SAKAGUCHI, N. ATABEY, *et al.* 1999. Heparan sulfate proteoglycan modulates KGF signaling through interaction with both ligand and receptor. Biochemistry **38:** 1765–1771.
34. RUBIN, J.S., R.M. DAY, D. BRECKENRIDGE, *et al.* 2001. Dissociation of heparan sulfate and receptor binding domains of hepatocyte growth factor reveals that heparan sulfate-c-Met interaction facilitates signaling. J. Biol. Chem. **276:** 32977–32983.
35. CHEN, G., T. USHIDA & T. TATEISHI. 1999. Fabrication of PLGA-Collagen hybrid sponge. Chem. Lett. **7:** 561–562.
36. CARBONETTO, S.T., M.M. GRUVER & D.C. TURNER. 1982. Nerve fiber growth on defined hydrogel substrates. Science **216:** 897–899.
37. PLANT, G.W., A.R. HARVEY & T.V. CHIRILA. 1995. Axonal growth within poly(2-hydroxyethyl methacrylate) sponges infiltrated with Schwann cells and implanted into the lesioned rat optic tract. Brain Res. **671:** 119–130.
38. SANTIN, M. *et al.*, 1996. Synthesis and characterization of a new interpenetrated poly(2-hydroxyethylmethacrylate)-gelatin composite polymer. Biomaterials **17:** 1459–1467.
39. SCHWARTZ, Z. & B.D. BOYAN. 1994. Underlying mechanisms at the bone-biomaterial interface. J. Cell. Biochem. **56:** 340–347.
40. KOLLER, M.R., M.A. PALSSON, I. MANCHEL, *et al.* 1998. Tissue culture surface characteristics influence the expansion of human bone marrow cells. Biomaterials **19:** 1963–1972.
41. QUI, Q., M. SAYER, M. KAWAJA, *et al.* 1998. Attachment, morphology, and protein expression of rat marrow stromal cells cultured on charged substrate surfaces. J. Biomed. Mater. Res., **42:** 117–127.
42. CURTIS, A. & C. WILKINSON. 1997. Topogrpahical control of cells. Biomaterials **18:** 1573–1583.
43. SINGHVI, R., G. STEPHANOPOULOS & D.I.C. WANG. 1994. Review: effects of substratum morphology on cell physiology. Biotechnol. Bioeng. **43:** 764–771.
44. SCHAMBERGER, P.C. & J.A. GARDELLA, JR. 1994. Surface chemical modifications of materials which influence animal cell adhesion: a review. Colloids & Surfaces B: Biointerfaces **2:** 209–223.
45. HAN, D.K. & J.A. HUBBELL. 1997. Synthesis of polymer network scaffolds from L-lactide and poly(ethylene glycol) and their interaction with cells. Macromolecules **30:** 6077–6083.
46. RUOSLATHI, E. 1991. Integrins. J. Clin. Invest. **87:** 1–5.
47. SCHWARTZ, M.A. 1997. Integrins, oncogenes, and anchorage independence. J. Cell Biol. **139:** 575–578.
48. CYBULSKY, A.V., A.J. MCTAVISH & M.D. CYR. 1994. Extracellular-matrix modulates epidermal growth-factor receptor activation in rat glomerular epithelial cells. J. Clin. Invest. **94:** 68–78.
49. SCHWARTZ, M.A. & C. LECHENE. 1992. Adhesion is required for protein kinase-C-dependent activation of the Na+/H+ antiporter by platelet-derived growth-factor. Proc. Natl. Acad. Sci. USA **89:** 6138–6141.
50. MIYAMOTO, S., H. TERAMOTO, J.S. GUTKIND & K.M. YAMADA. 1996. Integrins can collaborate with growth factors for phosphorylation of receptor tyrosine kinases and MAP kinase activation: roles of integrin aggregation and occupancy of receptors. J. Cell Biol. **135:** 1633–1642.
51. MCNAMEE, H.P., D.E. INGBER & M.A. SCHWARTZ. 1993. Adhesion to fibronectin stimulates inositol lipid-synthesis and enhances PDGF-induced inositol lipid breakdown. J. Cell Biol. **121:** 673–678.
52. KOYAMA, H., E.W. RAINES, K.E. BORNFELDT, *et al.* 1996. Fibrillar collagen inhibits arterial smooth muscleproliferation through regulation of Cdk2 inhibitors. Cell **87:** 1069–1078.
53. JACKS, T. & R.A. WEINBERG. 1998. The expanding role of cell cycle regulators. Science **280:** 1035–1036.

54. PINES, J. 1995. Cyclins and cyclin-dependent kinases: theme and variations. Adv. Cancer Res. **66:** 181–212.
55. SHERR, C.J. 1995. D-type cyclins. Trends Biochem. Sci. **20:** 187–190.
56. KING, R.W., P.K. JACKSON & M.W. KIRSCHNER. 1994. Mitosis in transition. Cell **79:** 563–571.
57. SHERR, C.J. & J.M. ROBERTS. 1995. Inhibitors of mammalian G(1) cyclin-dependent kinases. Genes & Devel. **9:** 1149–1163.
58. HUNTER, T. & J. PINES. 1994. Cyclins and cancer II: cyclin-D and Cdk inhibitors come of age. Cell **79:** 573–582.
59. HALL, M., S. BATES & G. PETERS. 1995. Evidence for different modes of action of cyclin-dependent kinase inhibitors: P15 and P16 bind to kinases, P21 and P27 bind to cyclins. Oncogene **11:** 1581–1588.
60. HARPER, J.W. *et al.*, 1995. Inhibition of cyclin-dependent kinases by P21. Mol. Biol. Cell **6:** 387–400.
61. DEMORA, J.F., A. UREN, M. HEIDARAN & E. SANTOS. 1997. Biological activity of P27(Kip1) and its amino- and carboxy-terminal domains in G2/M transition of *Xenopus* Oocytes. Oncogene **15:** 2541–2551.
62. GU, W. *et al.* 1993. Interaction of myogenic factors and the retinoblastoma protein mediates muscle-cell commitment and differentiation. Cell **72:** 309–324.
63. CHEN, P.L., D.J. RILEY, Y.M. CHEN & W.H. LEE. 1996. Retinoblastoma protein positively regulates terminal adipocyte differentiation through direct interaction with C/Ebps. Genes Devel. **10:** 2794–2804.
64. DI CUNTO, F. *et al.*, 1998. Inhibitory function of P21(Cip1/Waf1) in differentiation of primary mouse keratinocytes independent of cell cycle control. Science **280:** 1069–1072.
65. ASSOIAN, R.K., 1997. Anchorage-dependent cell cycle progression. J. Cell Biol. **136:** 1–4.
66. CHIEN, S., S. LI & J.Y.J. SHYY. 1998. Effects of mechanical forces on signal transduction and gene expression in endothelial cells. Hypertension **31:** 162–169.
67. CHEN, K.D. *et al.*, 1999. Mechanotransduction in response to shear stress: roles of receptor tyrosine kinases, integrins, and Shc. J. Biol. Chem. **274:** 18393–18400.

Cell–Cell and Cell–Extracellular Matrix Adhesion Receptors

CAROLINE DAMSKY

University of California, San Francisco, San Francisco, California 94143, USA

KEYWORD: extracellular matrix adhesion receptors

Cell–cell and cell–extracellular matrix (ECM) adhesion receptors play critical roles in both anchorage and signal transduction. A key feature of their function is their ability to organize signaling complexes at sites of contact with their extracellular environment. Depending on other aspects of the environment, including the nature and organization of the ECM, the presence of growth and differentiation factors, and the presence of mechanical stimuli, these signals can promote or restrain cell proliferation, promote differentiation, trigger matrix remodeling, or promote enhanced tissue organization. Understanding how to manipulate signaling through adhesion receptors to promote the desired end-points for specific tissue engineering problems is a critical key to successful tissue repair and reconstruction.

Two particularly important key issues need further development:

1. Fundamental research on how cells respond to three-dimensional extracellular non-rigid matrices. Most work with cells *in vitro* has involved culture on two-dimensional rigid substrates. For many cell types, this is not a biologically relevant environment. Studies from labs that have worked with three-dimensional matrices show that cells behave very differently under such conditions. Mechanisms by which signals from ECM, neighboring cells, and growth/differentiation factors synergize to regulate cell growth and survival, affect commitment to tissue-specific differentiation programs, and regulate tissue remodeling need to be re-evaluated at the molecular level from the point of view of these more relevant environments. Greater knowledge at the molecular level will make the process of designing effective strategies and biomaterials more efficient and less a matter of trial and error.

2. Input from mechanical stimuli. Cells in tissues constantly experience mechanical stimuli. Even cells in static culture experience the effects of gravity. It can be considered a kind of fifth dimension in the environment (with time being the fourth dimension). Stimuli such as shear-stress, fluid-flow, compression, stretch,

Address for correspondence: Caroline Damsky, University of California, San Francisco, Box 0512, Room HSW 604, 513 Parnassus, San Francisco, CA 94143. Voice: 415-476-8922; fax: 415-302-7338.

damsky@cgl.ucsf.edu

Ann. N.Y. Acad. Sci. 961: 154–155 (2002). © 2002 New York Academy of Sciences.

etc., not only alter the organization and distribution of structural elements and organelles within cells, but also become transduced into biochemical input that modulates signaling networks within and between cells.

SELECTED READING

1. DAMSKY, C.H. 1999. Extracellular matrix-integrin interactions in osteoblast function and tissue remodeling. Bone **25:** 95–96.

Convergence of Molecular Signaling and Tissue Engineering

MOHAMMAD A. HEIDARAN

BD Technologies, Research Triangle Park, North Carolina 27709, USA

KEYWORDS: tissue engineering; molecular scaffolding

Tissues and organs consist of specialized living cells arrayed within a complex structural and functional framework known generally as extracellular matrix (ECM). The great diversity observed in ECM composition contributes enormously to the properties and function of each organ and tissue: the rigidity and tensile strength of bone, the resilience of cartilage, the flexibility and hydrostatic strength of blood vessels, and the elasticity of skin are examples of how different ECM compositions contribute to tissue function. Equally important is the role of ECM during growth, development, and wound repair, where it provides a reservoir for soluble signaling molecules, and through its own dynamic composition, a source of additional signals to migrating, proliferating, and differentiating cells.

Artificial substitutes for ECM, called scaffolds, can consist of natural or synthetic polymers, or both, and have been used successfully alone and in combination with cells and soluble factors to induce tissue formation or promote tissue repair. Cells are also central to many tissue engineering strategies, and significant efforts have been made to identify and propagate pluripotent stem cells, to identify signaling events important for proper differentiation, and to identify ideal microenvironments for maximum cellular function. These efforts that have led to a convergence of research in bioengineering, biomaterials, ECM, cell growth and differentiation, and soluble factors that control cell fate.

Powerful recent developments in the multidisciplinary field of tissue engineering have yielded a novel set of tissue replacement parts and implementation strategies. Scientific advances in biomaterials, stem cells, growth and differentiation factors, and biomimetic environments have created unique opportunities to fabricate tissues in the laboratory from combinations of engineered extracellular matrices ("scaffolds"), cells, and biologically active molecules.

Among the major challenges now facing tissue engineering is the need for complex functionality, as well as biochemical stability in laboratory-grown tissues destined for transplantation. The continued success of tissue engineering, and the

Address for correspondence: Mohammad A. Heidaran, BD Technologies, P.O. Box 12016, 21 Davis Drive, Research Triangle Park, NC 27709. Voice: 919-597-6345; fax: 919-313-6400.
mheidara@bd.com

Ann. N.Y. Acad. Sci. 961: 156–157 (2002). © 2002 New York Academy of Sciences.

eventual development of true human replacement parts will grow from the convergence of molecular signaling principles combined with engineering and basic research advances in tissue, matrix, growth factor, stem cell, and developmental biology.

SELECTED READING

1. LITTLEWOOD, T. & G. EVAN. 1998. A matter of life and death. Science **281:** 1317–1321.
2. HELDIN, C.H. 2001. Signal transduction: multiple pathways, multiple options for therapy. Stem Cells **19:** 295–303.
3. LIOTTA, L.A. & E.C. KOHN. 2001. The microenvironment of tumour-host interface. Nature **411:** 375–379.

The Role of Extracellular Matrix Heparan Sulfate Glycosaminoglycan in the Activation of Growth Factor Signaling Pathways

DONALD P. BOTTARO

EntreMed, Inc., Rockville, Maryland 20850, USA

KEYWORDS: heparan sulfate glycosaminoglycan; growth factor signaling pathways

Much insight has been gained from the study of growth factor (GF) signaling using simple, well-characterized cultured cell models. An important challenge at present is understanding the overall impact of GF signaling in more complex systems where multiple biochemical and physical stimuli present in the extracellular milieu modify intracellular signaling and, in turn, biological responses. Heparan sulfate glycosaminoglycan (HS) is a ubiquitous extracellular matrix component that is important for intercellular communication mediated by a wide variety of GFs. By means of GF immobilization, HS participates in the spatial and temporal regulation of GF distribution and turnover, acts as a GF reservoir creating locally increased concentrations, and protects GFs from denaturation and/or degradation. In addition, HS participates directly in the formation of stable GF–cell surface receptor complexes, influencing receptor activation and downstream signaling events. The acquisition of high-resolution three-dimensional HS/protein complex structures and the discovery of specificity among HS compositions imparting high-affinity protein binding are at the forefront of research in this area. Ultimately this information should facilitate the development of complex artificial scaffolds for tissue regeneration, repair, and replacement.

SELECTED READING

1. BOTTARO, D.P. & M.A. HEIDARAN. 2001. Engineered extracellular matrices: a biological solution for tissue repair, regeneration and replacement. e-biomed 2: 9–12.

Address for correspondence: Donald P. Bottaro, EntreMed, Inc., 9640 Medical Center Drive, Rockville, MD 20850. Voice: 240-864-2777; fax: 240-864-2601.
donb@entremed.com

Ann. N.Y. Acad. Sci. 961: 158 (2002). © 2002 New York Academy of Sciences.

Cell Adhesion Molecules Activate Signaling Networks That Influence Proliferation, Gene Expression, and Differentiation

KATHRYN L. CROSSIN

The Scripps Research Institute, La Jolla, California 92037, USA

KEYWORD: neural cell adhesion binding

After the discovery of several different families of cell-to-cell adhesion molecules (CAMs), each with many members, it became clear that many of these molecules might have signaling functions that were as important as their adhesive abilities. Recent studies on molecules of the immunoglobulin superfamily have revealed that the binding of these molecules at the cell surface can lead to alterations in cell states. In particular, our studies on the neural CAM, N-CAM, have shown that N-CAM binding decreases cell proliferation and alters gene expression. Gene expression changes are mediated in part by the activation of two transcription factors, the glucocorticoid receptor and nuclear factor kappa B (NFκB). Activation of NFκB can be partially prevented by blockade of several different signal pathways, indicating that CAM binding initiates multiple signals to bring about changes in gene expression.

Our recent studies indicate that N-CAM binding increases the number of neurons differentiating from neural stem cells in culture, a finding that may have practical value in the control of neuron production after transplantation into the brain. These studies will be discussed briefly in the context of the multiplicity of adhesion molecules on any given cell that provide multiple inputs to similar signaling events.

It is becoming critical in the field of cellular signaling to consider the effects of cell adhesion as part of a network of interactions among not only CAMs, but also extracellular matrix molecules, integrins, growth factors, and their receptors. The challenge is to determine how these networks interact to yield particular stable differentiated cell states. The ability to manipulate these variables to control the balance between proliferation and maintenance of the differentiated state for particular cell types is essential in designing effective cellular replacement therapies.

Address for correspondence: Kathryn L. Crossin, The Scripps Research Institute, Room SR302, Mail Stop SBR 14, 10550 North Torrey Pines Road, La Jolla, CA 92037. Voice: 858-784-2623; fax: 858-784-2646.
    kcrossin@Scripps.edu

Ann. N.Y. Acad. Sci. 961: 159–160 (2002). © 2002 New York Academy of Sciences.

SELECTED REFERENCES

1. CHOI, J., L.A. KRUSHEL & K. CROSSIN. 2001. NF-B activation by N-CAM and cytokines in astrocytes is regulated by multiple protein kinases and redox modulation. Glia **33:** 45–56.
2. CROSSIN, K.L. & L.A. KRUSHEL. 2000. Cellular signaling by neural cell adhesion molecules of the immunoglobulin superfamily. Dev. Dyn. **218:** 260–279.
3. GREENSPAN, R.J. 2001. The flexible genome. Nat. Rev. Genet. **2:** 383–387.
4. HAZAN, R., L.A. KRUSHEL & K.L. CROSSIN. 1995. EGF receptor-mediated signals are differentially modulated by concanavalin A. J. Cell. Physiol. **162:** 74–85.

Molecular Signaling

PATRICIA DUCY

Department of Molecular and Human Genetics, Baylor College of Medicine, Houston, Texas 77030, USA

KEYWORD: molecular signaling

For a cell to respond to its environment and, moreover, to go on to divide, to migrate, or to differentiate, signals coming from the extracellular compartment need (1) to be sensed, (2) to reach the nucleus, and (3) then have to trigger the specific expression/ repression of particular factors. Our knowledge of this transmission of information is still often fragmented. For instance, it is a common situation to know which ligand/receptor interaction is going to activate a particular kinase pathway but to lack specific insights on the transcriptional events and target genes that will be eventually involved. On the other end, numerous nuclear factors are known to control specific gene expression, for example in cell differentiation programs, but rare are the signaling cascades that have been defined as controlling their activity. Thus, an important goal of the coming years will be to fill these gaps of knowledge by establishing continuous molecular bridges between cell physiology, signal transduction, and gene expression regulation.

Another field of opportunity would be to translate our knowledge of the molecular events and molecules known to be important during embryonic development to the understanding of adult physiology and physiopathology. It is unlikely that these three genetic programs involve totally different sets of regulatory genes, and, indeed, a still low but growing number of genes are shown to fulfill roles before and after birth. Thus, analyzing systematically the pattern of expression and the function of developmental "master genes" not only postnatally but also in aging models could become a useful strategy to define novel pathways amenable to therapeutic interventions.

SELECTED REFERENCES

1. DUCY, P. 2000. Cbfa1: a molecular switch in osteoblast biology. Dev. Dynam. **219:** 461–471.
2. DUCY, P., T. SCHINKE & G. KARSENTY. 2000. The osteoblast: a sophisticated fibroblast under central surveillance. Science **289:** 1501–1504.

Address for correspondence: Patricia Ducy, Baylor College of Medicine, One Baylor Plaza, Room 8911, Houston, TX 77030. Voice: 713-798-7954; fax: 713-798-1530.
pducy@bcm.tmc.edu

Ann. N.Y. Acad. Sci. 961: 161 (2002). © 2002 New York Academy of Sciences.

Mechanical Signaling

DONALD INGBER

Departments of Pathology and Surgery, Harvard Medical School/Children's Hospital, Boston, Massachusetts 02115, USA

KEYWORDS: mechanical signaling; integrins; tissue engineering

In order to determine design criteria for tissue engineering, it is necessary to understand how complex physiological pathways function within the physical context of real tissues. It is recognized that tissue patterning and architecture are influenced by sequential activation of gene expression and metabolic pathways; however they are also significantly affected by mechanical factors. Mechanical stresses applied to cell surface adhesion receptors (e.g., integrins) can activate intracellular signaling pathways and induce gene transcription. Mechanical stresses that produce cell distortion also can switch cells between growth, differentiation, motility, and apoptosis programs.

Tissue growth and development are controlled through interplay between soluble cytokines, insoluble adhesion molecules, and mechanical forces. Importantly, all three types of signals converge on cell surface adhesion receptors that couple extracellular anchoring scaffolds (extracellular matrix, other cells) to the intracellular cytoskeleton. Recent work from many laboratories shows, for example, that integrins mediate transmembrane transfer of mechanical signals and that some forms of mechanochemical transduction occur within the specialized cytoskeletal complex known as the focal adhesion that forms at the site of integrin binding. However, while individual signaling cascades may integrate within the focal adhesion, the whole cell appears to function as the "mechanosensor" when it comes to behavioral control at the cell level. For example, application of mechanical stresses to integrins can activate specific signaling pathways and turn on gene expression equally well in round and spread (mechanically distorted) cells. Yet, round cells integrate those signals with other cues and turn on a cellular suicide program (apoptosis), whereas spread cells take in the same inputs and translate them into a growth response.

Taken together, these results suggest that when it comes to the engineering of artificial tissues, cell shape and extracellular matrix mechanics must be regarded as equally important contributors to developmental regulation as chemical signals. In fact, surgeons already make use of mechanical forces to achieve desired *in vivo*

Address for correspondence: Donald Ingber, Children's Hospital/Harvard Medical School, Enders 1007, 300 Longwood Avenue, Boston, MA 02115. Voice: 617-355-8031; fax: 617-232-7914.

donald.ingber@tch.Harvard.edu

Ann. N.Y. Acad. Sci. 961: 162–163 (2002). © 2002 New York Academy of Sciences.

responses (as in distraction osteogenesis and skin expansion). Thus, it is critical that future approaches to tissue engineering incorporate relevant aspects of cellular mechanoregulation. Factors that may be important for tissue engineering design include level and direction of mechanical strain as well as use of dynamic versus static force regimens. Selection of the appropriate oscillation frequency, amplitude, and work cycle form (e.g., sinusoidal versus step function) for force application also may be critical for the tissue remodeling response. But equally important will be the material properties and form of the microengineered tissue scaffolds (artificial extracellular matrices) on the micron scale because this is the physical microenvironment the cell experiences and hence, the design feature that has a most direct impact on cell behavior.

SELECTED READING

1. HUANG, S. & D.E. INGBER. 1999. The structural and mechanical complexity of cell growth control. Nat. Cell Biol. 1: E131–E138.
2. MEYER, C.J., F.J. ALENGHAT, P. RIM, et al. 2000. Mechanical control of cAMP signaling and gene transcription through activated integrins. Nat.Cell Biol. 2: 666–668.
3. CHEN, C.S. & D.E. INGBER. 1999. Tensegrity and mechanoregulation: from skeleton to cytoskeleton. Osteoarthritis Cartilage 7/1: 81–94.
4. CHICUREL, M., C.S. CHEN & D.E. INGBER. 1998. Cellular control lies in the balance of forces. Curr. Opin. Cell Biol. 10: 232–239.
5. CHICUREL, M.E., R.H. SINGER, C. MEYER & D.E. INGBER. 1998. Integrin binding and mechanical tension induce movement of mRNA and ribosomes to focal adhesions. Nature 392: 730–733.
6. CHEN, C.S., M. MRKSICH, S. HUANG, et al. 1997. Geometric control of cell life and death. Science 276: 1425–1428.
7. SINGHVI, R., A. KUMAR, G. LOPEZ, et al. 1994. 1994. Engineering cell shape and function. Science 264: 696–698.
8. INGBER, D.E. 1993. The riddle of morphogenesis: a question of solution chemistry or molecular cell engineering? Cell 75: 1249–1252.
9. INGBER, D.E. 1991. Integrins as mechanochemical transducers. Curr. Opin. Cell Biol. 3: 841–848.

Molecular Signaling

Breakout Session Summary

Moderators

CAROLINE DAMSKY, *University of California, San Francisco*
MOHAMMAD HEIDARAN, *BD Technologies*

Panelists

DONALD BOTTARO, *EntreMed, Inc.*
KATHRYN CROSSIN, *Scripps Research Institute*
PATRICIA DUCY, *Baylor College of Medicine*
DONALD INGBER, *Harvard Medical School/Children's HospitaL*

BROAD STATEMENT

Tissues and organs consist of specialized living cells arrayed within a complex structural and functional framework generally known as the extracellular matrix (ECM). ECM composition and mechanics are important factors that contribute to the function and characteristics of each organ and tissue, such as the rigidity and tensile strength of bone, the resilience of cartilage, the flexibility and hydrostatic strength of blood vessels, and the elasticity of skin. Also important is the role of the ECM during growth, development, and wound repair, where it serves as a reservoir for soluble signaling molecules and, through its own dynamic composition and mechanics, a source of additional signals to migrating, proliferating, and differentiating cells.

Artificial substitutes for the ECM, called scaffolds, consisting of natural and/or synthetic polymers, have been used successfully alone or in combination with cells and soluble factors to induce tissue formation and promote tissue repair. Cells are also central to many tissue engineering strategies, and significant efforts have been made to identify and propagate pluripotent stem cells, to identify signaling events important for proper differentiation, and to identify ideal microenvironments for maximum cellular function. These efforts have led to a convergence of research in bioengineering, biomaterials, ECM, cell growth and differentiation, and soluble factors that control cell fate.

Recent developments in the multidisciplinary field of tissue engineering have provided a novel set of tissue replacement parts and implementation strategies. Scientific advances in biomaterials, stem cells, growth and differentiation factors, and biomimetic environments have created unique opportunities to fabricate tissues in the laboratory from combinations of engineered ECMs (scaffolds), cells, and biologically active molecules.

Ann. N.Y. Acad. Sci. 961: 164–167 (2002). © 2002 New York Academy of Sciences.

VISION

The goals of reparative medicine are to enhance normal tissue regeneration and to engineer artificial tissues for use as replacements for damaged body parts. While significant advances have been made in the development of prosthetic devices that can repair structural defects (e.g., vascular grafts) and even replace complex mechanical behaviors (e.g., artificial joints), the challenge for the future is to develop therapies and devices that restore the normal biochemical functions of living tissues in addition to their structural features. To accomplish this objective, precise design criteria must be established to guide developmental efforts. These criteria must be based on a thorough understanding of the molecular signaling networks, cellular interactions, and biophysical aspects of tissue formation. Such understanding will accelerate our abilities to promote optimal tissue reconstruction and repair.

OBJECTIVES

To facilitate this vision, future research should address areas associated with growth factors (i.e., hormones, cytokines, and other factors that regulate cellular transduction and control cell behaviors), mechanical and stress-induced signaling, ECMs and scaffolds (including cell-to-matrix interactions), cell adhesion receptors and molecules, and temporal (aging) effects. Multidisciplinary and multiorganizational approaches to addressing these research needs are expected to be fruitful.

CHALLENGES

Among the major challenges for tissue engineering are the needs for complex functionality and biochemical stability in laboratory-grown tissues destined for transplantation. Realization of the potential benefits offered by tissue engineering in the development of true human replacement parts will require convergence of molecular signaling principles with research advances in tissue, matrix, growth factor, stem cell, and developmental biology. Specific challenges associated with developing a thorough understanding of molecular signaling and tissue regeneration include the following:

1. Understanding how to manipulate signaling through cell adhesion receptors and molecules to promote the desired endpoints for specific tissue engineering problems. Cell–cell and cell–ECM adhesion receptors and molecules play critical roles in both anchorage and signal transduction. A key feature of receptors is their ability to organize signaling complexes at sites of contact with their extracellular environment. Depending on other aspects of the environment including the nature and organization of the ECM, the presence of growth and differentiation factors, and the presence of mechanical stimuli, these signals can promote or restrain cell proliferation, promote differentiation, trigger matrix remodeling, or promote enhanced tissue organization. The ability to manipulate these variables to control the balance between proliferation and maintenance of the differentiated state for particular cell types is essential for designing effective cellular replacement therapies.

2. Establishing continuous molecular bridges between cell physiology, signal transduction, and gene expression. For a cell to respond to its environment, divide, migrate, or differentiate, signals from the extracellular compartment need to be sensed, reach the nucleus, and then trigger expression or repression of specific factors. Understanding of this information transmission process is currently incomplete. For example, while it is common to know which ligand/receptor interaction will activate a specific kinase pathway, specific insights into the transcriptional events and target genes that will eventually be involved are not known. Also, although numerous nuclear factors are known to control expression of specific genes (e.g., cell differentiation programs), very few signaling cascades have been defined as controlling their activity. Filling these gaps by establishing continuous molecular bridges is key to understanding cell transduction and differentiation.

3. Understanding the impact and mechanisms of growth factor signaling in complex systems where multiple biochemical and physical stimuli modify intracellular signaling and biological responses. Although much insight has been gained from studies of growth factor signaling using simple, well-characterized cultured cell models, studies involving more complex systems are necessary. The acquisition of high-resolution, three-dimensional extracellular matrix component/protein complex structures and the discovery of specificity among ECM components that impart high-affinity protein binding are currently at the forefront of research in this area. Information resulting from related research will facilitate the development of complex artificial scaffolds for tissue regeneration, repair , and replacement.

4. Translating knowledge of molecular events and molecules known to be important during embryonic development to the understanding of adult physiology and physiopathology. It is unlikely that these genetic programs involve totally different sets of regulatory genes. In fact, a small but growing number of genes have been shown to fulfill roles after as well as before birth. A systematic analysis of expression patterns and functions of developmental "master genes" in postnatal and aging models could provide useful information to define novel pathways and strategies for therapeutic intervention and regenerative repair.

5. Understanding how cells sense mechanical forces and integrate them with signals from other tissue control elements (e.g., growth factors and ECM). Cells in tissues constantly experience mechanical stimuli. Even cells in static culture experience the effects of gravity. Stimuli such as shear-stress, fluid-flow, compression, stretch, etc. not only alter the organization and distribution of structural elements and organelles within cells, but also become transduced into biochemical input that modulates intra- and intercellular signaling networks and in turn, gene expression. Understanding the importance of mechanical stresses and microarchitecture in cell signaling is important for the design of medical devices (i.e., for promoting wound healing) as well as the engineering and manufacture of artificial replacement tissue.

6. Fundamental research on cell response to three-dimensional, nonrigid extracellular matrices. Most work with cells *in vitro* has involved culture on two-dimensional rigid substrates. For many cell types, this is not a biologically relevant environment. Studies with three-dimensional matrices have shown that cells behave very differently under such conditions. Mechanisms by which signals from the ECM, neighboring cells, and growth/differentiation factors synergize to regulate cell growth and survival, affect commitment to tissue-specific differentiation programs, and regulate tissue remodeling need to be evaluated at the molecular level with regard to the more relevant three-dimensional microenvironments. Increased knowledge of the impact of three-dimensional microenvironments at the molecular level will expedite and improve tissue engineering strategies and biomaterials design.

ACKNOWLEDGMENTS

The Molecular Signaling Breakout Session was coordinated by Dr. Richard Swaja, NIBIB, NIH, and Ms. Mollie Sourwine, NIBIB, NIH.

The Provant® Wound Closure System Induces Activation of p44/42 MAP Kinase in Normal Cultured Human Fibroblasts

TERRI L. GILBERT,[a] NICOLE GRIFFIN,[b] JOHN MOFFETT,[c] MARY C. RITZ,[a] AND FRANK R. GEORGE[a]

[a]Regenesis Biomedical Inc., Scottsdale, Arizona 85257, USA

[b]Department of Microbiology, Arizona State University, Tempe, Arizona, USA

[c]Harrington Arthritis Research Center, Phoenix, Arizona 85006, USA

KEYWORDS: Provant; cell proliferation; fibroblasts; MAP kinase

INTRODUCTION

Healing of wounds to closure can be broken down into four phases: acute inflammation, granulation tissue formation, epithelialization, and tissue remodeling.[1] Chronic wounds—such as pressure ulcers—arise when one of these phases is interrupted. The Provant® Wound Closure System, based on Cell Proliferation Induction® (CPI®) technology, was developed as a novel bioactive approach to the serious need for effective wound treatment. Clinical trials show that Provant decreases time to closure for chronic wounds by 50%.[2]

CPI technology makes use of a specific radiofrequency stimulus to induce proliferation of fibroblasts, epithelial cells,[3] and lymphocytes.[4] These effects occur under isothermal conditions.[4,5] Because the response to CPI is mitogenic, mitogen-activated protein (MAP) kinase pathways might play a role. These pathways mediate many cellular processes, such as cell replication, transcription, and programmed cell death.[6–8] There are three major pathways, all characterized by the final kinase in the respective cascades: p38 MAP kinase, p44/42 MAP kinase, and the c-Jun N-terminal kinase (JNK). The p44/42 MAP kinase pathway is known for its mitogenic effects while the other kinases have been implicated in stress-induced pathways.[9]

Address for correspondence: Frank R. George, Regenesis Biomedical Inc., 1435 N. Hayden Road, Scottsdale, AZ 85257. Voice: 480-970-4970; fax: 480-970-8792.
george@regenesisbiomedical.com

Ann. N.Y. Acad. Sci. 961: 168–171 (2002). © 2002 New York Academy of Sciences.

METHODS

Proliferation Assay

Normal human fibroblast (HDF) cells were plated in 96-well plates at a density of 1,000–40,000 cells/well. Cells were Provant-treated for 30 min, and grown overnight. Media was aspirated and plates were stored at −80°C until quantified. Cell density was quantified using CyQuant GR dye (Molecular Probes), which excites at 480 nm and emits at 520 nm.

MAP Kinase Assay

Cells were treated with CPI signals for the stated times and then lysed with Laemmli sample buffer. Lysates were run on SDS–PAGE, transferred via semi-dry apparatus to PVDF membrane and immunoblotted for phosphorylated p44/42 MAP kinase. Protein bands were labeled using the ECF (Amersham) kit and detected by Phosphoimager.

RESULTS

When normal human fibroblasts (HDFs) in culture are treated with Provant, the cells are clearly more confluent after 16 hours than are sham-treated controls (FIG. 1). Cell growth of Provant-treated cells was 130–220% greater than controls. The effect is more robust in low serum concentrations. To determine whether p44/42

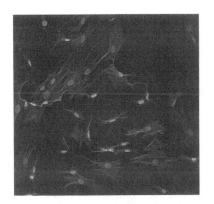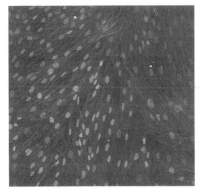

Sham Treated   Provant Treated
100X magnification

FIGURE 1. HDFs were treated with Provant, then allowed to grow 16 hours. Cells were fixed and stained with DAPI (nuclear-specific) and phalloidin labeled with Alexa Fluor 488 (actin-specific).

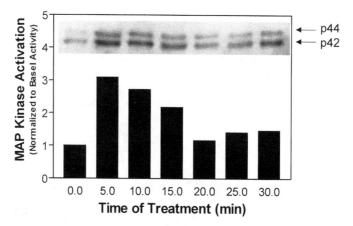

FIGURE 2. Using Western blotting techniques, lysates generated after initiation of treatment were probed for activated p44/42 MAP kinase. The graph is normalized to basal activity.

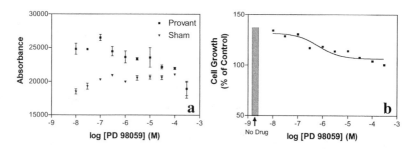

FIGURE 3. (a) PD 98059-induced decrease in growth of Provant-treated cells compared to controls. (b) Normalized dose–response curve. *Bar* denotes cell growth in the absence of drug.

MAP kinase was activated, cell lysates from Provant-treated cultures were immunoblotted for phosphorylated p44/42 MAP kinase (FIG. 2). The enzyme that phosphorylates p44/42 is referred to as MEK. To confirm that this cascade was activated, we treated cells with PD 98059, a specific inhibitor of MEK, prior to treatment with CPI signals. FIGURE 3a shows that the CPI effect is blocked as PD 98059 concentration increases. A normalized curve is shown in FIGURE 3b.

CONCLUSION

Treatment of normal human fibroblasts with Provant increases cell proliferation by up to two-fold within 24 hours. This effect is most pronounced when cells are grown in serum-poor media. This model system most likely mirrors the environment encountered in chronic wounds. This suggested to us that Provant is acting as an exogenous, nonmolecular mitogen. When we searched for the activation of p44/42 MAP kinase, a typical mitogenic signaling molecule, in response to Provant treatment, we saw activation of p44/42 during the treatment. MAP kinase activation within the first 30 minutes of treatment suggests that CPI signals may be activating the p44/42 MAP kinase cascade. The ability to induce cells to start proliferating, especially in chronic wounds, makes the Provant® Wound Closure System a valuable addition to the field of reparative medicine.

REFERENCES

1. SCHAFFER, C.J. & L.B. NANNEY. 1996. Cell biology of wound healing. Int. Rev. Cytol. **169:** 151–181.
2. GEORGE, F.R., R. GALLEGOS, W.D. DACHMAN, *et al.* 2002. Clinical efficacy of cell proliferation induction (CPI): a novel biophysical stimulus for accelerated wound Healing to closure. Wounds. In press.
3. GEORGE, F.R., R.J. LUKAS, J. MOFFETT, *et al.* 2002. In vitro mechanisms of cell proliferation induction (cpi): a novel bioactive treatment for accelerating wound healing. Wounds **14:** 107–115.
4. CLEARY, S.F., L.M. LIU & R.E. MERCHANT. 1990. In vitro lymphocyte proliferation induced by radio-frequency electromagnetic radiation under isothermal conditions. Bioelectromagnetics **11:** 47–56.
5. CLEARY, S.F., L.M. LIU & R.E. MERCHANT. 1990. Glioma proliferation modulated in vitro by isothermal radiofrequency radiation exposure. Radiat. Res. **121:** 38–45.
6. DENT, P., W.D. JARVIS, M.J. BIRRER, *et al.* 1998. The roles of signaling by the p42/p44 mitogen-activated protein (MAP) kinase pathway: a potential route to radio- and chemo-sensitization of tumor cells resulting in the induction of apoptosis and loss of clonogenicity. Leukemia **12:** 1843–1850.
7. PELECH, S.L. & D.L. CHAREST. 1995. MAP kinase-dependent pathways in cell cycle control. Prog. Cell Cycle Res. **1:** 33–52.
8. DAVIS, R.J. 1995. Transcriptional regulation by MAP kinases. Mol. Reprod. Dev. **42:** 459–467.
9. COBB, M. H. 1999. MAP kinase pathways. Prog. Biophys. Mol. Biol. **71:** 479–500.

p38 MAP Kinase Regulation of AP-2 Binding in TGF-β1-Stimulated Chondrogenesis of Human Trabecular Bone-Derived Cells

R. TULI,[a] M.R. SEGHATOLESLAMI,[b] S. TULI,[a] M.S. HOWARD,[b] K.G. DANIELSON,[b] AND R.S. TUAN[a]

[a]Cartilage Biology and Orthopaedics Branch, National Institute of Arthritis and Musculoskeletal and Skin Diseases, National Institutes of Health, Bethesda, Maryland 20892, USA

[b]Department of Orthopaedic Surgery, Thomas Jefferson University, Philadelphia, Pennsylvania 19107, USA

ABSTRACT: Collagenase-treated, explanted human trabecular-bone chips are an excellent source of osteoblast-like cells. We have recently shown the multiple differentiation potential of these cells; in addition to osteogenesis and adipogenesis, these cells also undergo chondrogenesis when maintained as high-density pellet cultures (250,000 cells/pellet) in a serum-free, chemically defined medium stimulated with TGF-β1 (10 ng/mL). In this investigation, we have analyzed how transactivating nuclear transcription factors, specifically AP-2 and SP-1, may interact with common cis-acting elements found in the regulatory region of cartilage-specific genes as part of the signal transduction mechanism of TGF-β1 and p38 during chondrogenesis of human trabecular bone–derived multipotential cells. Both TGF-β1 stimulation and p38 MAP kinase activation affect the binding of AP-2 as well as SP-1 to oligonucleotides with sequence similarity to the overlapping AP-2/SP-1 sites found in the putative 52-bp immediate upstream regulatory region and the 5′-untranslated region of the human aggrecan gene. Electrophoretic mobility shift assays show that TGF-β1 treatment of the bone-derived cells inhibits AP-2 DNA binding but enhances the DNA binding ability of SP-1. Additionally, treatment of these TGF-β1–stimulated cells with p38 MAP kinase inhibitor, SB203580, rescued the AP-2 DNA binding but did not affect SP-1 DNA binding. These findings indicate that AP-2 DNA binding is the target of both TGF-β1 and p38 MAP kinase signaling pathways and suggest a possible signal transduction cascade whereby TGF-β1 induction of chondrogenesis involves the activation of p38 MAP kinase and the subsequent inhibition of DNA binding by AP-2, thereby preventing the transcriptional repression of the aggrecan gene.

KEYWORDS: mesenchymal progenitor cell; human trabecular bone; chondrogenesis; p38 mitogen-activated protein (MAP) kinase; activated protein–2 (AP-2)

Address for correspondence: Rocky S. Tuan, Ph.D., Cartilage Biology and Orthopaedics Branch, National Institute of Arthritis and Musculoskeletal and Skin Diseases, National Institutes of Health, 50 South Drive, Building 50, Room 1503, Bethesda, MD 20892-8022. Voice: 301-451-6854; fax: 301-402-2724.

tuanr@mail.nih.gov

Ann. N.Y. Acad. Sci. 961: 172–177 (2002). © 2002 New York Academy of Sciences.

INTRODUCTION

Collagenase-treated, explanted human trabecular-bone chips are an excellent source of osteoblast-like cells.[1] We have recently shown the multiple differentiation potential of these cells.[2–4] These cells are capable of undergoing chondrogenesis when maintained as high-density pellet cultures (250,000 cells/pellet) in a serum-free, chemically defined medium and stimulated with TGF-β1 (10 ng/ml).[5–7] In this study, we have analyzed the role of p38 MAP kinase as a potential component of the TGF-β1-stimulated signaling cascade involved in human osteoblast (hOB) chondrogenic gene expression and differentiation. Recent work by Valmhu *et al.* points to the importance of the proximal promoter (700 bp) and the untranslated region encompassing the first exon of the human aggrecan gene in its high expression during chondrogenesis.[8] Since numerous closely positioned Sp-1 and AP-2 response elements are identified in these regulatory regions, we have examined the possible alterations in binding and levels of these transcription factors as potential targets for the TGF-β- and/or P38-induced chondrogenic signaling activity. Here we report that the changes in the levels of cellular binding of Sp-1- and AP-2-responsive DNA sequences are indeed some of the targets of TGF-β1-induced chondrogenic activity in hOB pellet cultures. In addition, we report that TGF-β1-regulated alterations of AP-2 binding, but not Sp-1, response element is mediated by the activity of p38 MAP kinase.

MATERIALS AND METHODS

Chondrogenic Differentiation of High-Density Pellet Cell Cultures

For chondrogenic differentiation, cells were cultured for 7 days as high-density cell pellets in a serum-free, chemically defined medium as described previously.[5–7] Control cell cultures were maintained without TGF-β1. For inhibition studies, p38 inhibitor (SB203580) was used at a final concentration of 5 mM. The culture medium was changed every 3 days.

RNA Isolation and RT-PCR

Total RNA from the pellet cultures was extracted using the methodology for TriZol reagent RNA extraction (Gibco BRL, Life Technologies). RNA (2 μg) was reverse-transcribed using the Superscript First-Strand Synthesis System for RT-PCR (Gibco BRL, Life Technologies). The following primers were used to amplify human aggrecan and glyceraldehyde 3-phosphate dehydrogenase (GAPDH; internal control) genes. Aggrecan: sense, TGAGGAAGGGCTGGAACAAGTACC; antisense, GGAGGTGGTAATTGCAGGGAACA. GAPDH: sense, GGGCTGCTT-TTAACTCTGGT; antisense, TGGCAGGTTTTTCTAGACGG.

Preparation of Nuclear Extracts

For nuclear extraction, 12 pellet cultures were used per treatment. On day 7, pellets were washed with cold PBS, resuspended in 400 μl of buffer A (10 mM Hepes-

KOH, pH 7.9; 1.5 mM $MgCl_2$; 10 mM KCl; 0.5 mM DTT; 10 µM glycerol phosphate), containing 50 µl/ml protease inhibitor cocktail for use with mammalian cell and tissue extracts (Sigma) and phosphatase inhibitor cocktail II (Sigma), then incubated on ice for 30 minutes and homogenized using a Dounce homogenizer (30 strokes). The nuclear fraction was collected by centrifugation at $1000 \times g$ for 10 min at 4°C, resuspended in 50 µl of buffer C (20 mM Hepes-KOH, pH 7.9; 25% glycerol; 420 mM NaCl; 1.5 mM $MgCl_2$; 0.2 mM EDTA; 0.5mM DTT; 10 µM glycerol phosphate) containing the same protease and phosphatase inhibitors as in solution A, and then incubated on ice for 90 min and harvested by centrifugation at $3000 \times g$ for 2 min. Aliquots of nuclear extract were stored at −70°C until use. Protein concentration was determined using the Bradford protein assay system (Pierce).

Gel Mobility Shift Analysis

Complementary oligo deoxynucleotides (ODN) corresponding to the aggrecan gene sequences from −232 to −249 (5′-TCCGCGCCGCCCCGGGAG-3′, 5′-CTCCCGGGGCGGCGCGGA-3′), and from −4 to +15 (5′-ATGCCCGCCCGC-CCGCCCAC-3′, 5′-GTGGGCGGGCGGGCGGGCAT-3′), were annealed by ramping from 94°C to 45°C at a rate of 0.5°C per 10 seconds. Labeling of the double-stranded oligo deoxynucleotides (ODN), binding, and gel shift assays were done according to the protocol provided by the Gel Shift Assay System kit obtained from Promega. In these binding assays, 2–4 µg of nuclear extracts and 500,000 cpm radioactive labels were used.

RESULTS AND DISCUSSION

The cell signaling events required for cartilage-specific core protein (aggrecan) gene expression in multipotential mesenchymal hOB cells isolated from trabecular bone are initiated and maintained once these cells are aggregated as pellet cultures and treated with TGF-β1. Our present work shows that this TGF-β1 regulation of aggrecan gene expression is mediated by the activity of p38 MAP kinase. Inhibition of p38 activity diminishes the aggrecan gene expression induced by TGF-β1 by 30% (FIG. 1), thereby suggesting possible cross-talk between Smad and MAP kinase signaling pathways, or direct activation of the MAP kinase pathway by TGF-β1. On the basis of previous work by Valmhu et al.,[8] who have identified numerous Sp-1 and AP-2 binding sites in the regulatory regions of the human aggrecan gene, we have examined the effects of TGF-β1 induction and p38 activity on the binding of Sp-1, a transcription factor shown to cooperate with TGF-β1-activated Smad proteins to induce collagen type I gene expression in human glomerular mesangial cells,[11] and AP-2, a transcription factor shown to have a major role in the joint development of Drosophila.[12]

Our results indicate that the nuclear extract from chondrogenic TGF-β1-stimulated hOB pellet cultures binds at a much lower efficiency to both upstream and downstream overlapping AP-2/Sp-1-responsive elements found in the regulatory regions of the aggrecan gene, as compared to the nuclear extracts from untreated cultures (FIG. 2). To determine whether the binding of both AP-2 and Sp-1 was individually

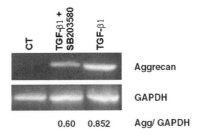

FIGURE 1. Effect of p38 MAP kinase inhibition on aggrecan gene expression in hOB pellet cultures induced with TGF-β1. Aggrecan gene expression is induced with TGF-β1, and the addition of p38 MAP kinase inhibitor SB203580 diminishes (30%) the effect of TGF-β1-induced aggrecan gene expression.

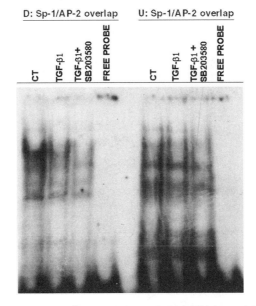

FIGURE 2. Effects of TGF-β1 stimulation and p38 inhibition on hOB nuclear binding to the aggrecan promoter regions containing AP-2/Sp-1 overlapping response elements. Both oligos (D: downstream, −4 to +15; U: upstream, −232 to −249) bind with less affinity to the TGF-β1-induced and SB203580-inhibited hOB cell extracts.

affected in these cell extracts, mobility gel shift assays were performed using separate consensus binding elements responsive to AP-2 or Sp-1 nuclear factors (FIG. 3). Nuclear extract from TGF-β1-induced hOB pellet cultures had less binding capacity (28%) to the AP-2-responsive ODN as compared to uninduced cultures (CT). This binding pattern was consistent with the endogenous AP-2/Sp-1 overlapping sequences. However, the TGF-β1-induced nuclear extract bound at a higher level (37%) to a Sp-1 consensus binding element, as compared to the uninduced extract

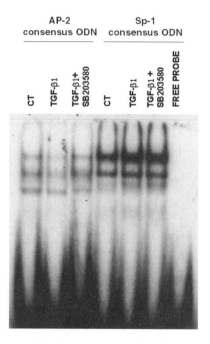

FIGURE 3. Inhibition of p38 MAP kinase affects TGF-β1 stimulated hOB nuclear binding to the AP-2 but not Sp-1 consensus binding element. Consensus sequences for oligodeoxynucleotides (ODN) AP-2: 5′-GATCGAACTGACCGCCCGCGGCCCGT-3′, 3′-TAGCTTGACTGGCGGG-CGCCGGGCA-5; Sp-1:5′-ATTCGATCGGGGCGG-GGCGAC-3′,3′-TAAGCTACGGCCGCCCCGCTG-5′.Binding reactions were done at room temperature for 20 min. In competition experiments using 1X concentrations of the respective cold probes, full competition of the banding patterns of both AP-1 and Sp-1 consensus ODNs was seen (data not shown).

(CT). Inhibition of p38 MAP kinase in the TGF-β1–induced cultures resulted in higher binding of the AP-2 ODN (15%, compared to CT), and had no effect on the Sp-1 ODN when compared to the TGF-β1-induced extract. These differences in binding patterns of consensus sequences versus endogenous AP-2/Sp-1 overlapping sequences suggest the individual AP-2 and Sp-1 binding to be influenced by the specificity of its binding sequences as well as possible interactions with other DNA-dependent binding nuclear factor(s) yet to be identified.

Therefore, treatment of cultures with SB203580, a potent p38 MAP kinase inhibitor, substantially downregulates TGF-β1-induced aggrecan gene expression, as evidenced by RT-PCR in our pellet culture system. These results implicate p38 MAP kinase as one of the common regulators mediating the signals initiated by TGF-β1 for chondrogenic progression in our pellet culture system. We have consistently found that, when cultured with both TGF-β1 and SB203580, AP-2 binding increases significantly, whereas Sp-1 binding is unaffected. This suggests a possible signal transduction cascade whereby TGF-β1 activation of p38 MAP kinase results in the inhibition of AP-2 DNA binding, which in turn leads to increased aggrecan gene expression. Taken together, these data suggest that chondrogenic induction of hOB cells by TGF-β1 can affect the binding of AP-2 as well as Sp-1 directly, and that AP-2 might act as a negative regulator of aggrecan gene expression.

ACKNOWLEDGMENTS

This study was supported in part by Grants AR44501, DE12864, AR39740, AR45181, and AR47396 from the National Institutes of Health.

REFERENCES

1. ROBEY, P.G. & J.D. TERMINE. 1985. Human bone cells in vitro. Calcif. Tissue Int. **37:** 453–460.
2. NOTH, U., A. OSYCZKA, R. TULI, *et al.* 2000. Trabecular bone derived human mesenchymal stem cells. Jefferson Orthopaedic J. **24:** 65–73.
3. NUTTALL, M.E., A.J. PATTON, D.L. OLIVERA, *et al.* 1998. Human trabecular bone cells are able to express both osteoblastic and adipocytic phenotype: implications for osteopenic disorders. J. Bone Miner. Res. **13:** 371–382.
4. NUTTALL, M.E. & J.M. GIMBLE. 2000. Is there a therapeutic opportunity to either prevent or treat osteopenic disorders by inhibiting marrow adipogenesis? Bone **27:** 177–184.
5. JOHNSTONE, B., T.M. HERRING, A.I. CAPLAN, *et al.* 1998. In vitro chondrogenesis of bone marrow-derived mesenchymal progenitor cells. Exp. Cell Res. **238:** 265–272.
6. YOO, J.U., T.S. BARTHEL, K. NISHIMURA, *et al.* 1998. The chondrogenic potential of human bone-marrow-derived mesenchymal progenitor cells. J. Bone Joint Surg. **80–A:** 1745–1757.
7. MACKAY, A.M., S.C. BECK, J.M. MURPHY, *et al.* 1998. Chondrogenic differentiation of cultured human mesenchymal stem cells from marrow. Tissue Eng. **4:** 415–428.
8. VALMHU, W.B., G.D. PALMER, J. DOBSON, *et al.* 1998. Regulatory activities of the 5'- and 3'-untranslated regions and promoter of the human aggrecan gene. J. Biol. Chem. **273:** 6196–6202.
9. PONCELET, A.C. & W.H. SCHNAPER. 2001. Sp1 and Smad proteins cooperate to mediate transforming growth factor-β1-induced collagen expression in human glomerular mesangial cells. **276:** 6983–6992.
10. KERBER, B., I. MONGE, M. MUELLER, *et al.* 2001. The AP-2 transcription factor is required for joint formation and cell survival in *Drosophila* leg development. Genes Devel. **128:** 1231–1238.

TGF-β1-Stimulated Osteoblasts Require Intracellular Calcium Signaling for Enhanced α5 Integrin Expression

LEON J. NESTI, [a,b] E.J. CATERSON,[a] MARK WANG,[a] RICHARD CHANG,[a] FELIX CHAPOVSKY,[a] JAN B. HOEK,[b] AND ROCKY S. TUAN[a]

Departments of [a]Orthopaedic Surgery and [b]Pathology, Anatomy, and Cell Biology, Thomas Jefferson University, Philadelphia, Pennsylvania 19107, USA

ABSTRACT: The osteoactive factor, transforming growth factor β1 (TGF-β1), influences osteoblast activity and bone function. We recently characterized a Smad-independent TGF-β1-induced Ca^{2+} signal in human osteoblasts (HOB) and demonstrated its importance in cell adhesion. Here, we further elucidate the role of the TGF-β1 Ca^{2+} signal in the mechanics of HOB adhesion. Osteoblast interaction with fibronectin (FN) through α5β1 integrin is principally responsible for osteoblast-substrate adhesion. Our results show that the TGF-β1 intracellular Ca^{2+} signal is responsible, in part, for stimulation of α5 integrin expression, but not β1 integrin or FN expression. Increased α5 integrin protein and mRNA expression was seen as early as 12 h after TGF-β1 treatment, but was inhibited by cotreatment with nifedipine, a Ca^{2+} channel blocker. TGF-β1 increased both FN and β1 integrin protein production within 48 h, independent of nifedipine cotreatment. Immunofluorescence observations revealed that TGF-β1 increased α5 integrin staining, clustering, and colocalization with the actin cytoskeleton, effects that were blocked by nifedipine. The TGF-β1 Ca^{2+} signal, a pathway crucial for HOB adhesion, enhances α5 integrin expression, focal contact formation, and cytoskeleton reorganization. These early events are necessary for osteoblast adhesion; thus they determine the fate of the cell and ultimately affect bone function.

KEYWORDS: TGF-β signaling; orthopedic tissue engineering

INTRODUCTION

Substantial information is available indicating that the osteoactive factor, transforming growth factor-β1 (TGF-β1), which is crucial for bone growth and development, influences osteoblast adhesion, proliferation, differentiation, and maturation. These properties are important in considering TGF-β1 for orthopedic tissue engineering applications such as fracture repair, implant fixation, and structural defects. The direct effect of TGF-β1 on osteoblast activity has been well documented; however, little is known about the mechanisms through which TGF-β1 exerts these ef-

Address for correspondence: Leon J. Nesti, Department of Orthopaedic Surgery, Thomas Jefferson University, 501 Curtis Building, 1015 Walnut Street, Philadelphia, PA 19107. Voice: 215-955-4321; fax: 215-955-4317.

leonnesti@yahoo.com

Ann. N.Y. Acad. Sci. 961: 178–182 (2002). © 2002 New York Academy of Sciences.

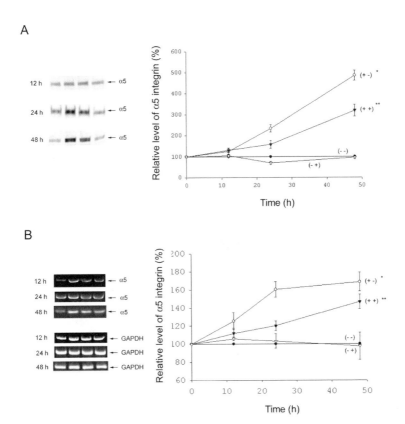

FIGURE 1. TGF-β1 effect on α5 integrin expression in HOBs analyzed by immunoblot (**A**) and RT-PCR (**B**). (**A**) TGF-β1 stimulation significantly increases α5 integrin protein level (* = P <0.005 compared to control). (**B**) TGF-β1 stimulation significantly increases the level of α5 integrin mRNA (* = P <0.005 compared to control). Blocking the TGF-β1 Ca^{2+} signal significantly inhibits α5 integrin protein (* = P <0.005) and mRNA (* *= P <0.05) levels. Nifedipine treatment alone showed no significant change compared to control. (– – = control; + – = TGF-β; + + = TGF-β1 and nifedipine; – + = nifedipine)

fects. Recently, we have described a novel TGF-β1 Ca^{2+} signaling pathway in primary human osteoblasts (HOBs) and have demonstrated its importance in osteoblast adhesion (unpublished data). Here, we further elucidate the mechanism through which the TGF-β1 Ca^{2+} signal influences osteoblast adhesion.

METHODS

Cell Preparation

Cultured primary human osteoblasts were isolated using the protocol of Robey and Termine.[1,2]

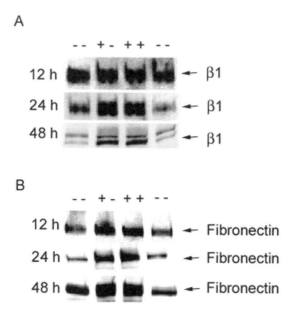

FIGURE 2. TGF-β1 effect on β1 integrin and fibronectin analyzed by immunoblot. TGF-β1 stimulation of HOBs increases β1 integrin (**A**) and fibronectin (**B**) levels at 12 hours, 24 hours, and 48 hours. Densitometric analysis normalized to control shows that blocking the Ca^{2+} signal with nifedipine does not affect the TGF-β1-stimulated increase in protein levels of fibronectin or β1 integrin. Nifedipine treatment alone shows no significant change compared to control. (− − = control, + − = TGF-β, + + = TGF-β1 and nifedipine, − + = nifedipine).

$[Ca^{2+}]_i$ Measurement

TGF-β1-induced $[Ca^{2+}]_i$ response was measured in Fura-2-loaded HOBs by fluorescence imaging using a previously described set-up (204 image sets, 5-sec intervals, 200-msec shutter speed).[3,4] TGF-β1 (R&D Systems, 10 ng/ml) was added, and $[Ca^{2+}]_i$ was measured for 15 minutes.

Immunoblot Analysis

TGF-β1 (10 ng/ml) and/or nifedipine (10 mM) were added to HOBs prior to protein extraction, and untreated cells served as control. Extracts were separated, incubated with antibodies to α5 integrin, β1 integrin, or fibronectin (Trans Labs), and immunoreactive protein was detected by HRP-chemiluminescence.

RT-PCR Analysis

Total RNA was isolated from control and treated HOB using a one step RT-PCR method (Titan) using primers for α5 integrin and GAPDH.

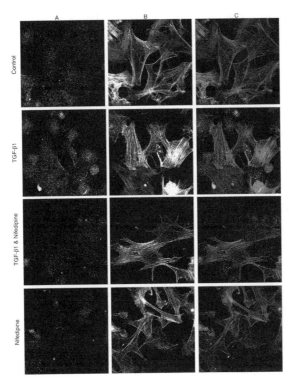

FIGURE 3. Effects of TGF-β1 on spatial distribution of α5 integrin and F-actin. TGF-β1-treated groups demonstrate increased α5 integrin staining, focal contact formation (*arrows*), and actin fiber formation and organization (*arrows*). Focal adhesion contacts are evidenced by overlap of the α5 integrin and F-actin images (*arrows*).

Immunohistochemistry/Cytoskeletal Staining

HOBs were treated with/without TGF-β1 and/or nifedipine for 48 hours, plated for 3 hours, and fixed, permeabilized, and immunostained with α5 integrin antibody (Trans Labs) and rhodamine-conjugated phalloidin (Molecular Probes). A confocal laser scanning microscope (CSLM) was used for two-channel acquisition.

RESULTS AND DISCUSSION

The mechanism of TGF-β1 signaling is thought to be mediated by receptor-associated Smad proteins.[5] We have recently characterized a TGF-β1-induced Ca^{2+} signal in osteoblasts and have demonstrated its importance in cell adhesion and independence from Smad signaling (unpublished data). Osteoblast interaction with the extracellular matrix protein fibronectin by α5β1 integrin on the cell surface is principally responsible for osteoblast substrate adhesion. Our results show that the TGF-β1-induced intracellular Ca^{2+} signal is responsible, in part, for the stimulation of α5 integrin expression, but not β1 integrin or fibronectin expression in HOBs (FIGS. 1 and 2). Increased α5 integrin protein and mRNA expression was seen as early as 12 hours after TGF-β1 treatment, but was inhibited by co-treatment of cells with nifedipine, an L-type Ca^{2+} channel blocker (FIG. 1 A and B). TGF-β1 treatment increased fibronectin and β1 integrin protein production within 48 hours in a manner unaffected by nifedipine treatment (FIG. 2 A and B). Immunofluorescence observations revealed increased α5 integrin staining, α5 integrin clustering, and colocalization with the actin cytoskeleton after TGF-β1 treatment, effects that were blocked by co-treatment with nifedipine (FIG. 3). TGF-β1-induced intracellular Ca^{2+} signal is thus important in the regulation of α5 integrin expression and distribution in HOBs. HOB adhesion is a crucial first step that may ultimately determine the fate of the cell. Understanding the mechanisms through with TGF-β1 affects this crucial first step, will allow us to more effectively use this growth factor for tissue engineering.

ACKNOWLEDGMENTS

We thank Paul Anderson, BME, EE, for imaging system and computer interface support. This work is supported in part by NIH Grants DE11327, DE16864, AR44501, AA08714, AA07186, and AA07215. L.J.N. is a recipient of NIH M.D-Ph.D. NRSA Grant F30AA05516.

REFERENCES

1. ROBEY, P.G. & J.D. TERMINE. 1985. Human bone cells *in vitro*. Calcif. Tiss. Int. **37:** 453–460.
2. SINHA, R.K. & R.T. TUAN. 1996. Regulation of human osteoblast integrin expression by orthopaedic implant material. Bone **18:** 451–457.
3. TSIEN, R. 1989. Fluorescent indicators of ion concentrations. Meth. Cell Biol. **30:** 127–156.
4. ROONEY, T. *et al.* 1989. Characterization of cytosolic calcium oscillations induced by phenylephrine and vasopressin in single fura-2-loaded hepatocytes. J. Biol. Chem. **264:** 17131–17141.
5. BLOBE, G.C. *et al.* 2000. Role of transforming growth factor-β in human disease. N. Engl. J. Med. **342:** 1350–1358.

Tissue Engineering

Functional Assessment and Clinical Outcome

STEVEN A. GOLDSTEIN

Orthopaedic Research Laboratories, Department of Orthopaedic Surgery, University of Michigan, Ann Arbor, Michigan 48109, USA

ABSTRACT: The issues that should be considered as part of the design and evaluation of tissue engineering constructs with respect to their targeted clinical application are reviewed. This paper provides a general framework for the process of bringing tissue-engineering constructs from the laboratory bench to the patient's bedside, rather than presenting a detailed review of the engineering or biologic principles or mechanisms that are necessary for successful tissue engineering. Many of the principles are animated by using examples from current studies being developed in my laboratory or those of my collaborators. In all likelihood, multiple solutions or approaches will be found that lead to successful tissue-engineering constructs. The focus here is on the identification of critical parameters to be considered rather than specific design solutions. The review is therefore organized to reflect feasible sequences of activities formulated to take tissue engineering from concept to clinical reality.

KEYWORDS: tissue engineering; bone graft substitute; gene-activated matrix

CONSTRUCT CONCEPTUALIZATION

Successful tissue engineering begins with a clear and precise definition of the clinical demand or problem being addressed. Regardless of whether the application is focused on replacing or augmenting failing heart tissue or on creating bone graft substitute materials, the design objective should begin with the delineation of the multiple specific attributes of the identified clinical problem. The incidence, prevalence, and demographics of the clinical problem need to be identified, and an appreciation for its complexity, with respect to a therapeutic intervention, needs to be developed. The current treatment options or standard of care must be accurately identified, since the new tissue-engineering solution must provide some advantage over the current therapy to merit investment in its development. It is also important to identify any unusual temporal factors associated with either the treatment or potential clinical evaluation. Most importantly, clarity in identifying the clinical problem is also the first step towards defining the required functional properties of the tissue-engineering solution. It is important to begin by identifying these targeted

Address for correspondence: Steven A. Goldstein, Ph.D., Orthopaedic Research Laboratories, Room G161, Mail Stop 0486, 400 North Ingalls, Ann Arbor, MI 48109-0486. Voice: 734-763-9674; fax: 734-647-0003.

stevegld@umich.edu

Ann. N.Y. Acad. Sci. 961: 183–192 (2002). © 2002 New York Academy of Sciences.

functional properties in a quantitative manner, since they will become the reference of comparison for outcome measures that will be used to qualify the tissue-engineering construct as being successful or unsuccessful. For example, if the tissue-engineering target is bone replacement, the required functional properties might be categorized as follows:

Biomechanical Behavior of Bone
Material properties
Architecture/geometry/morphology
Organ level mechanical function

Biologic Function
Regulatory maintenance
Participation in physiologic homeostasis
Capacity for repair, adaptation and remodeling

As will be noted below, the critical targeted functional properties identified for specific tissue-engineering construct designs are important for guiding the design phase or discovery phase, as well as for providing specific reference variables to use as outcome measures in preclinical and eventual clinical trials.

DESIGN AND FUNCTION HIERARCHY

Following a paradigm similar to that used to solve an engineering design problem, the next phase associated with successful tissue engineering is focused on characterizing the specific design attributes and technology that will serve as the backbone of the construct under consideration. I have found it particularly valuable to consider the design attributes and targeted functions within a hierarchical framework. This concept can be demonstrated in FIGURE 1, by reviewing the images used in designing a bone tissue-engineering construct.

The targeted clinical property is a bone that has mechanical properties and geometry that enable the patient to ambulate and participate in activities of daily living. While, at the organ level, the bone must provide these macroscopic properties, bone structure and function can be decomposed into multiple hierarchical scales. These scales range from tissue morphology through the extracellular matrix to the included cells, and finally the intracellular machinery responsible for orchestrating gene expression and protein synthesis. The only way to attain the targeted organ level properties is through appropriate interactions of all of the constituents within each architectural hierarchical level as well as the correct integration of signals or properties across scales. This perspective may be valuable in that it can provide a guide for not only design conceptualization, but also evaluation of the function of constructs through a series of logical sequential tests at multiple scales.

In practice, the ability to perform this segmentation can faciliatate a design optimization process. Since all tissue engineering constructs involve the use of cells, matrices, and biofactors, this hierarchical paradigm can be used to assess how the source or manipulation of cells (*in vivo* or *ex vivo*, for example), the design of the matrices, and the choice of biofactors can influence the behavior of the construct. At the conclusion of this design phase, the hierarchical analyses, both within scale and

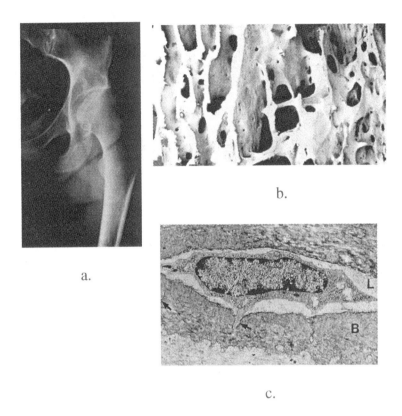

FIGURE 1. Simplified hierarchical pattern of bone. Clinical function is defined at the organ level (**a**) as depicted by the function of the whole femur. The function of the femur, however, is dependent on the properties/function of the bone tissue architecture and properties (**b**), which is dependent on the cell/matrix interactions (**c**).

across scale, provide the proof of concept for the technology. Most often, this proof of concept involves cellular, molecular, or physiologic analyses *in vitro* and, importantly, *in vivo*.

Most investigators begin the *in vivo* verification of their technology in small animals with a particular focus on those biomechanical or biologic outcome variables that are most critical with respect to the targeted functional properties that were defined as part of the clinical conceptualization. This is also the stage at which a determination can be made concerning the potential of the technology to progress towards clinical utility. With convincing proof of concept and early *in vivo* demonstration of functional efficacy, transition to pivotal preclinical studies is often warranted.

DESIGN OF PIVOTAL PRECLINICAL STUDIES

Typically, the development of a tissue engineering construct requires the performance of several preclinical studies prior to evaluation in human subjects. As discussed earlier, some preclinical studies are a part of the evaluation of proof of concept and are followed by expanded studies that are designed to evaluate efficacy in a relevant or well-accepted model. Usually, parallel studies are also performed to evaluate safety. These safety studies are frequently performed by third-party contract laboratories that have extensive experience with assays to evaluate toxicity and other safety-related issues.

For many tissue-engineering constructs, the pivotal preclinical studies designed to demonstrate efficacy involve a transition from the use of small animals to large-animal models. While some of the attributes of the small-animal models will be discussed below, the need for large animals may be dependent on a need to evaluate responses of the construct under conditions that better simulate a physiologic match with the human clinical condition. The preclinical studies typically include an evaluation of dose or utilization profile, and success is significantly dependent on choosing appropriate endpoint measures. The safety studies, in contrast, are focused on evaluating toxicity, either locally or systemically, bioavailability, and the relationship between bioavailability, toxicity, and dose. Often, the safety studies can also provide assessment of the effects of storage or sterilization on the properties of the construct.

The designs, as well as the success of preclinical studies, are dependent on a number of issues. These include:
(1) the experimental model;
(2) functional assays;
(3) dose response;
(4) correspondence of preclinical and clinical variables; and
(5) safety considerations and assessments.

Experimental Model

The appropriate choice of an experimental model is critical to the success of the preclinical studies. Part of the criteria associated with choice or design of the experimental model is related to the targeted properties and, often, the expected commercial market of the tissue-engineered construct. For example, if the tissue-engineered construct is designed for a very specific clinical application, and a market strategy includes a relatively narrowly defined use, then the experimental model should be designed to simulate the clinical condition that is targeted. On the other hand, if the goal is to seek approval of a construct to be used in a broad range of clinical applications, then the experimental model may not precisely resemble any one specific clinical condition, but instead be general enough such that the results can be translated to this broad set of clinical applications. Examples of these two design objectives are illustrated as follows:

Specific Clinical Application

An example of a specific clinical application is the development of a tissue-engineered bone graft substitute for use in augmentation of spine fusions. The expected clinical use of the graft substitute might also include incorporation within a specialized device to enhance the spinal fusion. Given the specificity of this clinical target, the experimental design would likely utilize an animal model of spine fusion at a location that might be similar to that of human clinical need and would also use a device that would be similar to the device used in human patients.

Broad Clinical Application

In contrast, development of a bone-tissue-engineered bone graft substitute material for enhancing fractures or filling bony defects might be intended for use in any defect in long bones. This broad application requires an animal model that includes a defect that supports a healing cascade that simulates a response expected to occur in many different sites.

In addition to a careful choice of animal model based on simulating clinical conditions, it is perhaps even more critical to consider the biologic and physiologic compatibility of the chosen animal to human physiology. Using the example of the bone graft substitute, the animal model chosen should be evaluated for its ability to simulate human bone chemistry, metabolic turnover, and morphology.

Quantitative Functional Assays

Quantitative and reproducible functional assays must be developed and utilized in the pivotal preclinical study. The quantitative aspects of these functional assays cannot be underestimated, since without highly quantitative assays, the ability to statistically evaluate the efficacy of the tissue-engineering construct may be compromised. In fact, the overall statistical robustness of both the experimental model and the experimental design of the study that utilizes the model must be considered carefully. It is recommended that investigators seek consultation with biostatistical experts.

Preclinical versus Clinical Studies

As noted several times earlier in this text, the final evaluation of a tissue-engineering construct involves clinical studies that will include a number of outcome measures to assess the efficacy and potential safety of the constructs. Considering that many of these tissue-engineering constructs are based on cellular or molecular mechanisms for incorporation and regeneration, the ability to have quantitative functional assays is limited to existing imaging modalities or minimally invasive or non-invasive measures. This is in stark contrast to preclinical studies in which the animals can be sacrificed and assays ranging from macroscopic to molecular in scale can be performed in highly quantitative ways. This limitation in translating studies into the clinical arena provides an incentive for additional considerations in the design of the preclinical studies. Investigators should consider developing surrogate assays as part of the preclinical studies. In other words, the preclinical studies should include outcome measures that are similar to those that are anticipated to be utilized in the clinical studies. For example, the evaluation of incorporating a bone graft substitute in a clinical setting might include radiographs or even CT scans. In the preclinical study, however, the bone can be extracted from the animal and much higher resolution imaging technologies and even histologic technologies can be utilized. In the preclinical studies, therefore, it would be very advantageous to develop correlations between microscopic-based measures that can be taken from excised bones and noninvasive imaging techniques such as radiographs and CT scans. If robust correlations are demonstrated, the radiographs or CT scans can be utilized as surrogate functional assays. This approach and specific attention paid towards the development and assessment of these surrogate assays in the preclinical phase may substantially enhance the ability to have effective assays for the eventual clinical studies. From a functional tissue-engineering perspective, this is particularly true when the mechanical properties of the constructs are considered. There are very few, if any, assays available to noninvasively characterize mechanical properties of tissue-engineering constructs. Therefore, a successful outcome depends on the development of surrogate assays for these mechanical properties during the preclinical studies.

The principles necessary to move tissue-engineering construct design from concept to the clinic have been summarized, albeit briefly. In an effort to more clearly demonstrate these principles, a more complete example will be provided below in a sequence of descriptions following the paradigm described earlier.

EXAMPLE: TISSUE ENGINEERING APPROACH FOR DEVELOPMENT OF A BONE GRAFT SUBSTITUTE

The purpose of the following section is to demonstrate the conceptual development of a bone tissue-engineered construct within the context of functional assessment during progression from the bench to the bedside.

Construct Conceptualization

There are many clinical conditions, ranging from severe trauma to reconstructive procedures, that necessitate the use of bone grafts for enhancing of healing or insuring that healing occurs. Two specific contrasting clinical needs can be defined. The first is the segmental loss of bone after substantial trauma, most often in long bones. Bone grafting has become a frequent and important part of treating these significant defects. A contrasting clinical need of interest involves procedures requiring fusion in the spine. There are a variety of clinical indications where spinal fusion becomes a therapeutic choice, but it has been demonstrated that the success rate of spinal fusion is limited and the use of bone graft to augment spine fusion is warranted. Without going into detail, investigations in the field have established a substantial need for development of bone graft substitutes. Prior work in the field has also established that the desired targeted properties of these substitutes include: (1) physical and chemical constituency that is compatible and, if possible, replicates normal bone; (2) a large surface-to-volume ratio for promoting cell recruitment and deposition; (3) the ability to fill large and potentially irregular volumes; (4) significant short-term mechanical integrity followed by an evolution of long-term properties that become equal to those of native bone; (5) a compatible environment for both osteoblast and osteoclast function such that normal bone formation and subsequent remodeling can occur; and (6) the capacity to support the eventual development of mechanical properties equal to those of normal bone, while re-establishing normal modeling and adaptational mechanisms for long-term maintenance.

Design of the Tissue-Engineering Approach

The strategic technology that was developed involves the local delivery of DNA in a therapeutic modality to stimulate the regenerative process of bone. More specifically, a three-dimensional matrix capable of supporting associated plasmid DNA encoding for a bone-promoting protein can be physically placed into a bone defect site or into a device being utilized to stimulate bone regeneration and repair. This technology includes the development of what has been named a gene-activated matrix (GAM), which includes the delivery matrix and encoded DNA. Upon insertion into a wound site, the construct promotes recruitment and migration of repair cells that have come into contact *in situ* with the incorporated DNA. Upon contact with DNA, the cells may be transfected and eventually synthesize and express the bone-promoting protein. In a sense, these repair cells are converted into internal bioreactors to create the local stimulating factor that enhances the bone regeneration.

This technology and its assessment has been described elsewhere.[1-3] While the details of these studies will not be presented, a brief summary of the findings will be presented.

As is described in the work cited above, cellular and molecular data supporting this concept have been presented and the proof of concept has been demonstrated. The proof of concept included not only *in vitro* studies, but also a series of small-animal studies demonstrating that the transfection can occur *in vivo* and can even promote the formation of bone. As a result, the studies demonstrated sufficient efficacy to warrant the design and implementation of pivotal preclinical studies.

Pivotal Preclinical Studies

Several specific, pivotal preclinical studies were made. A small-animal study in rats was designed that involved the creation of a defect in the femoral diaphysis, representing a complex and difficult fracture to heal. Placement of the gene-activated matrix in this defect was monitored during the life of the animal, post surgery, by radiographs similar to those that would be utilized in a clinical study. At sacrifice, the treated bones were extracted and analyzed mechanically, histologically, and by high-resolution imaging techniques. The results demonstrated that bone was formed by this technology and that its morphology, and even mechanical integrity, were indicative of efficacy.

The small-animal study set the stage for a larger-animal study that utilized a defect in the tibia of large dogs. The rationale for moving from the small animal to the large animal involved recognition that the physiology of rat bone differs from human bone in its nutritional makeup and its ability to be remodeled. The large-animal model (canine) involved the creation of a defect in the tibia that was very difficult to heal, therefore providing a robust test for the technology. Without going into details, the bone-generating delivery system was surgically placed in the defects, radiographs were taken to follow the progression of healing, and at sacrifice the bone was evaluated mechanically, histologically, and morphologically. Again, the data suggested great promise and are beginning to support necessary efficacy that may lead to clinical trials.

Finally, another clinical model was developed, in a study designed to test the efficacy of the constructs in modeled spinal fusion. This model involved the use of

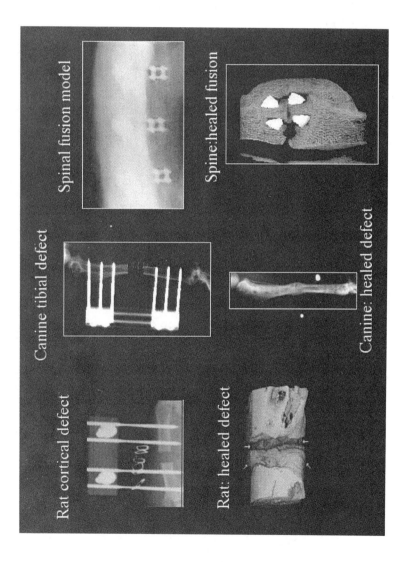

FIGURE 2. Experimental models for small- and large-animal studies. The defects that were created to simulate the clinically targeted properties were treated by surgical implantation of the gene-activated matrices. These kinds of studies may aid in evaluating the efficacy of a tissue-engineering foundation.

sheep, whose spines are similar to those of humans in size and morphology and, thus, a device needed to augment the spinal fusion could be used. This is a case of a very specific animal model that mimics a specific clinical need. The study actually simulated three spinal fusions in each animal. The evaluation occurred after sacrifice of the animal, when the experimental region could be removed and analyzed using histologic techniques as well as high-resolution imaging, including electron microscopy. Once again, the data were supportive of the move forward toward clinical trials as this very clinically specific model provided direct correspondence to a potential clinical use.

In the design of these models and their outcome measures, the targeted clinical utility as well as the framework of future clinical studies were considered. The functional assays ranged from biomechanical testing to sequential imaging. In illustration of this example, models and sample results are shown in FIGURE 2.

SUMMARY

This paper present some of the issues to be considered when designing experiments and methodologies to bring tissue-engineering constructs from concept to the clinic. Although they may be inferred at several points, the regulatory requirements have not been described, but regulatory issues provide a very important additional guide to experimental design and development of functional assays. In a very general sense, regulatory agencies such as the Food and Drug Administration have specific requirements for the outcome of both preclinical efficacy and safety studies, as well as requirements for establishing manufacturing composition, methods, and controls. The selection and design of the outcome variables for the clinical studies, as well as an assessment of risk benefit will also be of critical importance to the regulatory review process.

The principles that must be considered for successful tissue engineering include the following:

(1) comprehensive description and characterization of the properties of the native tissue;

(2) thorough design optimization based on clinical need and targeted functional properties;

(3) conceptualization of the tissue-engineering construct within a hierarchical paradigm that includes functional assessment across multiple scales from the cellular and molecular level, through the organ level;

(4) assessment of both biomechanical and biologic function and adaptation; and

(5) the development of preclinical studies that include considerations of characterizing surrogate variables in the preclinical studies that may aid in functional assessment during the clinical phase.

ACKNOWLEDGMENTS

The bone-tissue engineering example using the gene-activated matrix is based on research supported by Selective Genetics, Inc. [The author has an affiliation with Selective Genetics and discloses that the results on continued studies using the gene-activated matrices may result in the potential for financial gain.]

REFERENCES

1. BONADIO, J., E. SMILEY, P. PATIL & S.A. GOLDSTEIN. 1999. Localized, direct plasmid-gene delivery *in vivo*: prolonged therapy results in reproducible tissue regeneration. Nat. Med. **5:** 753–759.
2. FANG, J., Y.-Y. ZHU, E. SMILEY, *et al.* 1996. Stimulation of new bone formation by direct transfer of osteogenic plasmid genes. Proc. Natl. Acad. Sci. USA **93:** 5753–5758.
3. GOLDSTEIN, S.A. 2000. *In vivo* nonviral delivery factors to enhance bone repair. CORR **379S:** S113–S119.

Functional Tissue Engineering

The Role of Biomechanics in Reparative Medicine

FARSHID GUILAK

Department of Surgery, Duke University Medical Center,
Durham, North Carolina 27710, USA

KEYWORDS: functional tissue engineering; biomechanics

"Tissue engineering" is a new but rapidly growing field that uses implanted cells, scaffolds, and biologically active molecules to replace or repair injured or diseased tissues and organs. Despite its early success, tissue engineers have faced challenges in repairing tissues that serve a predominantly biomechanical function. In order for tissue engineers to effectively repair or replace these load-bearing structures, it will be necessary first to address a number of significant questions on the interactions of engineered constructs and repair tissues with mechanical forces, both *in vivo* and *in vitro*. An evolving discipline called "functional tissue engineering" (FTE) seeks to address these challenges.

The United States National Committee on Biomechanics (USNCB) formed a sub-committee in 1998 that adopted the concept of functional tissue engineering with the following specific goals:

(1) increasing awareness among tissue engineers about the importance of restoring "function" when engineering tissue constructs;

(2) identifying the critical structural and mechanical requirements needed for each tissue-engineered construct; and

(3) encouraging tissue engineers to incorporate these functional criteria in design and manufacturing processes to optimize the overall success of engineered tissues.

What constitutes "success" will be expected to differ among tissues. For example, tissues or systems that are designed to prolong life may tolerate a lower margin for error than those that are designed to improve the quality of life. The difficulty in performing a surgical procedure, and the duration of a specific treatment will influence the cost-effectiveness of a procedure and therefore may also factor into its perceived success. For example, therapies of replacement or regeneration of blood vessels or bone might be expected to last the lifetime of the individual, while replacement of cartilage may be considered successful if it delays total joint replacement for five to ten years.

Address for correspondence: Farshid Guilak, Duke University Medical Center, Room 375 MSRB, Box 3093, Durham, NC 27710. Voice: 919-684-2521; fax: 919-681-8490.
guilak@duke.edu

Ann. N.Y. Acad. Sci. 961: 193–195 (2002). © 2002 New York Academy of Sciences.

The following principles of functional tissue engineering have recently been proposed in a general format that is meant to apply to a variety of tissues.[1] These principles are not considered to be complete, but are proposed as an initial set of guidelines that are expected to evolve along with the fields of tissue engineering and regenerative medicine. It is hoped that the answers to these questions will significantly influence the quality of the implants that tissue engineers design and the repair outcome after surgery.

PRINCIPLES OF FUNCTIONAL TISSUE ENGINEERING

Defining standards of success for tissue repair. A critical step in achieving success in tissue-engineered repair and regenerative medicine will be the development of appropriate standards of clinical success and the completion of prospective outcomes studies to compare the safety and efficacy of different procedures. Inherent in this process will be the further development of minimally invasive methods for the assessment of tissue function (e.g., imaging, biomarkers).

Understanding the biomechanical properties of native tissues. A thorough understanding of the sub-failure and failure properties of native tissues will facilitate the design and engineering of repair tissues that provide the appropriate functional properties.

Prioritization of specific biomechanical properties as design parameters. The relative importance of the different material properties in the design of an engineered repair tissue is not known. Given the difficulties in matching all of the material properties and structure of native tissues in an engineered construct, a key issue in the success of engineered repairs will be the prioritization of different biomechanical properties as design parameters.

Measurement of *in vivo* stresses and strains in native and repair tissues. In attempting to define design parameters for the biomechanical function of repair tissues, knowledge of the mechanical context in which normal and repair tissues will serve for different *in vivo* activities will be required to establish patterns of activity and the limits of expected usage.

Investigation of the effects of mechanical factors on tissue repair *in vivo*. Once implanted, tissue-engineered constructs will be subjected to significant loads and deformations *in vivo*. Both the mechanical and biological consequences of *in vivo* loading must be understood to improve the success of engineered repairs and to develop appropriate rehabilitation protocols for regenerative medicine.

The use of physical factors to enhance tissue regeneration *in vitro*. Mechanical stress is an important modulator of cell physiology, and there is significant evidence that physical factors may be used to improve or accelerate tissue regeneration and repair *in vitro*. A more thorough understanding of cell and tissue response to mechanical stress and other physical factors within native and artificial extracellular matrices would improve the ability to form functional tissue replacements *in vitro*.

There is clearly a need to establish both functional criteria as well as design parameters for tissue regeneration. Other rapidly evolving new technologies may have a significant impact on functional tissue engineering, and it is important to consider these principles of functional tissue engineering in light of the role of novel growth factors, new biomaterials, gene therapy, and other changing technologies. Incorporating such principles in the design of functional tissue replacements, from the molecular scale to the macroscopic scale, will, it is hoped, result in safer and more efficacious tissue repairs and replacements.

REFERENCE

1. BUTLER, D., S.A. GOLDSTEIN & F. GUILAK. 2000. Functional tissue engineering: the role of biomechanics. J. Biomech. Eng. **122:** 570–575.

Fluorescence Imaging and Engineered Biosensors

Functional and Activity-Based Sensing Using High Content Screening

RAVI KAPUR

Cellomics, Inc., Pittsburgh, Pennsylvania 15219, USA

KEYWORDS: biosensors; fluorescence imaging

Living cells are the targets of candidate drug compounds, active biomolecules, and toxic agents comprising organisms, macromolecules, organic molecules, or inorganic ions. Synthetic and natural polymers, such as used in engineering tissues, also affect these cells. Cellomics, Inc. high content screening (HCS) technology transforms living cells into biosensors that screen compound efficacy and can additionally interrogate the environment for presence of organisms, pathogens, toxins and other cellular irritants.

The use of cells as biosensors exploits the exquisitely sensitive and specific molecular detection and amplification system developed by cells to sense minute changes in their external milieu. Our high content screening technology couples the evolved sensing capabilities of cells with the use of novel reagents that convert intracellular molecular and chemical events into spatial and temporal fluorescent light signals.

The unique advantages of high content screening are manifold. High content screens extract information on a cell-by-cell basis rather than providing an average population-response measurement. Furthermore, subpopulations (e.g., transfected cells, viable cells) or individual cells of interest can be distinguished from a population of mixed cells. Interactions between drug candidates, biomolecules, and multiple cellular targets, as well as downstream events, can be monitored in a single HCS assay via multicolor fluorescence. High content screens can be performed using fixed or live cell formats to yield the temporal-spatial dynamic information necessary to determine the role of selected targets in cell functions. Thus, cell physiology, including cytotoxicity, can be quantified in HCS, providing multifaceted functional information. Another key component of HCS is the ability to manage the enormous volumes of raw data and images generated in screening. This is currently address-

Address for correspondence: Ravi Kapur, Cellomics, Inc., 100 Technology Drive, Pittsburgh, PA 15219. Voice: 412-826-3600; fax: 412-826-3850.
rkapur@cellomics.com

Ann. N.Y. Acad. Sci. 961: 196–197 (2002). © 2002 New York Academy of Sciences.

able using Cellomics Store, which additionally incorporates tools to visualize this biologically rich data, not only from the ArrayScan System, but also from other screening systems.

Quantitative, automated intracellular multicolor fluorescence imaging can be a useful tool for assessment of biological efficacy of hybrid devices comprising abiotic (synthetic polymer) and biotic (cells and active biomolecules) elements.

SELECTED REFERENCES

1. GIULIANO, K.A. & D.L. TAYLOR, D. 1998. Fluorescent-protein biosensors: new tools for drug discovery. Trends Biotechnol. **16:** 135.
2. GIULIANO, K.A., R.L. DeBIASIO, R.T. DUNLAY, et al. 1997. High-content screening: a new approach to easing key bottlenecks in the drug discovery process. J. Biomol. Screen. **2:** 249.
3. GIULIANO, K.A., P.L. POST, K.M. HAHN & D.L. TAYLOR. 1995. Fluorescent protein biosensors: measurement of molecular dynamics in living cells. Annu. Rev. Biophysics Biomolec. Struct. **24:** 405.
4. KAPUR, R., K.A. GIULIANO, M. CAMPANA, et al. 1999. Streamlining the drug discovery process by integrating miniaturization, high throughput screening, high content screening, and automation on the CellChip^{TM} system. Biomed. Microdevices **2:** 99–111.
5. COWARD, P., S.D. CHAN, H.G. WADA, et al. 1999. Chimeric G proteins allow a high-throughput signaling assay of Gi-coupled receptors. Analyt. Biochem. **270:** 242.
6. CONWAY, B.R., L.K. MINOR, J.Z. XU, et al. 1999. Quantification of G-protein coupled receptor internalization using G-protein coupled receptor-GFP conjugates with the ArrayScan™ high content screening system. J. Biomol. Screen. **4:** 75.
7. DING, G.J.F., P.A. FISCHER, R.C. BOLTZ, et al. 1998. Characterization and quantitation of NF-κB nuclear translocation induced by interleukin-1 and tumor necrosis factor-alpha. J. Biol. Chem. **273:** 28897.

Functional Considerations in Tissue-Engineering Whole Organs

MICHAEL V. SEFTON

Institute of Biomaterials and Biomedical Engineering, University of Toronto, Toronto, Ontario M5S 3G9, Canada

KEYWORDS: whole-organ tissue engineeering; heart replacement

The LIFE Initiative, a multi-institutional affiliation of researchers, was created to use tissue engineering and regeneration to produce an endless supply of human vital organs for transplantation.[1] With an unlimited supply of vital organs, replacing a damaged or failed organ could become not substantially different than any other surgical procedure. Of more significance, a large number of patients may benefit from the implantation of smaller structures—heart muscle, valves, vascular grafts—so that the "spin-off" benefits of a tissue-engineering effort may become even more important than the whole organ. Consider a heart as an example of a large three-dimensional organ. The cells of the heart are arranged in precise architectures, and the correct organization and functional coordination of these cells are requisite for efficient contractile activity. The possibility that diminished cardiac function may be recovered through the implantation of cells or biosynthetic constructs has great appeal.[2,3]

From working with microencapsulated cells and immunoisolation systems for many years,[4] we have learned that successful implementation of a tissue-engineering construct requires (1) an adequate, viable cell mass; (2) the appropriate behaviour of the cells; and (3) sufficient durability of the function *in vivo*. The specific requirements are determined by the application, the nature of the cells, the implantation site, and the biocompatibility of the device.

There are numerous critical issues to growing a heart, some of which apply to all large constructs and some which are more specific to the heart. The generic issues reflect the fundamental nature of how an organ is different from a tissue and how a heart (or liver or kidney) is different from a blood vessel. By definition, engineering an organ requires multiple cell types, each of which phenotype must be maintained for long periods of time and in a form that permits complete intercellular communication. More problematic may be the intrinsic nature of large cell-based constructs (tissues or organs) and the corresponding difficulty of supplying cells deep within the construct with nutrients.

Address for correspondence: Michael V. Sefton, University of Toronto, 4 Taddle Creek Road, Room RS 407, Toronto, ON M5S 3G9, Canada. Voice: 416-978-3088; fax: 416-978-4317.
sefton@chem-eng.utoronto.ca

Ann. N.Y. Acad. Sci. 961: 198–200 (2002). © 2002 New York Academy of Sciences.

As if these issues were not enough, one can also easily identify problems specifically related to the functional requirements. While obtaining a source of human cardiomyocytes or a functioning equivalent cell may not be insurmountable, getting the cells to function as an organ is problematic. The cells need to form the appropriate subcellular structures, intercellular connections, and matrix arrangements required for functional coordination and directed beating. The alignment of myocytes in register and the proper formation of the intercalated disks between myocytes are also critical in enabling electrical pulses to be transmitted in the correct direction at normal speeds and in allowing suitable force transmission between myocytes. In addition, the heart contains specialized cells that participate in the electrical conduction routes found throughout the heart (e.g., the sino-atrial (SA) node, the atrio-ventricular (AV) node, and Purkinje fibers). These specialized cells are crucial to the coordination of the heart's contractile effort, and including them in the proper places in a biosynthetic substitute may be critical. Obviously, there are clear differences between the rhythmic twitching of cultured cardiac cells *en masse* and the organized, efficient, regulated beating of the heart; only the latter will generate the force required to pump blood at systolic pressure levels.

Thus a critical feature of a heart is both its movement and the complexity of the electrical conduction pathways. Neither problem has received much attention in the tissue-engineering literature. However, given the variety of electrical conduction–related diseases in a normal myocardium, there is good reason to suspect that simple mimicry of heart muscle may fall short of the goal. Thus, functional tissue engineering in the context of a heart has specific and well-established attributes.[5] The heart must deliver 5 to 15 L/min of blood at pressures from 100 to 150 mm Hg or 1 to 8 watts (pressure volume [PV] work only) with a fatigue life of $>10^8$ cycles over 10 years. Presumably it must also observe Frank–Starling mechanics and have the nonlinear, viscoelastic properties of normal myocardium. Heart muscle must stretch in response to capillary filling pressure and eject a volume of blood that varies with demand. The latter requires a uniform and well-coordinated contraction that generates the required power. There is a long way to go from the small number of cardiomyocyte-like cells beating in a culture dish to the complex architecture of the myocardium and the duplication of its nonlinear viscoelastic properties.

Nonetheless, there is much that can be done without reference to the whole heart and is the basis for a number of projects working towards a cardiac patch or *in situ* cardiac repair. Thus the cardiac patch can be considered a stepping stone towards a whole heart, or, looking at it from the opposite direction, the patch is a spin-off benefit of working towards a heart. The road to a full heart is obviously long and conditional on these issues being resolved. At the same time the various components—cardiomyocytes, elastomeric scaffolds, immune control strategies—will be valuable both commercially and clinically in their own right.

Tissue engineering whole organs such as a functional heart is clearly an ambitious undertaking. But the problem of growing whole organs can be broken down into more manageable components and interim milestones. Furthermore, reaching an interim goal such as a cardiac patch or a strategy for vascularizing thick slabs of tissue has an intrinsic therapeutic value on its own. In fact, these spin-off benefits may be more valuable and may benefit more patients than would the whole organ. Thus the LIFE initiative has been created to tissue-engineer complex three-dimensional, mechanically robust and dynamic organs like the heart.[1] But the underlying goal is to advance the science and art of tissue engineering, to create novel arrangements for collaborative research on a global scale, and to foster a new industry based on the capacity for treating human disease with replacement tissues and organs.

REFERENCES

1. AKINS, R. & M.V. SEFTON. 2001. Tissue engineering the human heart. New Surg. **1:** 26–32.
2. AKINS, R.E. 2000. Prospects for the use of cell implantation, gene therapy, and tissue engineering in the treatment of myocardial disease and congenital heart defects. *In* Medizinische Regeneration und Tissue Engineering. K. Sames, Ed. :1–16. EcoMed. Landsberg, Germany.
3. REINLIB, L. & L. FIELD. 2000. Cell transplantation as future therapy for cardiovascular disease? Circulation **101:** e182–e187.
4. SEFTON, M.V., M.H. MAY, S. LAHOOTI & J.E. BABENSEE. 2000. Making microencapsulation work: conformal coating, immobilization gels and in vivo performance. J. Contr. Rel. **65:** 173–186.
5. BURTON, A.C. 1972. Biophysical Basis of the Circulation. Yearbook Medical Publishers. Chicago, IL.

Functional Assessment and Tissue Design of Skeletal Muscle

HERMAN H. VANDENBURGH

Brown University School of Medicine, Providence, Rhode Island 02906, USA

KEYWORD: **bioartificial skeletal muscle**

Tissue engineering human skeletal muscle *ex vivo* for structural use *in vivo* will require the capability to generate adult myofibers with the appropriate size and packing density to generate the directed forces necessary to perform functional work. Human skeletal muscle stem cells (myoblasts or satellite cells) are easily isolated from adult muscle and expanded in tissue culture to billions of cells. They can be reconstituted in three dimensions with natural or artificial scaffolding material and formed into organized tissues containing partially differentiated contractile muscle fibers. When electrically stimulated to contract, these bioartificial muscles (BAMs) generate forces that are only 1 to 2 percent of the specific force generated by normal skeletal muscle. This is due to high extracellular matrix content, low myofiber packing density, and less than fully differentiated myofibers in the BAMs. Computerized tissue-engineering hardware and software developed in our lab allows real time μN to mN force measurements for weeks to months *in vitro* on the tissue-engineered constructs. Manipulation of the tissue-culture medium, growth factor supplements, perfusion conditions, electrical stimulation, and repetitive mechanical loading with this equipment should eventually allow the production of stronger, more *in vivo*-like constructs. In addition, this equipment allows the material testing of BAM viscoelastic and failure properties *ex vivo* before implantation. The development of the tools for real-time, long-term functional assessment of tissue-engineered constructs *in vitro* is one of the most critical areas for the field.

While the structural use of BAMs is a distant goal, several near-term goals are feasible and will contribute to our knowledge of BAM performance once implanted *in vivo*. This includes their use as implantable protein delivery "devices" following genetic engineering of the myoblasts, and their tissue engineering into BAMs. We have successfully engineered BAMs from small- and large-adult animal myoblasts to secrete therapeutic levels of recombinant proteins such as growth hormone, insulin-like growth factor-1, erythropoietin, and vascular endothelial growth factor that

Address for correspondence: Herman H. Vandenburgh, Brown University School of Medicine, 164 Summit Avenue, Room 227, Providence, RI 02906. Voice: 401-793-4273; fax: 401-454-3157.

herman_vandenburgh@brown.edu

Ann. N.Y. Acad. Sci. 961: 201–202 (2002). © 2002 New York Academy of Sciences.

can survive for 4 months *in vivo*. These studies provide useful information on the survival and integration of BAMs when implanted into various muscle and nonmuscle host sites. Future technological developments in this near-term area will provide useful information for the long-term goal of engineering skeletal muscle implants as structural substitutes with native force generation capacity.

ACKNOWLEDGMENTS

This work was carried out in collaboration with Courtney Powell, Paul Kosnik, and Martin Nackman of the Department of Pathology, Miriam Hospital, Brown University School of Medicine, and Cell Based Delivery, Inc., Providence, Rhode Island.

SELECTED REFERENCES

1. VANDENBURGH, H.H., J. SHANSKY, M. DEL TATTO & J.CHROMIAK. 1998. Organogenesis of skeletal muscle in tissue culture. *In* Methods in Molecular Medicine, Vol. 18: Tissue Engineering Methods and Protocols. J.R. Morgan & M.L. Yarmush, Eds.: 217–227. Humana Press. Totowa, NJ.
2. POWELL, C., J. SHANKSY, M. DEL TATTO & H.H. VANDENBURGH. 2001. Bioartificial muscles in gene therapy. *In* Methods in Molecular Medicine, Vol. 69: Gene Therapy Protocols. J.R. Morgan, Ed.: 149–160. Humana Press. Totowa, NJ.
3. KOSNIK, P.E., R.G. DENNIS & H.H. VANDENBURGH. 2002. Tissue engineering skeletal muscle. *In* Functional Tissue Engineering: The Role of Biomechanics. F. Guilak *et al.*, Eds. In press.

Functional Assessment of Tissues with Magnetic Resonance Imaging

ALAN P. KORETSKY

National Institute of Neurological Disorders and Stroke, NIH,
Bethesda, Maryland 20892, USA

KEYWORDS: magnetic resonance imaging; tissue function

Over the past ten years, magnetic resonance imaging (MRI) has rapidly progressed from a purely anatomical imaging technique to one that reports on a wide variety of tissue functions. While most of these techniques have been developed for clinical problems or animal models, it is clear that they will be useful for studying engineered tissues both *in vitro* and *in vivo*. The goal of this presentation is to introduce the wide range of anatomical and functional information that can be assessed with magnetic resonance imaging.

ANATOMY WITH MRI

The major impact of MRI on the clinical sciences has derived from its ability to obtain millimeter resolution of tissue non-invasively.[1] Indeed for almost every known pathological condition, there is an anatomical MRI technique that can give contrast to distinguish pathologic from normal tissue. Recent increases in magnetic fields and image-acquisition strategies are pushing the resolution that can be obtained with MRI: 50–100-micron resolution with *in vitro* samples and rodents in very high magnetic fields (> 7T) can be routinely obtained in a few minutes and applications to engineered tissues are appearing.[2] In the human brain, resolution approaching 300 microns can be expected in the next few years.

MRI TO MONITOR MOTION

Since its early days, MRI has been used to monitor motion in biological samples. This range of motion goes from the macroscopic (such as ventricular dynamics of the heart) to the microscopic (e.g., diffusion of water). Specialized MRI tagging ap-

Address for correspondence: Alan P. Koretsky, National Institute of Neurological Disorders and Stroke, NIH, Building 10, Room B1S317, 9000 Rockville Pike, Bethesda, MD 20892. Voice: 301-402-9659; fax:301-402-0119.
koretsky@ninds.nih.gov

Ann. N.Y. Acad. Sci. 961: 203–205 (2002). © 2002 New York Academy of Sciences.

proaches enable detailed analysis of contraction of muscle at roughly millimeter resolution.[3] Diffusion MRI offers sensitivity to cellular volume and can probe the extent of anisotropy in complex media. These properties have made diffusion MRI useful for evaluating tissue damage due to stroke[4] and to trace white matter tracts in the brain.[5] Recent successes at labeling cells with strong, iron oxide–based contrast agents show much progress in enabling MRI to monitor cell migration in tissues.[6]

MRI TO MONITOR HEMODYNAMICS

Techniques that allow MRI to monitor tissue hemodynamics are having a large impact in assessing the function of tissues. Indeed, since assessing vascularization of engineered tissues is of major importance, these MRI techniques should have broad application. Blood volume is readily measured using MRI contrast agents that are restricted to the blood.[7] Oxygenation of hemoglobin can be indirectly assessed using the paramagnetic properties of hemoglobin to give so-called blood oxygenation level–dependent MRI. This has been the primary tool for functional MRI of the brain.[8] Regional blood flow can be quantitated using techniques that spin-label arterial blood.[9] All of these techniques are being applied to a broad range of problems in physiology including tumor angiogenesis, cardiac function, and brain function. It is exciting to apply them to engineered tissues both *in vitro* and in *vivo*.

MRI TO MONITOR METABOLISM

Most MRI is based on detection of hydrogen atoms in water, although ^1H in other molecules and other nuclei such as ^{31}P (to study cellular energetics), ^{13}C (to study metabolic fluxes), ^{19}F (to study drug metabolism), and ^{23}Na (to study ion homeostasis) can be readily detected *in vivo*.[1] The major disadvantage of these techniques is their relatively limited spatial resolution (about 1 cc). There is a history of using MR spectroscopic techniques to study cellular metabolism in a variety of bioreactors that hold much potential for studying engineered tissues.[10]

FUNCTIONAL AND MOLECULAR SPECIFIC CONTRAST AGENTS

Contrast agents have played a large role in the clinical usefulness of MRI. Function- and molecule-specific contrast agents that are sensitive to a number of physiologic functions are being developed. These can be quite simple substances or ones that require significant chemical synthesis. For example, it is well known that manganese ion can enter excitable cells through voltage gated calcium channels and is an excellent MRI contrast agent. These properties have been combined to use manganese to mark active regions in the rodent brain.[11] Interestingly, manganese also is transported in an anterograde fashion in the nervous system, making it useful for tracing neuronal connections.[12] New, so-called "smart contrast agents" are being developed whose contrast changes with a change in calcium concentration as occurs during an action potential.[13] Finally, peptides or antibodies that bind specific molecular targets can be attached to a variety of MRI contrast agents for imaging the distribution of specific membrane receptors.

Numerous other promising areas of MRI development may also have an impact on tissue engineering. For example, current density imaging holds promise for looking directly at electric fields in tissues. MRI elastography holds promise to measure the physical properties of tissues. A number of strategies for monitoring gene expression by MRI are awaiting widespread application. The rapid growth of functions that MRI can measure indicates that it should play a major role in assessing engineered tissues, *in vitro*, in animals, and in humans.

REFERENCES

1. GADIEN, D.G. 2000. NMR and Its Applications to Living Systems. Oxford University Press. New York.
2. POTTER, K. *et al.* 2000. Response of engineered cartilage tissue to biochemical agents as studied by proton magnetic resonance microscopy. Arthritis Rheum. **43:** 1580–1590.
3. MOORE, C.C. *et al.* 2000. Top. Magn Reson. Imaging **11**: 359–371.
4. HACKE, W. & S. WARACH. 2000. Diffusion-weighted MRI as an evolving standard of care in acute stroke. Neurology **54:** 1548–1549.
5. BASSER, P.J. *et al.* 2000. In vivo fiber tractography using DT-MRI data. Magn. Reson. Med. **44:** 625–632.
6. BULTE, J. *et al.* 1999. Neurotransplantation of magnetically labeled oligodendrocyte progenitors: magnetic resonance tracking of cell migration and myelination Proc. Natl. Acad. Sci. USA **21:** 15256–15261.
7. ZAHARCHUK, G. *et al.,* 1998. Continuous assessment of perfusion by tagging including volume and water extraction (CAPTIVE): a steady-state contrast agent technique for measuring blood flow, relative blood volume fraction, and the water extraction fraction. Magn. Reson. Med. **40:** 666–678.
8. OGAWA, S. *et al.* 2000. An approach to probe some neural systems interaction by functional MRI at neural time scale down to milliseconds. Proc. Natl. Acad. Sci USA **97:** 11026–11031.
9. BARBIER, E. *et al.* 2001. Perfusion imaging using dynamic arterial spin labeling (DASL) Magn. Reson. Med. **45:** 1021–1029.
10. GILLIES, R.J. *et al.* 1993. Design and application of NMR compatible bioreactor circuits for extended perfusion of high-density mammalian-cell cultures. NMR Biomed. **6:** 95–104.
11. LIN, Y.-J. & A.P. KORETSKY. 1997. Manganese ion enhances T-1-weighted MRI during brain activation: an approach to direct imaging of brain function. Magn. Reson. Med. **38:** 378–388.
12. PAUTLER, R.G. *et al.* 1998. In vivo neuronal tract tracing using manganese-enhanced magnetic resonance imaging. Magn. Reson. Med. **40:** 740–748.
13. LI, W-H. *et al.* 1999. A calcium-sensitive magnetic resonance imaging contrast agent. J. Am. Chem. Soc. **121:** 1413–1414.

Determination of Responsiveness in Biological Systems Is Dependent on Highly Sensitive Quantitative Measures

REGIS J. O'KEEFE

Center for Musculoskeletal Research, University of Rochester, Strong Memorial Hospital, Rochester, New York 14642, USA

KEYWORD: tissue engineering outcome

The engineering of musculoskeletal tissues is an active area of research, with tremendous potential to have an impact on the healing and regeneration of bone, cartilage, and soft-tissues, including tendon, ligament, meniscus, and disc. The design of these various tissues requires consideration of both material and biological science and potentially involves a complex interplay of a scaffolding material, cells, and growth factors or genes. Selection of the optimal combination of these materials is dependent upon sensitive and quantitative measures of performance. Paradigms for the evaluation of performance in animals have been established and rely on harvesting the tissues and assessing histologic, biochemical, biomechanical, and molecular features. While animal studies will limit and focus design parameters, assessment will ultimately be dependent upon the performance of these materials in human trials. Thus, paradigms for the evaluation of tissue engineering constructs in humans need to be developed. Similar to animal studies, the outcome measures should be sensitive and quantitative. While validated clinical outcome measures are valuable, direct assessment of the materials will rely on noninvasive methods of assessment. Imaging modalities have tremendous potential to have an impact on the development of engineered tissues and will involve the design of innovative methods of image processing. As an example, we have presented elsewhere the development of a novel computerized imaging methodology which has been essential for the development of a human clinical trial to assess the efficacy of a bioengineered molecule, Enbrel, to prevent peri-prosthetic bone loss. We have established the ability of this agent to prevent particle-mediated inflammatory bone loss in animal models using histological measures after animal sacrifice. In order to establish the utility of Enbrel in humans, we have developed and used sensitive and reproducible imaging modalities for functional analysis of the drug effect.

Address for correspondence: Regis O'Keefe, Strong Memorial Hospital, Rochester, NY 14642. Voice: 716-275-3100; 716-756-4727.

Regis_okeefe@urmc.Rochester.edu

Ann. N.Y. Acad. Sci. 961: 206 (2002). © 2002 New York Academy of Sciences.

Functional Assessment of Engineered Tissues and Elements of Tissue Design

Breakout Session Summary

Moderator

FARSHID GUILAK, *Duke University Medical Center*

RAVI KAPUR, *Cellomics, Inc.*

Panelists

MICHAEL V. SEFTON, *University of Toronto*

HERMAN H. VANDENBURGH, *Brown University School of Medicine*

ALAN P. KORETSKY, *National Institutes of Health*

ANDRES KRIETE, *Tissue Informatics, Inc.*

REGIS J. O'KEEFE, *University of Rochester*

Rapporteur

FARSHID GUILAK, *Duke University Medical Center*

BROAD STATEMENT

Tissue engineering is a rapidly growing field that seeks to repair or replace tissues and organs by delivering combinations of cells, biomaterials, and/or biologically active molecules. Tissue engineering merges several aspects of engineering, biology, and medicine, and many rapid achievements in this field have arisen from significant advances in the integration of these fields. Despite early successes, however, few functional tissue-engineered products are currently available for clinical use. The development and application of rational design criteria and technologies for the assessment of tissue function would be expected to improve the success of engineered products.

VISION

Tissue engineering has the potential to dramatically alter the treatment of numerous diseases by enabling the repair of injured or diseased tissue with living replacements. The overriding vision of this field is to improve the speed, extent, and duration of tissue repair over currently available methods. Most tissue-engineered products must serve multiple, complex functions, including inter-related metabolic and structural (i.e., biomechanical) demands. Many challenges still remain in the ability to restore the native function of various organs and tissues. In the next 5–10

Ann. N.Y. Acad. Sci. 961: 207–209 (2002). © 2002 New York Academy of Sciences.

years, advances in two important and related areas are presented as mechanisms to improve the outcome of engineered tissue replacements: (1) the assessment of function in engineered tissues, and (2) the application of rational design principles. Incorporation of these approaches must span multiple hierarchical scales, from the macroscopic level, directed at satisfying the clinical requirements of the product, to the microscopic level, directed at satisfying the cell and molecular requirements for long-term functional success.

OBJECTIVES

The ultimate goal of tissue engineering is to restore the function of injured or diseased tissues. However, many challenges remain, particularly with respect to functional assessment and the design of tissue replacements. Many of the complex structural and metabolic functions of tissues and organs are not fully understood, even for native tissues. In this respect, what constitutes "success" needs to be defined *a priori* and will be expected to differ among tissues. For example, tissues or systems that are designed to prolong life may tolerate a lower margin for error than those that are designed to improve the quality of life. The difficulty in performing a surgical procedure, and the duration that a specific treatment lasts will influence the cost-effectiveness of a procedure and therefore may also factor into its perceived success. For example, therapies of replacement or regeneration of blood vessels or cardiac muscle might be expected to last the lifetime of the individual, while replacement of cartilage may be considered successful if it delays total joint replacement for five to ten years.

OBSTACLES AND CHALLENGES

The following obstacles and challenges were identified: (1) A primary need in this field is the development of fundamental principles and standards that cross multiple disciplines. In this respect, researchers with different expertise and backgrounds will be more likely to seek common goals in the development of engineered tissues. (2) As it is unlikely for an engineered replacement to meet all of the functional demands of native tissue or organ, it is critical to prioritize the needs and to determine the relationship between specific functional properties and the overall clinical success of product. The definitions of "function" should be broadened to include biomechanical, electrophysiological, and metabolic properties, where appropriate. (3) Trial-and-error approaches have led to rapid advances in the field of tissue engineering, but are inherently limited in many cases. A balance of "rational" design approaches with trial-and-error methods is recommended. Such approaches will require further basic science studies in tissue engineering systems in addition to translational research. (4) Interdisciplinary collaborations are the foundation of tissue engineering and will be necessary for significant advances in the field. Such approaches must be fostered from the student level to that of research partnerships through increased interaction and communication among disciplines.

The breakout panel identified the following specific areas as critically in need of investigation: (1) minimally invasive imaging in tissue engineering; (2) high-throughput cellular activity profiling and molecular characterization; (3) modeling of biological systems in reparative medicine; (4) the influence of biophysical factors on engineered tissues; (5) and controlled, prospective, and randomized outcomes studies in reparative medicine.

ACKNOWLEDGMENTS

The Functional Assessment of Engineered Tissues and Elements of Tissue Design breakout session was coordinated by Dr. Tracy E. Orr, CSR, NIH and Dr. James Panagis, NIAMS, NIH.

Bioreactors and Bioprocessing for Tissue Engineering

ANTHONY RATCLIFFE[a] AND LAURA E. NIKLASON[b]

[a]Advanced Tissue Sciences, La Jolla, California 92037, USA

[b]Departments of Anesthesia and Biomedical Engineering, Duke University, Durham, North Carolina 27708, USA

ABSTRACT: Bioreactor design in tissue engineering is complex, and at the early stages of its development. Design of biologically effective, yet scalable, devices requires intimate collaboration between engineers and biologists to ensure that all aspects are considered fully. Growth conditions, harvesting time, scale-up, storage, and sterility issues all need to be considered and incorporated into the design of bioreactors. Each tissue-engineered product will likely require individualized bioreactor design. However, without a comprehensive understanding of each of these components, bioreactor design and tissue growth to manufacture product will remain at a relatively rudimentary and limited level. Increased fundamental understanding of the issues can have a dramatic impact on the ability to generate tissue-engineered product safely, economically, and in the numbers that are required to fully address the patient populations in need.

KEYWORDS: tissue engineering; bioreactors; organ culture

INTRODUCTION

In the field of reparative medicine, methods are rapidly being developed to generate cell and tissue-based constructs that can function clinically in repair and replacement.[1] These research successes will need to be translated into products that can be used successfully to treat the intended patient populations (often large numbers per year).[2] A major challenge is to translate these research-scale product designs into large-scale production of biologically based products that are reproducible, safe, clinically effective, and are economically acceptable and competitive. This requires substantial understanding of the biology and function of the tissue-engineered product. In addition, sophisticated technical skill in design and manufacture of the bioreactors and growth process are required, as is an understanding of the specific challenges encountered in scale-up of the process, and storage and delivery of the final product.

Address for correspondence: Laura E. Niklason, M.D., Ph.D., Departments of Anesthesia and Biomedical Engineering, Duke University, Room 136 Hudson Hall, Research Dr. @ Science Dr., Durham, NC 27708. Voice: 919-660-5149.

nikla001@mc.duke.edu

Ann. N.Y. Acad. Sci. 961: 210–215 (2002). © 2002 New York Academy of Sciences.

The overall goal is to have bioreactors and process monitoring that reliably and reproducibly form, store, and safely deliver complex and perfusable tissues or organs that can sustain function *in vivo*. This chapter provides an overview of the field today, and identifies key technical challenges and opportunities in bioprocessing of tissue-engineered products.

BIOREACTORS IN USE TODAY

Dishes and Flasks

The simplest and most widely used bioreactor today is the culture dish.[3] This provides an environment that is sterile, easy to use, simple, and economical to manufacture. These features have made simple dishes and flasks the universal bioreactors in use today. However, the fact that the culture dish requires individual manual handling for medium exchange, cell seeding, etc., ultimately limits its usefulness when large numbers are required. Scale-up of the culture dish only goes so far, as, for example, having multi-well plates. To move into large-scale production, dish or flask systems would be required wherein sophisticated robotics could be used, and this has not been accomplished to date.

Spinner Flasks

Cells in monolayer culture are not generally nutrient-limited. This is because passive diffusion is more than adequate to supply a 10-micron tissue thickness. In contrast, the supply of oxygen and soluble nutrients becomes critically limiting for the culture of tissues that are thicker than 100–200 microns.[4] This diffusion limitation is partly alleviated by stirring of the culture medium. Continuous stirring in a spinner flask provides continuous exposure of the tissue surface to fresh nutrients as well as to (controllable) shear forces.[5] Spinner flasks may be fitted with ports and filters for gas exchange, and hence may constitute a more "closed system" than conventional dishes and flasks. However, spinner flasks suffer from similar drawbacks in terms of requirements for individual handling and difficulty of scale-up. Furthermore, the conditions in simple spinner flasks are not tailored to mimic the milieu of specific tissues in the body.

Tissue-Specific Bioreactor Systems

Cartilage

Hyaline cartilage lines the surfaces of skeletal joints and is subject to injury that can lead to osteoarthritis. Engineering of cartilage *in vitro* has been accomplished using simple dish and spinner flask systems, and also in more complex systems that mimic the periodic compressive stresses of the native joint.[6,7] Bioreactors for cartilage have been custom-designed to provide unconfined compression, with a strain amplitude of 10%. Under conditions of periodic compressive strain, the mechanical properties of engineered cartilage are significantly improved. In addition, glycosaminoglycans and collagen, both of which are components of the secreted extracellular matrix, are significantly enhanced by dynamic loading. Engineered cartilage provides some of the most compelling evidence that mimicking the native mechanical environment *in vitro* can be beneficial for producing functional tissues.

Ligament

Bioreactor design can provide sometimes complex mechanical cues to cultivated tissues. For the tissue engineering of ligament from precursor cells, the application of multidimensional mechanical strains may be important.[8] Bioreactors that control translational and rotational strain to cells embedded in a collagen gel have been shown to produce ligament-like structures. Increased fibroblast phenotype, with expression of collagen and tenascin-C, as well as cellular alignment, are produced by multidimensional strains. Hence, bioreactor stimuli can direct the differentiation lineage of uncommitted precursor cells *in vitro*. This area of investigation will doubtless expand in the near future.

Cardiovascular

The cardiovascular system is home to the most costly and debilitating diseases of the Western world.[9] As such, there exist powerful medical and economic incentives to develop replacement blood vessels and myocardial tissues. Functional autologous arteries can be cultured using pulsatile perfusion bioreactors.[10,11] Autologous vascular cells are cultured on tubular, biodegradable scaffolds contained within closed bioreactor systems. The bioreactors apply pulsatile radial distensions to developing vessels, at a controlled frequency and radial strain. Studies in several animal species have shown that pulsatile culture conditions promote collagen deposition, and significantly improve mechanical properties of engineered blood vessels.[12]

Other investigators have shown that mechanical stimulation may alter the differentiation of cardiomyocytes *in vitro*.[13] Unidirectional mechanical stretch causes cardiomyocyte organization into parallel arrays, as well as cellular hypertrophy. In addition, accumulation of both fetal and adult myosin heavy chain isoforms may point to alterations in cellular differentiation state. Since the circulatory system is subject to continuous pulsatile inputs, is not surprising that cells derived from the cardiovascular system respond to bioreactor-delivered mechanical forces with changes in cellular behavior.

Other Bioreactor Stimuli

Mechanical stimuli are not the only physiologically and developmentally relevant physical inputs that may be harnessed in bioreactors. Several groups have shown a phenotypic and differentiation effect of electrical currents on neuronal cells.[14] Effects of electrical stimuli on osteoblast behavior have also been well documented.[15] Our laboratory has also shown a biological effect of pulsatile electrical current on the phenotype of skeletal muscle–derived cells (Niklason, unpublished observations). Although the field of tissue engineering has not yet fully exploited electrical signals in bioreactor-based systems, this is likely to become an area of increasing interest.

BIOREACTORS IN MANUFACTURING

Closed bioreactor systems offer major advantages for manufacturing, since sterility can be assured while maintaining viability of the tissue product. This approach has been used successfully in the manufacture of two tissue-engineered products, TransCyte™ and Dermagraft®, both manufactured by Advanced Tissue Sciences. TransCyte™ is made using a sealed bioreactor designed as a cassette made from polycarbonate, containing a scaffold (Biobrane) onto which dermal fibroblasts are seeded and allowed to attach.[16] By maintaining a fluid flow within the bioreactor system, the cultures are able to form a three-dimensional matrix representative of a human dermis. These bioreactors are designed such that groups of eight can be placed side by side, with the necessary manifold tubing to provide the cells and culture media in a uniform manner. Within a manufacturing facility, these 8-bioreactor units are stacked within specially designed incubators, to ensure a uniform and consistent environment where the tissue can be grown.

At the end of the growth process, the cassettes can be sealed by closing the inlet and outlet tubing, and the tissues can be stored frozen. These cassettes are also used as the delivery packages for the product, and are only opened at the clinical site for immediate application. This bioreactor system therefore allows for automated cell seeding, media change, in-process monitoring of growth, storage and delivery, and at the same time provides a scaled-up system.

A similar approach was used in the design of bioreactors for Dermagraft®, with some significant modifications.[17] A simpler bioreactor was made using flexible, gas-permeable plastic. The soft-walled bioreactor contains 8 units per bag. To achieve increased scale-up, twelve bags are connected by an injection-molded header system, again allowing simultaneous seeding and media change. This modular unit therefore grows 96 pieces of tissue-engineered skin replacement. Multiple modular units may be connected and simultaneously processed to achieve a 1000:1 scale-up. These bags can be sealed, packaged individually, stored frozen, and delivered to the site, where they are only opened for application, thereby ensuring sterility.

FACTORS AFFECTING TISSUE GROWTH IN MANUFACTURING

The growth of tissue-engineered products is dependent on the immediate environment that the constructs are subjected to. The factors that can modulate growth include temperature, culture medium, chemical factors, and mechanical environment, including fluid flow and perfusion. Each of these individually can have a dramatic impact on the growth of the tissue, although if controlled they can be used as major positive modulators.

Temperature is an obvious factor, and at the research level cultures are routinely grown in well-designed and controlled incubators. However, in a manufacturing system, very large incubators or even entire rooms are required, and these are more difficult to control. To ensure minimal temperature modulation, medium changes should preferably be performed within the temperature-controlled environment, but this may not be possible. If so, the time that the bioreactors are out of the temperature-controlled environment must be uniform and minimized.

Providing the three-dimensional tissues with nutrients may rely on passive diffusion, or may be more actively delivered by direct perfusion. Tissues that have been manufactured to date have relied on diffusion, although tissues envisioned for future products will require a more active delivery process. This has been done successfully in cartilage tissue engineering, where perfusion bioreactors have been designed and used to make cartilage constructs.[18,19] However, direct perfusion introduces a new level of complexity when scale-up is encountered, and the engineering challenges may be significant.

MONITORING TISSUE GROWTH

Minimizing variability of conditions does not necessarily result in perfectly uniform growth between systems. In the manufacturing environment, the time at which to harvest the tissue-engineered constructs so that they meet the release criteria may vary. Therefore, it is necessary to monitor growth during the culture process, to ensure that the harvest time is optimal for each batch. The monitoring method is likely to be individualized for each product, although the monitoring of glucose uptake during culture has been used successfully in the manufacture of Dermagraft® and TransCyte™. For products that may subserve an important mechanical function, such as blood vessel or cartilage, monitoring the mechanical properties of the constructs during the growth process may be necessary. This will require substantial design features to be included into the bioreactor, and will certainly increase complexity of the bioreactor design and growth process. However, direct testing of individual constructs noninvasively will allow for a high degree of certainty in harvesting product within the pre-determined manufacturing specifications.

REFERENCES

1. LYSAGHT, M.J. & J. REYES. 2001. The growth of tissue engineering. Tissue Eng. 7: 485–493.
2. LANGER, R. & J.P. VACANTI. 1993. Tissue engineering. Science 260: 920–926.
3. MORGAN, J.R. & M.L. YARMUSH. 1999. Tissue Engineering Methods and Protocols. Humana Press. Totowa, NJ.
4. LIGHTFOOT, E.N. 1974. Transport Phenomena and Living Systems. John Wiley. New York.
5. LANZA, R.P., R. LANGER & W.L. CHICK. 1997. Principles of Tissue Engineering. R.G. Landes. Austin, TX.
6. MAUCK, R.L., M.A. SOLTZ, C.C. WANG, et al. 2000. Furnctional tissue engineering of articular cartilage through dynamic loading of chondrocyte-seeded agarose gels. J. Biomech. Eng. 122: 252–260.
7. GOOCH, K.J., T. BLUNK, D.L. COURTER, et al. 2001. IGF-I and mechanical environment interact to modulate engineered cartilage development. Biochem. Biophysical. Res. Commun. 286: 909–915.
8. ALTMAN, G.H., R.L. HORAN, I. MARTIN, et al. 2002. Cell differentiation by mechanical stress. FASEB J. 16: 270–272.
9. ROSS, R. 1993. The pathogenesis of atherosclerosis: a perspective for the 1990's. Nature 362: 801–809.
10. NIKLASON, L.E., J. GAO, W.M. ABBOTT, et al. 1999. Functional arteries grown in vitro. Science 284: 489–493.

11. NIKLASON, L.E., W.A. ABBOTT, J. GAO, *et al.* 2001. Morphologic and mechanical characteristics of bovine engineered arteries. J. Vasc. Surg. **33**: 628–638.
12. SOLAN, A., S. MITCHELL, M. MOSES & L.E. NIKLASON. Effects of mechanical conditions on engineered blood vessels. Submitted for publication.
13. VANDENBURGH, H.H., R. SOLERSSI, J. SHANSKY, *et al.* 1996. Mechancial stimulation of organogenic cardiomyocyte growth in vitro. Am. J. Physiol. **270**: C1284–C1292.
14. BROSENITSCH, T.A. & D.M. KATZ. 2001. Physiological patterns of electrical stimulation can induce neuronal gene expression by activating N-type calcium channels. J. Neurosci. **21**: 2571–2579.
15. WIESMANN, H., M. HARTIG, U. STRATMANN, *et al.* 2001. Electrical stimulation influences mineral formation of osteoblast-like cells in vitro. Biochim. Biophys. Acta **1538**: 28–37.
16. NAUGHTON, G. 1999. Skin: the first tissue-engineered products: the advanced tissue sciences story. Sci. Am. **280**: 84–85.
17. PURDUE, G.F., J.L. HUNT, J.M. STILL, *et al.* 1997. A multicenter clinical trial of biosynthetic skin replacement, Dermagraft-TC, compared with cryopreserved human cadaver skin for temporary coverage of excised burn wounds. J. Burn Care Rehab. **18**: 52–57.
18. DUNKELMAN, N.A., M.P. ZIMBER, R.G. LEBARON, *et al.* 1995. Cartilage production by rabbit articular chondrocytes on polyglycolic acid scaffolds in a closed bioreactor system. Biotechnol. Bioeng. **46**: 299.
19. DAVISSON, R., R. SAH & A. RATCLIFFE. 2002. Perfusion increases cell content and matrix synthesis in chondrocyte three-dimensional culture. Tissue Eng. In press.

Tissue Engineering as a Subdivision of Bioprocess Engineering

Reparative Tissue Engineering as a Subspecialty of Tissue Engineering

KYUNG A. KANG

*Chemical Engineering Department, University of Louisville,
Louisville, Kentucky 40292, USA*

KEYWORDS: tissue engineering; bioprocess engineeering

INTRODUCTION

Bioengineering is one of the newest disciplines in engineering education, and tissue engineering is one of its newest subdivisions. Over the past decade, the field of tissue engineering (especially reparative tissue engineering) has been rapidly expanding, due to advances in technologies for human DNA sequencing/gene manipulation, high demands for the products, and the increased availability of sophisticated instrumentation.

Tissue engineering may be expected to continue to expand and develop because of the human interest in longer and higher-quality life. Some important issues related to reparative tissue engineering that require consideration are: a clear definition and classification of its sub-areas; an optimized approach for problem definition and solution, given the highly multi-disciplinary nature of this field; regulatory issues related to the FDA approval process; national/international laws that regulate research; ethical considerations; and the appropriate education of tissue engineers.

DEFINITION

The recent rapid progress in tissue engineering has led to some confusion over a clear definition of the term "tissue engineering" itself, as well as over the research areas within tissue engineering. Therefore, the scope and boundary of tissue engineering need to be clearly defined.

Address for correspondence: Kyung A. Kang, Chemical Engineering Department, University of Louisville, Louisville, KY 40292. Voice: 502-852-2094; fax: 502-852-6355.
kyung.kang@Louisville.edu

Ann. N.Y. Acad. Sci. 961: 216–219 (2002). © 2002 New York Academy of Sciences.

TEAMWORK APPROACH TO SUCCESS

Bioengineering has usually been divided into two subdisciplines: bioprocess engineering and biomedical engineering, depending upon whether the engineering principle is applied without direct human contact or is directly related to the human body. For example, biopurification of therapeutic proteins falls within bioprocess engineering, while bioimaging is within biomedical. Tissue engineering, however, has characteristics of both bioprocess engineering and biomedical engineering. For instance, animal and plant tissue culture for biomaterial production is considered to be bioprocess engineering, while studying the effect of lactic dehydrogenase (LDH) on cardiac muscle is considered to be biomedical engineering. Further, skin tissue growth for transplantation may be both bioprocess and biomedical engineering.

Due to the delicate and complex nature of tissues, technically sophisticated tools are needed for process monitoring during production. Some examples are nanotechnology and/or MEMS technology for micromanipulation of tissue, and noninvasive micro biosensors for monitoring tissue growth rates without damaging the cells. Tissue engineering requires a very integrated, continuous relationship with basic, applied, and medical sciences. The commitment of large resources to achieve final products of reparative tissue engineering will require an effective teamwork approach. Legal and business expertise may be required in addition to engineers and scientists. The ideal team must be configured as early as possible to encourage fast progress and a successful outcome, while the nature of the final product dictates the structure of the team.

REGULATORY ISSUES

Currently, the main issues related to the governmental regulations for the products from reparative tissue engineering are:

- Regulations for the product quality in terms of FDA approval: Since the history of products associated with reparative tissue engineering is relatively short, it is unclear whether the currently applied FDA approval process is applicable or if the regulation is still under development.

- The allowable degree of gene manipulation and tissue repair, as well as the proper law enforcement for violations: Due to the current ethical conflicts in the type of tissues and organs and in the user selection procedures, there have been very frequent disputes on what, how, and who are allowed. Are U.S. laws covering these issues in place and able to clearly resolve the conflict?

- The boundary of national and international laws: Even for existing bioproducts, there have been many instances where U.S. regulations are different from corresponding ones abroad. Such differences are likely to become more commonplace with products related to reparative tissue engineering, since they can involve major organs of the body and serious gene manipulations. Are the United States and other countries currently in discussion on how to resolve these issues?

ETHICS AND LAWS

As reparative tissue engineering products become more frequently used, there will likely be many instances where the law may allow the product's use, but an individual's ethics may not, or vice versa. When these two entities do not agree, and if there is frequent disagreement, then how might this have an impact on the stability of a society?

EDUCATION OF TISSUE ENGINEERS

Because of the unique but diverse nature of tissue engineering, the curriculum for educating tissue engineers must be different from that for traditional engineers. For example, traditional engineering education does not necessarily include teaching and communication with people from other disciplines, including those that must deal with FDA regulatory issues. This current educational practice forces companies to teach these invaluable skills to their new bioengineering employees after their graduation. But, the cost of this training is passed to the consumer by increased product costs. Unless the educational structure for bioengineers undergoes a modification, there will be a shortage of adequately trained tissue engineers to meet demand.[1]

IMPORTANCE OF THE BIOPROCESS ENGINEERING ASPECT OF TISSUE ENGINEERING

As results of tissue engineering research materialize as products, the role of bioprocess engineering becomes extremely important. This is because, as a product is applied to the human body, (1) tissue engineering production processes need to be designed for uniform product quality with extremely high standards of assurance; (2) the supply needs to be timely and supported by adequate mass production and storage; and (3) the costs need to be reasonable for the customers. Considering that products are intimately connected with human life, the quality assurance should be much more tightly regulated than any others. When the final product is in a living form, as many of the tissue engineering products are, then the mass production, an appropriate way of storage, and timely supply are tightly linked. Due to the needs for the process design optimization with these critical issues listed above and the economic analysis of these rather expensive products, tissue engineering requires a much more structured application of bioprocess engineering principles.

The history of manufacturing tissue-engineering products is not very long and bioprocesses for these products are still being established, probably more empirically than systematically. The time for the application of these products for the treatment of human diseases is upon us. Now may be the best time for researchers, research funding agents, regulatory agents, and educators to communicate and deliberate with bioprocess engineers to create an appropriate environment fostering bioprocess engineering for tissue-engineering products.

REFERENCE

1. BRULEY, D.F., K.A. KANG, F. MOUSSY & T.F. WIESNER. 1995. Symbiosis of biomedical and bioprocess engineering utilizing TQM for bioengineering education and research. *In* 1995 ASEE Annual Conference Proceedings. ASEE. Washington, DC.

Bioreactors and Bioprocessing

Breakout Session Summary

Moderators

LAURA E. NIKLASON, *Duke University*

ANTHONY RATCLIFFE, *Advanced Tissue Sciences, Inc.*

Panelists

KELVIN BROCKBANK, *Organ Recovery Systems, Inc.*

DUANE F. BRULEY, *University of Maryland Baltimore County*

KYUNG A. KANG, *University of Louisville*

BROAD STATEMENT

Bioprocessing involves both upstream (bioreactors) and downstream (separation and purification) elements to produce bioproducts. Bioprocessing is fundamental to the pharmaceutical and biotechnology fields for producing large quantities of products for clinical and research use at a reasonable cost. The health-care industry will become increasingly dependent on bioprocessing bioproducts that improve quality of life and economic competitiveness.

VISION

Biomedical engineers are rapidly developing procedures and techniques to produce functional cells and tissues at the laboratory bench. *The next challenge is to translate this research into large-scale production of safe, reproducible tissues that are clinically appropriate for reparative medicine.* This will require research to develop technologies to initiate cell proliferation under conditions that maintain functional phenotypes; to design bioprocessing systems that promote large-scale growth and histogenesis; and to develop efficient separation and purification systems. Progress in these areas is essential to ensure the availability of reparative cells, three-dimensional tissues, and tissue composites for clinical repair and regeneration treatments.

Ann. N.Y. Acad. Sci. 961: 220–222 (2002). © 2002 New York Academy of Sciences.

OBJECTIVES

Currently, the expansion of cell lines and the growth of three-dimensional tissues are major areas of focus in developing new bioprocessing techniques. Solving these important technical issues requires a better understanding of processes such as interactions between cells and biodegradable scaffolds, angiogenesis in large three-dimensional tissues, transport of materials within living tissues, maintenance of sterility from cell seeding to implantation, growth of cells and tissues under varying conditions of biomechanical loading, identification of defined culture media (lacking animal serum), and signaling between cells.

Much of the recent bioprocessing research has been done using bioreactor systems adapted for the production of reparative tissues for transplantation, such as vascular substitutes, skin, or cartilage. However, bioreactors are also being investigated for use as short-term extracorporeal cell-based therapies such as the bioartificial liver or kidney. Miniaturized bioreactors, or microencapsulation devices, have been adapted for either extracorporeal or implanted cell-based drug delivery, for hormones such as insulin or erythropoetin. These devices present other research challenges. One is the development of high flux immunoprotective membranes that do not inhibit the essential mass transport functions. Another is engineering processes to attach functional cells to durable biomaterials, rather than having cells seeded onto biodegradable materials.

Bioreactors are also used in the pharmaceutical industry, both for drug production and testing. This application would also benefit from three-dimensional human tissues that more accurately mimic *in vivo* function.

Further advances in separation and purification will be essential to the quality, reproducibility, and throughput of bioprocessing. Challenges include extending cell viability, preventing biological fouling, developing control biosensors, and implementing continuous processing and quality-control procedures. In addition, improved *in silico* computational modeling of flow distribution and mass transport within the porous media of bioreactor membranes would lead to more successful design for *ex vivo* and *in vivo* bioprocessing systems.

Production and distribution of reparative cells and tissues will require advances in several areas. These include storage technology, where engineering of improved cryopreservation procedures is of utmost importance. Improvements are also necessary in the isolation of pure cell populations, the guarantee of disease-free cell sources, and identification and isolation of stem/progenitor cells. Techniques for efficient gene transfection of large numbers of cells would also be of benefit.

Treatment of certain clinical conditions may require the patient to bioprocess the engineered cells and tissues *in situ* in order to provide healing and repair. Challenges include understanding how natural and synthetic pathways can be utilized to induce cell and biomechanical signaling, how matrices are formed, and how nano-machines can be adapted to build reparative tissues. Implanted biodegradable scaffolds might be engineered to guide self-processing or reparative tissues. Alternately, engineered products might induce existing tissues to resume their natural function.

Several nontechnical areas present challenges to bioprocessing. These include the development of standard reference materials, guidelines, and educational strategies. Achievement of all the identified goals will require strategies to stimulate and strengthen multidisciplinary research collaborations. The success of reparative medicine relates directly to rate of advances in bioprocessing capabilities to produce adequate quantities of engineered material.

CHALLENGES

In order to achieve the objectives, it will be necessary to: (1) design and develop a new generation of bioreactors, separation approaches, and purification methods, which are needed to fulfill the promise of reparative medicine. System design should include *in vitro*, paracorporeal, *in vivo* and *in situ* approaches; (2) promote administrative strategies that emphasize multi-disciplinary research projects, education, and collaborative review and funding; and (3) promote multidisciplinary bioprocessing research.

ACKNOWLEDGMENTS

The Bioreactors and Bioprocessing Breakout Session was coordinated by Dr. Loré Anne McNicol (NEI, NIH) and Dr. John Watson (NHLBI, NIH).

Vascular Assembly in Natural and Engineered Tissues

KAREN K. HIRSCHI,[a] THOMAS C. SKALAK,[b] SHAYN M. PEIRCE,[b] AND
CHARLES D. LITTLE[c]

[a]Departments of Pediatrics and Molecular & Cellular Biology, Center for Cell and
Gene Therapy and Children's Nutrition Research Center, Baylor College of Medicine,
Houston, Texas 77030, USA

[b]Department of Biomedical Engineering, University of Virginia, Box 800759,
Health System, Charlottesville, Virginia 22908, USA

[c]Department of Anatomy and Cell Biology, Kansas University Medical School,
3901 Rainbow Boulevard, Kansas City, Kansas 66160, USA

ABSTRACT: With the advent of molecular embryology and exploitation of genetic models systems, many genes necessary for normal blood vessel formation
during early development have been identified. These genes include soluble effectors and their receptors, as well as components of cell–cell junctions and mediators of cell–matrix interactions. *In vitro* model systems (2-D and 3-D) to
study paracrine and autocrine interactions of vascular cells and their progenitors have also been created. These systems are being combined to study the
behavior of genetically altered cells to dissect and define the cellular role(s) of
specific genes and gene families in directing the migration, proliferation, and
differentiation needed for blood vessel assembly. It is clear that a complex spatial and temporal interplay of signals, including both genetic and environmental, modulates the assembly process. The development of real-time imaging and
image analysis will enable us to gain further insights into this process. Collaborative efforts among vascular biologists, biomedical engineers, mathematicians, and physicists will allow us to bridge the gap between understanding
vessel assembly *in vivo* and assembling vessels *ex vivo*.

KEYWORDS: blood vessel assembly; neovascularization; vascular progenitors;
vascular patterning analysis

INTRODUCTION

The vasculature functions to deliver nutrients and oxygen to, as well as remove
waste products from, all tissues of the body. Proper tissue growth, repair, and regeneration require the establishment of a circulatory network that will meet the specific

Address for correspondence: Karen K. Hirschi, Department of Pediatrics and Molecular &
Cellular Biology, Baylor College of Medicine, One Baylor Plaza, N1030, Houston, TX 77030.
Voice: 713-798-7771; fax: 713-798-1230.
khirschi@bcm.tmc.edu

Ann. N.Y. Acad. Sci. 961: 223–242 (2002). © 2002 New York Academy of Sciences.

metabolic needs of the tissue. Tissues that are engineered *ex vivo* have the same demands to ensure survival and sustain function. Unfortunately, as yet, there are no blueprints for constructing a vascular network within three-dimensional (3-D) tissue structures. Defining such a blueprint is dependent upon (1) obtaining information about the molecular-, cellular-, and network-level contributors to this process using experimental models of vascular assembly and remodeling, and (2) integrating this information through the use of computational systems biology.

Studies conducted in a variety of experimental models have contributed to the understanding of the growth factors, mechanical forces, cell types, and signaling mechanisms integral to vascular assembly and remodeling. These include models in which integrated and efficient vascular networks are effectively established *in vivo*. In murine, avian, and fish embryos, for example, mechanisms of blood vessel and vascular network assembly can be experimentally dissected while the vasculature is established *de novo*. Adult models of vascular remodeling, such as rat or rabbit hindlimb ischemia, have provided insight into how biomolecular signals orchestrate this complex and coordinated process during tissue repair. *In vitro* models, in which vascular cells are stimulated to undergo certain aspects of microvascular remodeling, such as endothelial tube sprouting and mural cell recruitment, have suggested roles for specific growth factors in these processes. Furthermore, the development of *ex vivo* 3-D model systems has enabled the examination of the role of variables such as mechanical shear stress in blood vessel formation.

Obtaining genomic and proteomic data from such experimental models is critical to gaining insight into vascular formation and remodeling; however, the integration of this information over space and time is also necessary for achieving a complete understanding of this complex process. Recent work in systems biology and the advent of computational biology have provided tools capable of organizing vast amounts of data and feedback control mechanisms in order to study the dynamic interactions at the system level.[1] An integrative analysis based on molecular- and cellular-level information will thus result in a continuous spectrum of knowledge enabling manipulation of vascular remodeling in engineered and natural tissues.

With an eye toward this integrative approach, the current understanding of embryonic blood vessel assembly and applications to vascular tissue engineering will be discussed, as will recent advances in imaging of embryonic blood vessel formation that lend further insight into this process. Furthermore, because the engineering of autologous tissues for adult patients, and the therapeutic manipulation of vascular remodeling in natural tissues, are the ultimate goals, aspects of postnatal neovascularization that differ from embryonic processes and have implications for *in vivo* remodeling of engineered constructs will also be overviewed. We will also discuss the availability of vascular progenitor cells, which reside within adult tissues, and their potential to contribute to the remodeling process *in vivo* and in implanted engineered tissue constructs.

The application of emerging biological information, obtained from two-dimensional (2-D) and 3-D model systems to the construction of different types of vessels (arteries, veins, capillaries, lymphatics), and circulatory networks may be facilitated by the development of quantitative and predictive models of vascular assembly and remodeling. Such "blueprints" could then direct the fabrication of scaffolds for engineered constructs and match cell source with desired vascular function. The challenges for the development of integrative vascular systems design and analysis will be discussed in light of current technology and future needs.

VASCULAR STRUCTURE

The circulation of blood requires functional arterial, venous, capillary, and lymphatic systems. The cellular composition and structure of vessels in these systems differ, reflecting in part differences in biological function. The arterial system transports oxygenated blood away from the heart to all parts of the body, and must withstand high-pressure flow. Arteries are typically composed of three layers: the tunica intima, tunica media, and tunica adventitia. The innermost intimal layer is made up of a single layer of endothelial cells, which has direct contact with blood constituents. The media is composed of multiple layers of smooth muscle cells, and is separated from the intima by an internal elastic lamina, which collectively provide structural integrity and control flow rates. The outermost vessel layer, or adventitia, consists of loose connective tissue containing smaller blood vessels and nerves that sustain the vessel wall.[2,3]

The venous system is responsible for return of blood to the heart, and usually contains 75–80% of the total blood volume. The larger caliber and thin-walled veins and venules are responsible for regulating shifts in blood distribution between central and peripheral circulation. The capillary system has enormous surface area, which facilitates passage of substances into and out of blood circulation. Capillaries are composed of small-caliber endothelial tubes (one to three endothelial cells in diameter), surrounded by a basement membrane and a single layer of pericytes.[4] The adventitia of capillaries is a thin layer of connective tissue that is contiguous with that of the surrounding tissue. Blood flow in capillaries is not uniform and is controlled by the contractile state of precapillary arterioles, as well as surrounding pericytes.

EMBRYONIC BLOOD VESSEL ASSEMBLY

The study of *de novo* formation of the circulation during embryonic development directly contributes to the understanding of blood vessel assembly in natural adult tissues and in engineered vessel grafts and vascularized constructs. During embryogenesis, the vasculature is the first organ system to develop and is essential for the development of all other organs. With the advent of molecular embryology and exploitation of malleable genetic model systems, many genes necessary for normal blood vessel formation during early development have been identified. These genes include soluble effectors and their receptors, as well as components of cell–cell junctions and mediators of cell–matrix interactions. To dissect and define the cellular role(s) of these distinct genes and gene families in directing the migration, proliferation, and differentiation of vascular cells and their progenitors during blood vessel assembly, simplified and well-controlled *in vitro* model systems have been established. Information collectively gained from such *in vivo* and *in vitro* systems forms the basis for our current knowledge of the paracrine and autocrine regulation of

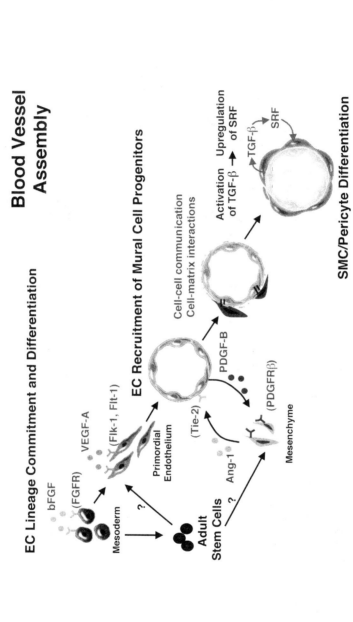

FIGURE 1. Blood vessel assembly. Blood vessel assembly involves the differentiation of multipotent mesodermal progenitors toward primordial endothelium. Further differentiation, cell migration, proliferation, and cell–matrix interactions are required for endothelial tube formation. Newly forming endothelial tubes recruit mural cell precursors from the surrounding mesenchyme, and via cell–cell and cell–matrix interactions, mural cell differentiation is promoted. Continued interactions between endothelial and mural cells are needed for cell growth control and vessel stabilization.

blood vessel assembly, as discussed below (summarized in FIG. 1), and provide the foundation for strategies to modulate vessel formation in natural and engineered tissues in adults.

Blood vessel formation in the embryo begins with the coalescence of mesodermal progenitors and their differentiation into endothelial and blood cells forming structures termed blood islands.[5] The fusion of blood islands gives rise to tube-like structures that define a primitive vascular plexus that is established before blood flow begins. Branching and remodeling of the initial plexus, combined with the onset of blood flow, lead to a well-defined circulatory network.[6–10]

The commitment of multipotent mesodermal cells to an endothelial lineage is influenced by basic fibroblast growth factor signaling,[11] and may involve other factors generated in adjacent tissue types. For example, Indian hedgehog (Ihh) is expressed in the yolk sac endoderm;[12] inhibition of its signaling in the adjacent mesoderm[13] results in disrupted endothelial cell differentiation and tube formation. The exact mechanisms by which Ihh modulates endothelial cell phenotype and blood vessel formation are not yet clear.

Other factors, such as Quaking[14] and vascular endothelial growth factor-A (VEGF-A[15]) are also produced in the yolk sac endoderm. Absence or malfunction of these genes also leads to disrupted vascular development,[14,16,17] but neither is thought to induce commitment of mesodermal cells to an endothelial-specific progenitor, or angioblast, phenotype. Although the precise role of Quaking in vessel assembly is still unknown, much has been learned about the role of VEGF receptor-mediated signaling is blood vessel formation over the last several years.

Embryonic loss-of-function and gain-of-function experiments indicate that VEGF-A signaling, mediated via at least two endothelial-specific receptor tyrosine kinases, VEGF-R2 (Flk-1) and VEGF-R1 (Flt-1), modulates the processes of endothelial cell migration and tube formation. Mice deficient in Flk-1[18] exhibit defective endothelial cell expansion and migration;[19] mice lacking Flt-1 have normal endothelial cells that do not assemble into tubes.[20] A murine deficiency of even one allele of VEGF-A results in defective remodeling of endothelial tube structures and embryonic lethality.[16,17] Embryos directly treated with exogenous VEGF-A exhibit enlarged vascular lumens, or vascular fusion;[21] in contrast, inhibition of endogenous VEGF(s) with soluble VEGF-R1 yields arrested vasculogenesis.[22] Hence, VEGF signaling is clearly essential for the initial steps of vessel formation,[23] but its production and bioavailability must be tightly controlled to ensure the normal formation of endothelial networks.

The newly formed endothelial tubes then govern the subsequent acquisition of mural cells (pericytes or smooth muscle cells) that make up the surrounding vessel wall. Proliferating endothelial cells secrete platelet-derived growth factor-B (PDGF-B) that acts as a chemoattractant and mitogen for mural cell precursors,[24–26] derived from the mesenchyme surrounding the endothelial tubes.[27,28] Thus, organ-specific mesodermal primordia contribute the mural cell layer(s) to developing vessels, which likely results in tissue-specific functional and regulatory properties of vascular smooth muscle cells and pericytes that are retained *in vitro*.[29] Furthermore, the anatomical position of developing vessels produces mural cell heterogeneity. Smooth muscle cells and pericytes in the head and neck region are derived from neural crest tissue, whereas mural cells in the lower portion of the body are derived from mesoderm,[30] as are all endothelial cells.[7,10,31]

The regulation of mural cell recruitment may involve genes other than PDGF-B, from diverse gene families, including angiopoietin-1,[32] tissue factor,[33] jagged-1, and COUP-TFII.[34] Mice lacking these factors have endothelial tubes, but no associated mural cells, and die *in utero*. Because in these animals, endothelial cells appear to differentiate and form tubes, faulty mural cell recruitment is proposed as the underlying defect.[35] The biological connection among these proteins, and with PDGF-B, remains to be determined.

Upon contact with endothelial cells, newly recruited mesenchymal cell progenitors are induced toward a mural cell fate,[28] in a process mediated by the activation of transforming growth factor-beta (TGF-β).[25,36] Although the process of TGF-β activation in response to heterocellular interactions is complex[37,38] and not completely understood, it is clear that activated TGF-β[39] and receptor signaling[40] play a critical role in vascular development.

Activated TGF-β is thought to directly induce mural cell-specific gene expression via TGF-β–control elements in the promoter regions of genes such as SM-α-actin.[41] TGF-β also induces mural cell differentiation via the upregulation of the transcription factor, serum response factor (SRF).[42] SRF binds to serum response elements in the promoter regions of mural cell-specific genes, including SM-γ-actin,[42-44] SM-α-actin,[45] SM22-α,[46,47] and calponin,[48,49] and induces their coordinated expression.[50] Although other transcription factors may serve as co-activators, SRF was found to be both necessary[42,50] and sufficient[42] to induce a smooth muscle cell phenotype.

Differentiated, contractile pericytes and smooth muscle cells not only modulate the flow of blood through endothelial tubes, but also serve to stabilize vessel structures.[51] Mesenchymal and mural cell contact suppresses endothelial cell growth,[26] and may sustain endothelial cell survival through the production of Ang-1[52] or VEGF-A.[53] Thus, it is clear that mural cells play a vital role in the establishment and maintenance of blood vessel structure and function; however, the exact mechanisms of endothelial-induced mural cell differentiation and growth control during vessel assembly are not yet clearly defined.

Observations of genetically altered mice suggest that gap junction communication between endothelial and mural cell progenitors may modulate these processes during blood vessel formation. Mice deficient for Cx43, although having survived through gestation, die within hours of birth secondary to malformations and malfunctioning of the subpulmonary outflow tract of the heart.[54] Mice deficient for Cx45 die early in gestation (E9.5–10.5 days) and exhibit abnormal vessel structures; endothelial tube structures form but are not invested with differentiated mural cells.[55] These *in vivo* studies suggest that gap junctions may be involved in the acquisition of mural cells during blood vessel formation. Recent studies using an *in vitro* model of vessel assembly provide more direct evidence that gap junction communication is required for endothelial-induced mural cell differentiation, and suggest that heterocellular communication is needed for the activation of TGF-β.[56]

Other requirements for normal blood vessel formation include cell–cell adhesion, establishment of a competent extracellular matrix, and appropriate cell–matrix interactions mediated via integrins such as $\alpha v \beta 3$ and $\alpha 5 \beta 1$. Thus, blood vessel assembly represents the culmination of many precisely timed events that are coordinated with the onset of blood flow. The recapitulation of this process *ex vivo* for the vascularization of tissue constructs, or the creation of vessel grafts, is a daunting task that will require careful monitoring and optimization of these many cellular events orchestrated in tandem. To achieve such a goal, it is necessary to not only understand the molecular regulation of these events, but to also acquire an appreciation for the space and time in which they occur, which will be discussed below.

POSTNATAL NEOVASCULARIZATION

Once the vasculature has formed, it is thought to be relatively stable postnatally, with infrequent cell turnover (i.e., human endothelial cell $T_{1/2}$ = 3.1 years[57]), although the contribution to preexisting endothelia by circulating progenitors is open to question (see below). However, in response to tissue injury or remodeling, and in the progression of prevalent pathological conditions such as atherosclerosis, retinopathy, and tumorigenesis, the formation of new blood vessels plays a central role.

Postnatal neovascularization was thought to occur solely via a process known as angiogenesis, which is largely driven by VEGF signaling.[35,58] Angiogenesis involves the proliferation and migration of differentiated endothelial cells from preexisting vessels to "sprout" new vessel tubes,[59] as well as intussusception,[60] a process involving the subdivision of a large caliber "mother" vessel into multiple daughter vessels.[61] Recent studies, however, indicate that multipotent stem cells found in circulating blood[62,63] or resident in adult tissues[64,65,70] also significantly contribute to newly forming blood vessels during tissue repair. Furthermore, this process involves the coalescence of multipotent progenitors and more closely resembles embryonic vasculogenesis rather than angiogenesis.[66]

Understanding the contribution of adult stem cells to injury-induced neovascularization in adult tissues will provide insights into *in vivo* remodeling and vascularization of implanted tissue constructs. Understanding the potential of distinct populations of such cells will enable the selection of appropriate progenitors for autologous vascular tissue engineering. Therefore, the source and potential of progenitors in adult tissues are discussed below.

Endothelial Cell Progenitors

Precursors to vascular endothelial cells in adults were first identified in circulating blood[62] and were recently shown to incorporate into vascular endothelium to a level of ~10% engraftment,[67] thereby enhancing blood flow in ischemic tissues.[63] The precursor population found in blood[62,68,69] is heterogeneous; thus, the true vascular cell progenitor(s) in circulation, if one exists, remains to be defined and therefore cannot be directly isolated for tissue engineering applications, but perhaps can be homed to sites of desired vascularization, as discussed below.

In contrast, a highly purified and homogeneous subpopulation of stem cells, termed SP cells, has been isolated from adult bone marrow (BM-SP)[70] and skeletal muscle[71](Majka *et al.*, manuscript submitted) tissues. Upon bone marrow transplantation into lethally irradiated recipients, SP cells from both tissue sources reconstitute all blood cell lineages[70,71] (McKinney-Freeman *et al.,* manuscript submitted). Given that developmental studies suggest a common progenitor for hematopoietic and endothelial cells, termed the "hemangioblast,"[72] the recent finding that SP cells, derived from adult bone marrow, also give rise to vascular endothelium[73] is intriguing. In fact, the phenotype of the BM-SP[73] is similar to that of embryonic hemangioblasts.[74] Reproducible isolation of this homogeneous population from clinically accessible tissues in adults, and thorough characterization of its molecular phenotype, will accelerate the development of directed, autologous vascular-tissue–engineering strategies. It is also possible that any mesodermal cell that displays multipotentiality can be, with sufficient data, reprogrammed to form all mesodermal tissues.

Vascular Smooth Muscle Cell Progenitors

It was previously thought that, during blood vessel assembly, vascular smooth muscle cells and pericytes were recruited, via endothelial-derived signals[25,26,28] from surrounding mesenchyme, or recruited along with endothelial cells from pre-existing vessels[24] to form the surrounding vessel wall. However, a recent finding suggests that endothelial and smooth muscle cells can also arise from the same progenitor. Yamashita and co-workers[75] found that embryonic stem cells expressing the VEGF receptor Flk-1[74] give rise to endothelial cells, as well as vascular smooth muscle cells, *in vivo* and *in vitro*. These studies suggest that endothelial cell precursors in adult tissues may serve as smooth muscle cell precursors as well. In fact, circulating progenitors of smooth muscle cells have been reported[76] and may arise from the same circulating cells that give rise to endothelial cells. Furthermore, BM-SP cells, which regenerate blood and vascular endothelium, have also been observed to give rise to vascular smooth muscle cells *in vivo* (Hirschi lab, unpublished data), providing further support for this idea. The vascular bipotential of such adult progenitors may enable their use in the creation of autologous vessel grafts.[77]

A vascular endothelium is by definition a mesothelium. In this regard, it is interesting to note that there is evidence in lower vertebrates that coelomic mesothelium can give rise to vascular precursor cells.[78] The possibility that this process occurs in warm-blooded vertebrates has not, to the best of our knowledge, been examined.

Molecular Profile of Adult Vascular Progenitors

Adult stem cells are, by definition, capable of becoming multiple differentiated cell types *in vivo*. Thus, directing their fate specifically toward vascular endothelial and smooth muscle cells for tissue-engineering purposes is dependent on defining their phenotype and understanding the molecular regulation of vascular development. As discussed above, we have learned much about the genes that regulate vascular cell commitment and differentiation (reviewed in Ref. 79); however, relatively little is known about the molecular profile of adult vascular progenitors. In fact, many blood- and marrow-derived populations studied are heterogeneous in nature, and the true progenitors in these populations have not been identified. More highly purified and homogeneous populations of adult stem cells such as BM-SP may, on the other hand, represent an ideal subpopulation of stem cells that can be molecularly defined more clearly and is more amenable to manipulation for autologous replacement therapies.

Toward this goal, the molecular profile of BM-SP cells, relative to the program of vascular cell commitment and differentiation, has been investigated.[73] It was determined, as expected, that such cells do not express genes reflective of mature endothelial cells (Flt-1, VE-cadherin, von Willebrand factor, factor VIII), nor do they express genes indicative of a smooth muscle cell phenotype (SM-α-actin, calponin, SM-22α). BM-SP cells do, however, express some genes shared by embryonic hemangioblasts, including PE-CAM-1[75] and Tal-1.[80] Although they do not express the tyrosine kinase receptor Flk-1,[74] BM-SP do express its ligand, VEGF-A,[73] which may act to direct the recruitment[81] or integration[82] of stem cells into existing endothelium. BM-SP also express Ang-1 and its receptor, Tie-2, which are involved in an important paracrine signaling pathway needed for the recruitment, or maintenance, of smooth muscle cells and pericytes to stabilize newly forming endothelial tubes.[32,83] The role of Ang-1 and Tie-2 in these adult stem cells is uncertain; Ang-1 signaling may modulate stem cell survival, as has been proposed for mature endothelial cells.[52] Much remains to be determined regarding the role of factors such as Ang-1 and VEGF in the modulation of stem cell survival, growth control, and interactions within the surrounding microenvironment. The surprising lack of VEGF receptors, which are critical for endothelial cell expansion and tube formation in the embryo, suggests that progenitors resident in adult tissues may require vascular inductive signals, or signals for migration, proliferation, and/or differentiation that differ from their embryonic counterparts.

An alternative, and more straightforward, explanation for the perplexing inconsistencies between embryonic and adult progenitors, is that cell types such as the "pure" hemangioblast, as specified by a battery of markers, may not exist in the adult. Rather, during formation of the primary mesoderm (gastrulation) a small percentage of mesoderm is set-aside and persists into adulthood. In this case, such cells would have the potential to form *all* mesodermal derivatives (endothelium, connective tissue, smooth, cardiac and skeletal muscle, bone, cartilage, and all leukocytes, phagocytes and other blood cell types). Additionally, such (hypothetical) primitive mesodermal cells would be free to migrate through loose connective tissues or to be carried by lymphatics or blood flow. Further, depending on local epigenetic factors (ECM, growth factors, cytokines) such cells would not, by necessity, exhibit any "specific" markers that would unequivocally characterize their potential to form only *one* mesodermal lineage, for example blood or endothelium.

Homing to Desired Sites of Vascularization

Engraftment of endothelial cell progenitors from blood circulation[62,68,69] or adult tissues[73] into blood vessels has been reported to reach levels of 3.5[73] to ~10%,[67] respectively. Although this level of engraftment has been reported to improve blood flow,[63] little is known about "optimal" levels of engraftment at injured sites, or whether engineered tissue constructs can be adequately vascularized *in vivo* via the same mechanisms. Developing strategies for improving engraftment of multipotent progenitors to specific sites *in vivo* could potentially accelerate the creation of functional tissue implants. Embryological studies suggest that VEGF-A may be useful for these purposes because it acts as a chemoattractant for hemangioblasts during developmental vasculogenesis.[81] Perhaps local expression of VEGF, provided as recombinant protein intravenously[84] or via gene transfer,[85] may serve to

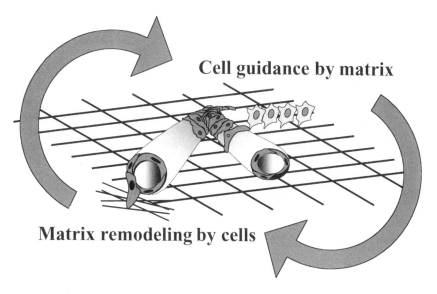

FIGURE 2. Blood vessel assembly *ex vivo*. Dynamic reciprocity of engineered tissue constructs involves cell guidance by environmental cues and active regulation of the extracellular matrix environment by implanted cells.

"home" endogenous progenitors that express the VEGF receptors to desired sites of engraftment, such as vascular grafts or tissue constructs.

FROM 2-D AND 3-D *IN VITRO* MODELS OF VASCULAR REMODELING TO ENGINEERED VASCULAR TISSUES

A thorough understanding of embryonic and postnatal vascular development and remodeling *in vivo*, as discussed above, will accelerate the design of vascularized systems. In addition, the study of vascular cell responses to environmental stimuli in 2-D and 3-D *in vitro* models is relevant to the dynamic adaptation of engineered tissue constructs to their scaffold and host environments. For example, models in which extracellular matrix environment is controlled and manipulated will further our understanding of a process termed dynamic reciprocity, in which the initial matrix geometry and composition guides cell-seeding patterns, while cells constituting the engineered tissue also modulate the tissue scaffold and host tissue (FIG. 2).

Previous studies of 2-D and 3-D *in vitro* models of the vasculature have focused on the interactions between cells and the surrounding cell matrix constituents,[86] the presence or absence of biochemical signaling factors and their local concentration gradients,[25] contact between cells,[56,87] and mechanical forces that impart stimuli to different vascular cell types.[88] These environmental and physiological cues are critical for initial acceptance of the engineered construct by the host tissue. Normal function of the implant requires the continual adaptation of engineered vessels to such stimuli. For example, the inosculation process, whereby the host vasculature connects to the vasculature of the engineered tissue construct, is a crucial step in establishing perfusion to the implanted engineered tissue. This process has been examined by implanting biodegradable polymer matrices containing human dermal microvascular endothelial cells into severe combined immunodeficient (SCID) mice.[89] In this study, the implanted cells formed functional human microvessels that anastomosed with the mouse vasculature.

Similarly, the development of a tissue-engineered small-diameter artery that is nonthrombogenic and resistant to infection and rejection requires an understanding of its mechanical properties, physiological transport capabilities, and functional and remodeling characteristics.[90] Whether the engineered vessel consists of cells seeded on a synthetic biodegradable polymer[91] or a tube of biopolymer composed of reconstituted collagen gel,[92] when implanted into the host tissue, the capability of the engineered vessel to undergo remodeling confers biocompatibility and ultimately prevents rejection. Various physical and functional features of engineered blood vessels that facilitate normal remodeling *in vivo*, such as mechanical burst strength and compliance, have been examined *in vitro*.[93] Thus, *in vitro* experimental models in conjunction with *in vivo* implantation studies are important tools for discovering, isolating, and examining environmental and physiological factors that contribute to vascular growth and remodeling. In turn, the proficiency of an engineered vascular construct to undergo normal remodeling after implantation into host tissue is a prerequisite for its normal long-term function.

DEVELOPING AN INTEGRATED APPROACH TO BLOOD VESSEL DESIGN AND ANALYSIS

Although the dawn of genomic research has led to the discovery of numerous genes and gene products, how these biomolecules work together to produce functional tissue assembly during development, maturation, and normal and pathological remodeling in the adult animal is currently unknown. The trend has been to focus on a particular molecular pathway and produce targeted intervention to create observable differences in organism functionality and/or architecture. Some of this work is highlighted in previous sections. However, because almost every known cell–cell and cell–matrix interaction involves multiple genes acting at specific tissue locations at specific times, understanding the entire complex system requires a detailed knowledge of gene functionalities and cellular behaviors as well as the spatial and temporal integration of this knowledge at the tissue level. Through a strategic combination of reductionist experimental approaches and integrative computer modeling, a global blueprint for vascular remodeling can be defined and utilized for engineering vascularized tissue constructs and advancing therapeutic vascularization strategies in natural tissues.

Cellular automata (CA) computer models are well suited for studying multifactorial complex systems, such as vascular remodeling, in which independent and discrete cell behavior is controlled by multiple variables changing in space and time. For example, CA models can integrate biochemical signaling events with cellular behaviors within assemblies of thousands of interacting individual cells in order to study vascular remodeling in terms of architectural patterning. FIGURE 3 shows a

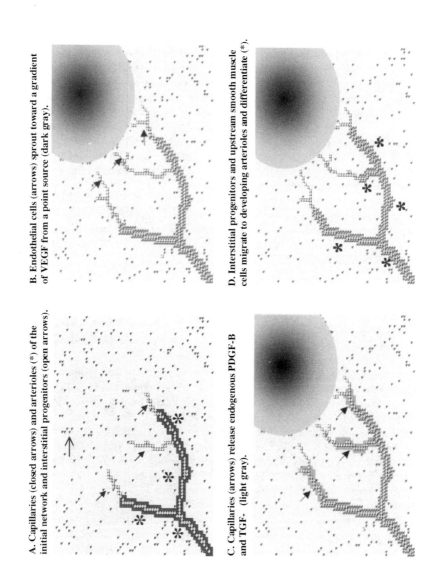

A. Capillaries (closed arrows) and arterioles (*) of the initial network and interstitial progenitors (open arrows).

B. Endothelial cells (arrows) sprout toward a gradient of VEGF from a point source (dark gray).

C. Capillaries (arrows) release endogenous PDGF-B and TGF- (light gray).

D. Interstitial progenitors and upstream smooth muscle cells migrate to developing arterioles and differentiate (*).

FIGURE 3. Computer modeling of blood vessel formation. Graphical results from a novel cellular automata computer simulation of capillary sprouting (**B**) and arteriolar remodeling (**C–D**).

simulated microvascular network generated by a CA model.[94] This network was stimulated to remodel by an initial exogenous application of VEGF-A, and the endothelial cell response to the biochemical stimulus is seen in the resulting capillary sprouting. This CA model is also capable of integrating physiological stimuli known to be important in the vascular remodeling process such as mechanical circumferential stretch,[95] environmental cues such as cell–matrix interactions,[96] metabolic states such as tissue hypoxia,[97] and gene expression characteristic of cell phenotypic states such as smooth muscle myosin expression by fully differentiated vascular smooth muscle cells.[98] Furthermore, specific molecular and cellular variables in the system can be modulated by a single keystroke, thus facilitating high-throughput analysis of the complex system. The approach can be tailored to incorporate new information so as to continually evolve with the body of knowledge and suggest experimental directions that may not be readily apparent. Finally, this new approach will also be useful in identifying bottlenecks or rate-limiting steps in the assembly process, thus facilitating development of therapeutic agents aimed at accelerating or suppressing functional blood vessel growth.

Vascular Pattern Analysis

CA models are also well suited to measuring and recording a vast number of variables that reflect patterns of spatial and temporal changes in vascular networks stimulated to undergo remodeling. In the example shown in FIGURE 3, an architectural pattern analysis, including the distribution and branching arrangement of arteries, capillaries, and venules, is indicative of overall network functionality.[99,100] The knowledge of architectural pattern formation obtained from the CA model can then be verified by *in vivo* vascular networks stimulated to remodel with microdelivery of therapy factors. Multisignal growth factor guidance of vascular pattern formation can in turn be translated to *in vivo* engineering of tissues. Additionally, CA models are capable of analyzing spatial and temporal gene expression patterns, which provides insight into gene function.[101] In summary, the systematic pattern analysis provided by the ability to store and organize data enables CA models to integrate numerous variables in time and space to produce patterns that can be deciphered to reveal integrated gene-, cellular-, and network-level functions.

MODERN MICROSCOPIC STUDIES ARE NEEDED TO PROVIDE THE STRATEGIES FOR INDUCTING VASCULARIZATION OF ENGINEERED TISSUES

Our understanding of vascular cell differentiation and vessel morphogenesis has been significantly advanced by the creative application of molecular genetics and by use of *in vivo* and *in vitro* model systems, as discussed above. Nevertheless, the complex motions that characterize tissue and organ formation in 3-D space are not found encoded in DNA and cannot be fully recapitulated in model systems. Thus, unfortunately, progress in understanding the dynamics of cell and tissue organization has moved slowly. Therefore, the formidable challenge now facing vessel biologists, and those interested in tissue engineering is, By what means will we gain the best understanding of vascular cell behavior within an interactive ensemble of living tissue?

That is, what is the *cell biological* basis of tissue and vessel organization? The answer to this question most likely lies in an integrative-approach for collecting and correlating dynamic data sets using light microscopy and eventually NMR micro-imaging.

The elucidation of vascular cell- and target-tissue dynamics will require the talents of biologists, physicists, engineers, and mathematicians. Creative new approaches to optics, photonics, and software design are now being employed to visualize and quantify the parameters associated with vascular morphogenesis in engineered tissues—to say nothing of embryos and normal organs. Time-lapse (motion) microscopy is now being combined with specifically tailored algorithms and dedicated operating software that facilitates computational studies of morphogenesis.[102-104] The ability to observe vascular cell behavior using four-dimensional microscopy (3-D and time) allows biologists to measure and quantify the behavior of individual cells, cellular cohorts and their surrounding extracellular matrix. This integrative approach will be important for monitoring vascularization of engineered tissues, for deciphering perturbation of vessel morphogenesis in experimental animals, and for recording vascular malformations in genetically altered animals.

Proper quantification of *dynamic* cell/tissue behavior can illuminate complex questions such as:

- What are the global cellular and tissue-level behaviors that occur during formation of a new, tissue specific, vascular pattern?

- How can we extract local vascular cell behavior from global tissue behavior during tissue vascularization, or while cells attempt to vascularize an engineered tissue?

- Should vessel primordia be part of the original architecture of engineered tissues or should bioengineers focus on inducing vascularization in partially formed tissue ensembles (e.g., the liver or muscle)?

- How do divergent vascular beds form anastomoses?

- How do the mechanical forces of blood flow influence vascular patterning and vessel wall architecture?

Despite the rich information base regarding ligand and receptors that influence vascular cell behavior discussed above, there is no rigorous, comprehensive understanding of tissue-level vascularization. Thousands of diagrams have been drawn depicting angiogenesis or vasculogenesis with very little, or no, microscopic data to validate the drawings. The key parameter that is virtually never depicted is that of relative motion. Whether vessels are forming in embryos or during postnatal organ growth, "real" growing (new) vessels must hit a moving target. It is not true that parenchymal tissues form and then become static while a capillary bed is assembled. Therefore, any serious understanding of vessel morphogenesis must account for the motion of both presumptive vessel cells and the global motion of surrounding tissue. This assertion is especially true when starting with small tissue ensembles (embryos) where extensive deformations and folding are inevitable. Thus, motion analysis of vascularization will have to fulfill at least two criteria. The first is the elucidation of global (tissue) deformations coincident with organ vascularization; the other is a detailed description of local cell motility, which specifies loop and curls in a capillary bed.

For example, by establishing an anatomical frame of reference, which itself may be in motion, it is possible to record and quantify large-scale tissue deformations and foldings associated with growth and/or assembly of engineered tissues. On the local level, and independent of global motions, it is possible, using time-lapse microscopy, to compute parameters that describe cell protrusive activity, response to mechanical stress and strain, velocity, and trajectory. Moreover, by making reasonable assumptions about the viscoelastic properties of tissues, it may soon be possible to predict tension fields within growing tissues (Little and colleagues, manuscript in preparation). It is possible that nascent endothelial cells respond in a predictable manner to tension fields as one component of vascular pattern formation. If so, tension could, hypothetically, be used to constrain or shape vascular bed morphogenesis.

FUTURE CHALLENGES

Future challenges in the field of vascular remodeling revolve around integrating the vast amounts of genetic, biochemical, cellular, and physiological information derived from this complex process into a comprehensive system-level understanding across multiple length and time scales. This systems-level goal cannot be reached in full without continuing to promote the "reductionist" approach that is needed to develop a fundamental understanding of each level of organization: gene, protein, cell, vessel, network, and tissue, through both hypothesis-driven and descriptive research. An effective marriage of the "holistic" and reductionist approaches will foster a deeper understanding of the vascular remodeling process than either approach could provide if pursued separately. In effect, computer models will suggest novel experimental approaches, while experimental and therapeutic approaches will provide the necessary governing principles and direct verification for computer simulations.

A number of new approaches are needed. Future efforts should focus on developing advanced computer models capable of multicomponent and multisignal simulation of vascular growth and development. Site- and cell-specific sampling methods for genetic components, such as DNA microanalysis and proteomics, will be necessary for characterizing new gene candidates and the functional circuitry of known genes. The advancement of experimental tools for spatial and temporal control of growth factor delivery/stimulation is also a necessity for understanding signaling interactions involved in vascular remodeling. In addition to the spatial and temporal characterization of the biochemical environment, monitoring mechanical factors at the cellular level will also be critical to forming a complete blueprint for vascular assembly and remodeling. Effective use of this blueprint also depends on improving fabrication methods for synthetic vascular tissues as well as continued development of noninvasive *in vivo* vascular imaging capabilities.

Addressing the above-mentioned issues is vital to meeting the challenge of creating vascular grafts and vascularized tissues; so, too, is the training of researchers/educators in interdisciplinary studies of complex living systems. Advances in vascular assembly will best be achieved by collaborative, integrative, and interdisciplinary efforts seeking to build a unified functional understanding from existing and developing knowledge bases.

ACKNOWLEDGMENTS

We thank Robert T. Tranquillo, Larry V. McIntire, Mark Post, and John Ranieri for their participation in an NIH discussion panel that stimulated and focused ideas emphasized in this review.

REFERENCES

1. VON DASSOW, G. *et al.* 2000. The segment polarity network is a robust developmental module. Nature **406:** 188–192.
2. SABIN, F.R. 1920. Studies on the origin of blood vessels and of red corpuscles as seen in the living blastoderm of the chick during the second day of incubation. Contrib. Embryol. **9:** 213–259.
3. MEIER, S. 1980. Development of the chick embryo mesoblast: pronephros, lateral plate, and early vasculature. J. Embryol. Exp. Morphol. **55:** 291–306.
4. MANDARINO, L.J. *et al.* 1993. Regulation of fibronectin and laminin synthesis by retinal capillary endothelial cells and pericytes *in vitro*. Exp. Eye Res. **57:** 609–621.
5. DOETSCHMAN, T.A., A. GOSSLER & R. KEMLER. 1987. Blastocyst-derived embryonic stem cells as a model for embryogenesis. *In* Future Aspects in Human *in Vitro* Fertilization. Feichtingen & P. Kemeter, Ed.: 187–195. Springer-Verlag. Berlin.
6. RISAU, W. *et al.* 1988. Vasculogenesis and angiogenesis in embryonic stem cell-derived embryoid bodies. Development **102:** 471–478.
7. COFFIN, J.D. & T.J. POOLE. 1988. Embryonic vascular development: immunohistochemical identification of the origin and subsequent morphogenesis of the major vessel primordia in quail embryo. Development **102:** 1–14.
8. POOLE, T.J. & J.D. COFFIN. 1989. Vasculogenesis and angiogenesis: two distinct morphogenetic mechanisms establish embryonic vascular pattern. J. Exp. Zool. **251:** 224–231.
9. NODEN, D.M. 1989. Embryonic origins and assembly of blood vessels. Am. Rev. Respir. Dis. **140:** 1097–1103.
10. PARDANAUD, L., F. YASSINE & F. DIETERLEN-LIÈVRE. 1989. Relationship between vasculogenesis, angiogenesis and haemopoiesis during avian ontogeny. Development **105:** 473–485.
11. RISAU, W., S.P. GAUTSCHI & P. BOHLEN. 1988. Endothelial cell growth factors in embryonic and adult chick brain are related to human acidic fibroblast growth factor. EMBO J. **7:** 959–962.
12. FARRINGTON, S.M., M. BELAOUSSOFF & M.H. BARON. 1997. Winged-helix, Hedgehog and Bmp genes are differentially expressed in distinct cell layers of the murine yolk sac. Mech. Dev. **62:** 197–211.
13. DYER, M.A. *et al.* 2001. Indian hedgehog activates hematopoiesis and vasculogenesis and can respecify prospective neurectodermal cell fate in the mouse embryo. Development **128:** 1717–1730.
14. NOVEROSKE, J.K. *et al.* 2002. Quaking is essential for blood vessel development. Genesis **32:** 218–230.
15. MIQUEROL, L., B.L. LANGILLE & A. NAGY. 2000. Embryonic development is disrupted by modest increases in vascular endothelial growth factor gene expression. Development **127:** 3941–3946.
16. CARMELIET, P. *et al.* 1996. Abnormal blood vessel development and lethality in embryos lacking a single VEGF allele. Nature **380:** 435–439.
17. FERRARA, N. *et al.* 1996. Heterozygous embryonic lethality induced by targeted inactivation of the VEGF gene. Nature **380:** 439–442.
18. SHALABY, F. *et al.* 1995. Failure of blood-island formation and vasculogenesis in Flk-1-deficient mice. Nature **376:** 62–66.
19. SCHUH, A.C. *et al.* 1999. *In vitro* hemaotpoietic and endothelial potential of flk-1(–/–) embryonic stem cells and embryos. Proc. Natl. Acad. Sci. USA **96:** 2159–2164.

20. FONG, G.-H. *et al.* 1995. Role of the flt-1 receptor tyrosine kinase in regulating the assembly of vascular endothelium. Nature **376:** 66–70.
21. DRAKE, C.J., D.A. CHERESH & C.D. LITTLE. 1995. An antagonist of integrin αvβ3 prevents maturation of blood vessels during embryonic neovascularization. J. Cell Sci. **108:** 2655–2661.
22. DRAKE, C.J. & C.D. LITTLE. 1999. VEGF and vascular fusion: implications for normal and pathological vessels. J. Histochem. Cytochem. **47:** 1351–1356.
23. DUMONT, D.J. *et al.* 1994. Dominant-negative and targeted null mutations in the endothelial receptor tyrosine kinase, tek, reveal a critical role in vasculogenesis of the embryo. Genes Dev. **8:** 1897–1907.
24. LINDAHL, P. *et al.* 1997. Pericyte loss and micro-aneurysm formation in platelet-derived growth factor B-chain-deficient mice. Science **277:** 242–245.
25. HIRSCHI, K.K., S.A. ROHOVSKY & P.A. D'AMORE. 1998. PDGF, TGF-β and heterotypic cell-cell interactions mediate the recruitment and differentiation of 10T1/2 cells to a smooth muscle cell fate. J. Cell Biol. **141:** 805–814.
26. HIRSCHI, K.K. *et al.* 1999. Endothelial cells modulate the proliferation of mural cell precursors via PDGF-BB and heterotypic cell contact. Cir. Res. **84:** 298–305.
27. NAKAMURA, H. 1988. Electron microscopic study of the prenatnal development of the thoracic aorta in the rat. Am. J. Anat. **181:** 406–418.
28. HUNGERFORD, J.E. *et al.* 1996. Development of the aortic vessel wall as defined by vascular smooth muscle and extracellular matrix markers. Dev. Biol. **178:** 375–392.
29. TOPOUZIS, S. & M.W. MAJESKY. 1996. Smooth muscle lineage diversity in the chick embryo. Two types of aortic smooth muscle cell differ in growth and receptor-mediated transcriptional responses to transforming growth factor-β Dev. Biol. **178:** 430–445.
30. KIRBY, M.L. & K.L. WALDO. 1995. Neural crest and cardiovascular patterning. Circ. Res. **77:** 211–215.
31. EKBLOM, P. *et al.* 1982. The origin of the glomerular endothelium. Cell Differ. **11:** 35–39.
32. SURI, C. *et al.* 1996. Requisite role of angiopoietin-1, a ligand for the TIE2 receptor, during embryonic angiogenesis. Cell **87:** 1171–1180.
33. CARMELIET, P. *et al.* 1996. Role of tissue factor in embryonic blood vessel development. Nature **383:** 73–75.
34. TSAI, S.Y. & M. TSAI. Chick ovalbumin upstream promoter-transcription factors (COUP-TFs): coming of age. Endocr. Rev. **18:** 229–240.
35. FOLKMAN, J. & P.A. D'AMORE. 1996. Blood vessel formation: What is its molecular basis? Cell **87:** 1153–1155.
36. ANTONELLI-ORLIDGE, A. *et al.* 1989. An activated form of transforming growth factor β is produced by cocultures of endothelial cells and pericytes. Proc. Natl. Acad. Sci. USA **86:** 4544–4548.
37. DENNIS, P.A. & D.B. RIFKIN. 1991. Cellular activation of transforming growth factor β requires binding to the cation-independent mannose 6-phosphate/insulin-like growth factor type II receptor. Proc. Natl. Acad. Sci. USA **88:** 580–584.
38. KOJIMA, S., K. NARA & D.B. RIFKIN. 1993. Requirement for transglutaminase in the activation of latent transforming growth factor-β in bovine endothelial cells. J. Cell Biol. **121:** 439–448.
39. DICKSON, M.C. *et al.* 1995. Defective haematopoiesis and vasculogeneisis in transforming growth factor-β1 knock-out mice. Development **121:** 1845–1854.
40. LARSSON, J. *et al.* 2001. Abnormal angiogenesis but intact hematopoietic potential in TGF-β type I receptor-deficient mice. EMBO J. **20:** 1663–1673.
41. HAUTMANN, M.B., C.S. MADSEN & G.K. OWENS. 1997. A transforming growth factor-beta (TGF-beta) control element drives TGF-β-induced stimulation of smooth-muscle alpha-actin gene expression in concert with 2 CARG elements. J. Biol. Chem. **272:** 948–956.
42. HIRSCHI, K.K. *et al.* 2001. Transforming growth factor-β induction of smooth muscle cell phenotype requires transcriptional and translational control of serum response factor. J. Biol. Chem. **277:** 6287–6295.
43. BROWNING, C.L. *et al.* 1998. The developmentally regulated expression of serum response factor plays a key role in the control of smooth muscle-specific genes. Dev. Biol. **194:** 18–37.

44. PHIEL, C.J. *et al.* 2001. Differential binding of an SRF/NK-2/MEF2 transcription factor complex in normal versus neoplastic smooth muscle tissues. J. Biol. Chem. **276:** 34637–34650.
45. BLANK, R.S. *et al.* 1992. Elements of the smooth muscle α actin promoter required in *cis* for transcriptional activation in smooth muscle: cell type-specific regulation. J. Biol. Chem. **267:** 984–989.
46. LI, L. 1997. Evidence for a serum response factor-mediated regulatory network governing SM22α transcription in smooth, skeletal, and cardiac muscle cells. Dev. Biol. **187:** 311–321.
47. KIM, S. *et al.* 1997. A serum response factor-dependent transcriptional regulatory program identified distinct smooth muscle cell sublineages. Mol. Cell. Biol. **17:** 2266–2278.
48. SAMAHA, F.F. *et al.* 1996. Developmental pattern of expression and genomic organization of the calponin-h1 gene. J. Biol. Chem. **271:** 395–403.
49. MIANO, J.M. *et al.* 2000. Serum response factor-dependent regulation of the smooth muscle calponin gene. J. Biol. Chem. **275:** 9814–9822.
50. LANDERHOLM, T.E. *et al.* 1999. A role for serum response factor in coronary smooth muscle differentiation from proepicardial cells. Development **126:** 2053–2062.
51. HIRSCHI, K.K. & P.A. D'AMORE. 1996. Pericytes in the microvasculature. Cardiovasc. Res. **32:** 687–698.
52. PAPAPETROPOULOS, A. *et al.* 1999. Direct actions of angiopoietin-1 on human endothelium:evidence for network stabilization, cell survival, and interaction with other angioges–nic growth factors. Lab. Invest. **79:** 213–223.
53. ZACHARY, I. 2001. Signaling mechanisms mediating vascular protective actions of vascular endothelial growth factor. Am. J. Physiol. **280:** C1375–C1386.
54. REAUME, A.G. *et al.* 1995. Cardiac malformation in neonatal mice lacking connexin 43. Science **267:** 1831–1834.
55. KRUGER, O. *et al.* 2000. Defective vascular development in connexin45-deficient mice. Development **127:** 4179–4193.
56. HIRSCHI, K.K. *et al.* Gap junction communication is necessary for endothelial-induced differentiation of progenitors toward a smooth muscle fate. Submitted.
57. SCHWARTZ, S.M. & E.P. BENDITT. 1976. Clustering of replicating cells in aortic endothelium. Proc. Natl. Acad. Sci. USA **73:** 651–653.
58. FERRARA, N. & T. DAVIS-SMYTH. 1997. The biology of vascular endothelial growth factor. Endocr. Rev. **18:** 4–25.
59. RISAU, W. 1995. Differention of endothelium. FASEB J. **9:** 926–933.
60. BURRI, P.H. & M.R. TAREK. 1990. A novel mechanism of capillary growth in the rat pulmonary microcirculation. Anat. Rec. **228:** 35–45.
61. PATAN, S. *et al.* 2001. Vascular morphogenesis and remodeling in a model of tissue repair: blood vessel formation and growth in the ovarian pedicle after ovariectomy. Circ. Res. **89:** 723–731.
62. ASAHARA, T. *et al.* 1997. Isolation of putative progenitor endothelial cells for angiogenesis. Science **275:** 964–967.
63. SCHATTEMAN, G.C. 2000. Blood-derived angioblasts accelerate blood flow restoration in diabetic mice. J. Clin. Invest. **106:** 571–578.
64. ORLIC, D. *et al.* 2001. Mobilized bone marrow cells repair the infarcted heart, improving function and survival. Proc. Natl. Acad. Sci. USA **98:** 10344–10349.
65. KOCHER, A.A. *et al.* 2001. Neovascularization of ischemic myocardium by human bone marrow-derived angioblasts prevents cardiomyocyte apoptosis, reduces remodeling and improves cardiac function. Nature Med. **7:** 430–436.
66. SPRINGER, M.L. *et al.* 1998. VEGF gene delivery to muscle: potential role for vasculogenesis in adults. Cell **2:** 549–558.
67. CROSBY, J.R. *et al.* 2000. Endothelial cells of hematopoietic origin make a significant contribution to adult blood vessel formation. Circ. Res. **87:** 728–730.
68. MUROHARA, T. *et al.* 2000. Tranplanted cord blood-derived endothelial procursor cells augment postnatal neovascularization. J. Clin. Invest. **105:** 1527–1536.
69. LIN, Y. *et al.* 2000. Origins of circulating endothelial cells and endothelial outgrowth from blood. J. Clin. Invest. **105:** 71–77.

70. GOODELL, M.A. *et al*. 1996. Isolation and functional properties of murine hematopoietic stem cells that are replicating *in vivo*. J. Exp. Med. **183:** 1797–1806.
71. GUSSONI, F. *et al*. 1999. Dystrophin expression in the mdx mouse restored by stem cells transplantation. Nature **401:** 390–404.
72. CHOI, K. *et al*. 1998. A common precursor for hematopoietic and endothelial cells. Development **125:** 725–732.
73. JACKSON, K.A. *et al*. 2001. Regeneration of cardiac muscle and vascular endothelium by adult stem cells. J. Clin. Invest. **107:** 1395–1402.
74. DRAKE, C.J. & P.A. FLEMING. 2000. Vasculogenesis in the day 6.5 to 9.5 mouse embryo. Blood **95:** 1671–1679.
75. YAMASHITA, J. *et al*. 2000. Flk1-positive cells derived from embryonic stem cells serve as vascular progenitors. Nature **408:** 92–96.
76. SHIMIZU, K. *et al*. 2001. Host bone marrow cells are a source of donor intimal smooth muscle-like cells in murine aortic transplant arteriopathy. Nature Med. **7:** 738–741.
77. NIKLASON, L.E. *et al*. 1999. Functional arteries frown *in vitro*. Science **284:** 489–493.
78. MUNOZ-CHAPULI, R. *et al*. 1999. Differentiation of hemangioblasts from embryonic mesothelial cells? A model on the origin of the vertebrate cardiovascular system. Differentiation **64:** 133–141.
79. CARMELIET, P. 2000. Mechanisms of angiogenesis and arteriogenesis. Nature Med. **6:** 389–395.
80. DRAKE, C.J. *et al*. 1997. Tal1/SCL is expressed in endothelial progenitor cells/angioblasts and defines a dorsal-to-ventral gradient of vasculogenesis. Dev. Biol. **192:** 17–30.
81. CLEAVER, O. & P.A. KRIEG. 1998. VEGF mediates angioblast migration during development of the dorsal aorta in Xenopus. Development **125:** 3905–3914.
82. SENGER, D. *et al*. 1996. Stimulation of endothelial cell migration by vascular permeability factor/vascular endothelial growth factor through cooperative mechanisms involving the $\alpha_v\beta_3$ integrin, osteopontin, and thrombin. Am. J. Pathol. **149:** 293–305.
83. SATO, T. *et al*. 1995. Distinct roles of the receptor tyrosine kinases Tie-1 and Tie-2 in blood vessel formation. Nature **376:** 70.
84. ASAHARA, T. *et al*. 1999. VEGF contributes to postnatal neovascularization by mobilizing bone marrow-derived endothelial progenitor cells. EMBO J. **18:** 3964–3972.
85. KALKA, C. *et al*. 2000. VEGF gene transfer mobilizes endothelial progenitor cells in patients with inoperable coronary disease. Ann. Thoracic Surg. **70:** 829–834.
86. CLARK, R.A.F. *et al*. 1996. Transient functional expression of $\alpha_v\beta_3$ on vascular cells during wound repair. Am. J. Pathol. **148:** 1407–1421.
87. COWAN, D.B., S.J. LYE & B.L. LANGILLE. 1998. Regulation of vascular connexin43 gene expression by mechanical loads. Circ. Res. **82:** 786–793.
88. WILSON, E. *et al*. 1993. Mechanical strain induces growth of vascular smooth muscle cells via autocrine action of PDGF. J. Cell Biol. **123:** 741–747.
89. NOR, J.E. *et al*. 2001. Engineering and characterization of functional human microvessels in immunodeficient mice. Lab. Invest. **81:** 453–463.
90. NEREM, R.M. 2000. Tissue engineering a blood vessel substitute: the role of biomechanics. Yonsei Med. J. **41:** 735–739.
91. NIKLASON, L.E. *et al*. 2001. Morphogenic and mechanical characteristics of engineered bovine arteries. J.Vasc. Surg. **33:** 628–638.
92. L'HEREUX, N. *et al*. 1993. *In vitro* construction of a human blood vessel from cultured vascular cells: a morphologic study. J. Vasc. Surg. **17:** 449–509.
93. GIRTON, T.S. *et al*. 2000. Mechanisms of stiffening and strengthening in media-equivalents fabricated using glycosylation. J. Biomech. Eng. **122:** 216–223.
94. PEIRCE, S.M., E. VAN GIESON & T.C. SKALAK. 2001. *In silico* cellular automata simulation of microvascular remodeling. Ann. Biomed. Eng. **29:** S75.
95. VAN GIESON, E.J., T.C. SKALAK & R.J. PRICE. Enhanced spatial smooth muscle cell coverage of microvessels exposed to increased hemodynamic stresses *in vivo*. Circ. Res. Submitted.
96. CLARK, R.A. *et al*. 1996. Transient functional expression of $\alpha_v\beta_3$ on vascular cells during wound repair. Am. J. Pathol. **148:** 1407–1421.
97. ADAIR, T.H. *et al*. 1987. Whole body structural vascular adaptation to prolonged hypoxia in chick embryos. Am. J. Physiol. **252:** H1228–H1234.

98. MANABE, I. & G.K. OWENS. 2001. CArG elements control smooth muscle subtype-specific expression of smooth muscle myosin in vivo. J. Clin. Invest. **107:** 823–834.
99. SKALAK, T.C. & R.J. PRICE. 1996. The role of mechanical stresses in microvascular remodeling. Microcirculation **3:** 143–165.
100. SKALAK, T.C., R.J. PRICE & P.J. ZELLER. 1998. Where do new arterioles come from? Mechanical forces and microvessel adaptation. Microcirculation **5:** 91–94.
101. MADSEN, C.S. *et al.* 1998. Smooth muscle-specific expression of the smooth muscle myosin heavy chain gene in transgenic mice requires 5'-flanking and first intronic DNA sequence. Circ. Res. **82:** 908–917.
102. RUPP, P.A., A. CZIROK & C.D. LITTLE. 2001. Computational biology of vascular cell behavior *in vivo*. Mol. Biol. Cell **12** (Suppl.)**:** 268a.
103. CZIROK, A., M. MATSUSHITA & T. VICSEK. 2001. Theory of periodic swarming of bacteria: application to *Proteus mirabilis*. Phys. Rev. E **63:** 31915–31926.
104. CZIROK, A. *et al.* 2002. Multi-field 3D scanning light microscopy of early embryogenesis. J. Microsc. Vol. 205, part I.

In Vivo and in Silico Approaches for Analysis and Design of Multisignal, Multicomponent Assembly Processes in Vascular Systems

THOMAS C. SKALAK

*Department of Biomedical Engineering, University of Virginia,
Charlottesville, Virginia 22908, USA*

KEYWORDS: vascular assembly; automata simulation; inosculation

A convergence of problems in reparative medicine today requires renewed attention to improving our basic understanding, developing predictive modeling techniques, and creating rational therapeutic methods of guiding vascular assembly in diseased natural tissues as well as in engineered tissues. Some of the high-profile applications today include therapeutic angiogenesis and collateralization in the ischemic heart and creating nutritive supply for engineered constructs. Each of these is currently being approached using products, reagents, and processes derived mainly from the past ten years of reductionist basic scientific thinking and experimental work. For example, in the field of applied angiogenesis, known as myocardial collateralization or revascularization, a wide variety of strategies based on application of growth factors and/or vascular precursor cells (including autologous bone marrow cells, circulating cells, and various types of fibroblasts) are currently being studied in animals and in clinical trials. Yet, it is already clear that a complex spatial and temporal interplay of signals, including both genetic and environmental cues, are needed for normal small vessel growth, remodeling, and maintenance. Thus, the next phase of research and development in this field will almost certainly require innovative new ways to study and design multicomponent, multisignal vascular systems, both in natural and engineered tissues. An emerging theme is that predetermination of final vascular or tissue structure in the microfabrication stage of engineered tissues is unlikely to succeed alone, but rather the harnessing of the natural, adaptive powers of the resident cells to guide functional vascular assembly and maintenance will provide the ability to supply large, complex tissues.

Biomedical engineers are ideally suited to help address the behavior of these complex systems. Consider the long list of components typically addressed today via the reductionist approach to studying therapeutic vascular assembly: hypoxia-

Address for correspondence: Thomas C. Skalak, Department of Biomedical Engineering, University of Virginia, Box 800759, Charlottesville, VA 22908. Voice: 804-924-0270; fax: 804-982-3870.

tskalak@Virginia.edu

Ann. N.Y. Acad. Sci. 961: 243–245 (2002). © 2002 New York Academy of Sciences.

inducible factor I (HIF-I); statins; ACE inhibitors; TGF-beta; PDGF; FGF-1, -2, and -4; VEGF; ephrins; angiopoietins; MMPs; intracellular signals, including NO; integrins; ECM proteins; and the receptors associated with these ligands. Each of these may be expressed either as a part of normal vascular development (vasculogenesis), later during vessel remodeling to a differentiated functional network with arterioles, capillaries, and venules, or during adult maintenance of a functional network. In early patterning, genetic and structural signals deriving from the extracellular matrix may strongly influence vascular network formation. In remodeling of a fully mature functional network, however, additional environmental cues, such as vascular wall cell strains due to intravascular flow and shear stress and/or blood pressure, may play an important role. Finally, vessels may grow by (a) primordial mesh formation followed by selective pruning and remodeling, (b) "sprouting" extensions from and enlarging existing vessels, (c) intussuception or splitting of existing capillaries, and (d) inosculation (anastomotic connection) with existing vessels. The implication of these observations is that several different design paths for engineering of *de novo* vascular networks in engineered constructs or remodeled networks *in vivo* can be envisioned.

One design path simulates the process of vasculogenesis, by seeding vascular precursors (endothelial cells, smooth muscle cells, fibroblasts, and stem cells) at sufficient densities within a parenchymal construct, allowing primordial mesh formation to occur, and subsequent remodeling to a functional network via naturally occurring molecular signals. A problem with this approach may turn out to be that the same diffusive constraints that limit cell seeding density in today's constructs will also limit initial seeding densities of vascular precursors in constructs larger than a few microliters, unless the initial geometric form factor is less than about 100 microns in at least one dimension. This might prove successful if later manipulations create the desired solid 3-D construct from the initial "2-D" seeding platform. Another design path would be to design for growth of the vascular system at a rate consistent with the growth of the engineered construct, so that nutrient delivery keeps pace with tissue demands. In this case, one could envision that the sprouting phase of vascular assembly would be critical, as this would supply the new vessels at an appropriate rate. A third design would be to deliver a vascular supply after a construct is formed, and this would require substantial capacity for either natural penetration (e.g., via MMPs) or mechanical penetration of the construct to guide the new vessels.

Any of the various designs for vascular assembly of functional networks could be optimized through the systematic study of the processes involved and their incorporation into predictive models of assembly. For example, if one chooses the first design path of primordial plexus formation, followed by remodeling, one might wish to design an optimal template for the plexus topology and dimensions. If one chooses to remodel existing vessels to collateralize a diseased tissue or to penetrate a newly implanted construct, the optimal spatial and temporal sequence of molecular and mechanical stimuli will need to be specified and implemented. Here, the central challenge will be to understand the genetic circuits (i.e., genes and interplay of gene products) controlling the assembly events and then apply the knowledge to optimize the patterning process. One approach under development (Skalak *et al.*, unpublished data) is an *in silico* cellular automata simulation method to predict cell motions, associations, and differentiation state based on the secretion, diffusion, binding, and functional effects of a complex interplay of molecular signals in a field of interacting discrete cells that include endothelial, smooth muscle, and fibroblast precursors. The method yields quantitative predictions of small vessel extension and wall phenotype/ structure that can be compared with experimentally observed vascular assembly patterns. The hope is that such approaches will place the myriad of putative players in the assembly process into a coherent picture that can guide design and fabrication of functional vessel networks. A complementary experimental approach is delivery of multiple growth factors via microcarriers in a prescribed temporal and spatial sequence to perform guided arterialization in areas demanding blood supply.

Other important issues requiring attention in the long-term are venous assembly, the inosculation process, and on/off switches for various steps in vascular assembly and maintenance. Venous assembly is little studied, but predictive methods to guide outflow channel assembly are essential to successful design. There is evidence for both predetermination of venous structures through ephrin expression and for hemodynamic effects on maintenance of the venous phenotype. Small venules exposed to elevated strains express smooth muscle differentiation markers and growth factors that suggest strain levels may act in part to maintain the venous/arterial structure differences. Inosculation is historically believed, by plastic surgeons particularly, to play an important role in reconnecting newly formed vessels in transplanted tissues to existing host tissue vessels. Yet, no study has successfully detailed how this process works at the large conduit level, and it may be the case that only microvessels have the ability to reconnect. If so, another design rule would be established, since growth of new small vessels *in situ* (after construct fabrication) would be necessary to establish host connections to implanted constructs.

New means to address these needs include computational simulation of vascular assembly, site- and cell-specific sampling methods, and DNA microarray analysis to identify new candidates and gene circuitry of existing candidates, proteomic description of components, experimental tools that allow spatial and temporal control of growth factor delivery/stimulation and cell-based therapies, fabrication methods allowing for creation of spatial gradients in ECM, signal molecule, and cell structures, and imaging techniques that permit chronic functional and phenotypic study of small vessel assembly in formative stages in intact, three-dimensional tissues.

ACKNOWLEDGMENTS

This work was carried out in collaboration with Richard J. Price, Eric Van Gieson, and Shayn M. Peirce, Department of Biomedical Engineering, University of Virginia, Charlottesville, VA 22908.

SELECTED REFERENCE

1. VAN GIESON, E.J. & T.C. SKALAK. 2001. Chronic vasodilation induces matrix metalloproteinase 9 (MMP-9) expression during microvascular remodeling in rat skeletal muscle. Microcirculation **8**: 25–31.

Vascular Assembly in Engineered and Natural Tissues

LARRY V. MCINTIRE

Rice University, Houston, Texas 77005, USA

KEYWORDS: angiogenesis; parenchymal cells; vascularization

Proper tissue repair or *de novo* regeneration requires formation of a patent extensive three-dimensional vascular network that meets the specific oxygen demands of the tissue involved. Network development involves a complex temporal and spatial interplay of cellular, chemical, and mechanical stimuli. Understanding these dynamics is required in order to adequately use the body's regenerative capacity to develop engineered tissue or to promote more rapid healing.

Research into the effects of tissue microenvironment on cell performance, growth factor expression, and vessel inosculation is essential to understanding and guiding microvascular formation. *In vitro* studies provide insight into cell responses to specific environmental factors. During angiogenesis, endothelial cells encounter various flow conditions. It is well documented that endothelial cell growth factor and motility factor expression vary greatly, depending on flow. The effects of flow and shear forces will indicate factors involved in initiating angiogenesis and detrimental effects of vessels not developed properly. Understanding the effect of other environmental factors, such as oxygen levels and mechanical strain, on network formation and stability will also lead to a greater understanding of mechanisms involved.

Parenchymal cells may provide a method for producing large quantities of vascular cells with drastically different properties from mature vascular cells. It is unknown if these cells can be used to improve tissue regeneration without resulting in uncontrolled and unregulated growth. Development of methods to isolate, study, and control development of multilineage cells could provide the greatest benefit to the progress of regenerative medicine.

Although individual growth factors are known to contribute to angiogenesis, the interaction of these factors in a temporal and spatial fashion is not known. Increased understanding of the angiogenic cascade will lead to better techniques for promoting proper angiogenesis in reparative biology. Research needs to be performed to determine the sequence and level of signals involved not only in vessel formation but in maintaining a network once it is established. The signals, matrix components, or

Address for correspondence: Larry V. McIntire, Rice University, G.R. Brown, Room W-100, 6100 Main Street, Houston, TX 77005. Voice: 713-348-4903; fax: 713-348-5154.
mcintire@rice.edu

Ann. N.Y. Acad. Sci. 961: 246–248 (2002). © 2002 New York Academy of Sciences.

cells necessary to promote and preserve vascular structure need to be determined. Because of the complex nature of angiogenesis and confounding parameters, much of this research could be addressed with accurate molecular and cellular computer models and simulations.

Because of the wealth of growth factors and cytokines involved in angiogenesis, unique analysis is required. DNA and protein microarrays have the potential to provide a method for rapid analysis of large libraries of compounds. Microarrays need to be developed and used to specifically study vascular cell responses to *in vitro* and *in vivo* chemical, physical, and molecular stimuli. Methods of analysis and well-defined metrics need to be developed that are comprehensive and standardized.

Clues to the desired performance of tissue-engineered products can be gained from studying typical reconstructive surgery techniques and wound healing environments. An ideal tissue-engineered construct would mimic a surgical graft, consisting of fully developed vessels and tissue. Anecdotal surgical evidence suggests the processes involved in graft take and survival, but it is not understood at a fundamental level. Knowledge about the spatial and temporal kinetics of vascularization and the interactions between donor and host vessels in surgery and healing will contribute to setting design goals and constraints for engineered tissue.

Ultimately, engineering a microvascular network will involve both *in vitro* and *in vivo* components. Although successful in small-scale animal models, *in vivo* scaffold prevascularization may not be applicable to larger defects. Research into methods of fabricating vascularized constructs *in vitro* is one of the most active areas of tissue engineering. Severe design limitations are encountered when scaling up from simple models to clinical-sized applications. Basic vessels have been formed, but more complex and extensive networks require more creative solutions. Once engineered, these networks would need to anastamose with host vasculature and remain patent. Essential to this task is an understanding of the molecular interactions and signals required for cell fusion as well as the feasibility of translating an *in vitro* network to an *in vivo* setting. This could lead to techniques for increasing the rates, of or guiding inosculation to, the engineered network.

The success of scaffolds implanted *in vivo* to form vascular networks must be understood completely. These networks may have characteristics that greatly deter oxygen exchange such as large avascular spaces, tortuous vessels, or unordered branching. This information cannot be gained using conventional angiogenic assays and imaging techniques. Methods are being developed that allow high resolution, three-dimensional quantitation of microvascular networks.

Vascularization is a complex process that, if understood, can be manipulated for improvements in clinical medicine. However, unregulated angiogenesis can result in severe consequences. For this reason it is essential that bioengineers, clinicians, and life scientists work together to elucidate the mechanisms involved in all aspects of angiogenesis so that it can be controlled, enhanced, or promoted for the greatest clinical benefit.

ACKNOWLEDGMENTS

This work was done in collaboration with Eric Brey (Institute of Biosciences and Bioengineering, Rice University; and Laboratory of Reparative Biology and

Bioengineering, University of Texas, M.D. Anderson Cancer Center, Houston, TX) and Charles W. Patrick, Jr. (Laboratory of Reparative Biology and Bioengineering, University of Texas, M.D. Anderson Cancer Center, Houston, TX).

SELECTED REFERENCES

1. STAMATAS, G.N. & L.V. MCINTIRE. 2001. Rapid flow-induced responses in endothelial cells. Biotechnol. Prog. **17:** 383–402.
2. MCCORMICK, S.M., S.G. ESKIN, L.V. MCINTIRE, *et al.* 2001. DNA microarray reveals changes in gene expression of shear stressed human endothelial cells. Proc. Natl. Acad. Sci. USA **98:** 8955–8960.
3. BREY, E., T.W. KING, C. JOHNSON, *et al.* 2002. A technique for quantitative 3D analysis of microvascular structure. Accepted for publication in Microvasc. Res.

Angiogenesis

Initiation and Maintenance

MARK J. POST

Associate Professor of Medicine and Physiology, Dartmouth Medical School, Hanover, New Hampshire 03755, USA

KEYWORDS: therapeutic angiogenesis; stem cell therapy; VEGF-A

Therapeutic angiogenesis has become a clinically feasible modality in the management of cardiovascular disease and has generated considerable interest as such. There is some evidence from clinical trials that the use of potent growth factors via a mostly systemic delivery mode is safe. It is less clear how growth factors can be applied to optimally support the growth of new, lasting blood vessels. In the trials, growth factors were generally given as single shots or as multiple repeated injections over the course of a couple of days, albeit in high dosages compared to the local levels needed for efficacy. From tracer studies, for instance, we found that with intracoronary FGF-2, less than 1% of the injected dose, can be retrieved from the heart and less than 0.05% will remain detectable 24 hours later. With a dose in the high μg/kg range, this still amounts to reasonable local concentrations that might be effective. The efficacy of this approach, however, is questionable, judging by the existing clinical data, and is most likely due to the short exposure times and inadequate initiation of angiogenesis.

Once growth of new blood vessels has been successfully initiated, it is necessary that they persist. It has been our experience in simple animal models of angiogenesis, such as the *in vivo* Matrigel model, that newly formed vessels regress rapidly after withdrawal of the growth factor. It has been suggested, but not shown convincingly *in vivo*, that timely application of the stabilizing factors angiopoietin-1 and PDGF-BB prolong the life of these vessels. The precise dynamics of regression and the required dwell time for angiogenic factors to prevent regression most likely depend on the model system/organ, the functionality of the blood vessels, and the type of growth factor or mixture of growth factors. The need for prolonged exposure to growth factors is not only a practical concern. We have shown that prolonged exposure to FGF-2 by a cell-based gene transfer induces a pathology reminiscent of a benign tumor. In contrast, cell-based gene transfer of VEGF-A turns an otherwise ineffective growth factor in the Matrigel model into a system of dose-dependent an-

Address for correspondence: Mark J. Post, Associate Professor of Medicine and Physiology, Dartmouth Medical School, Hanover, NH 03755, USA. Voice: 603-650-3903; fax: 603-650-6130.
marcus.j.post@dartmouth.edu

Ann. N.Y. Acad. Sci. 961: 249–250 (2002). © 2002 New York Academy of Sciences.

giogenesis. Thus, certain delivery modalities of growth factors may introduce new side effects.

Regression of blood vessels is a physiologic process in response to loss of function, such as in the uterine system after pregnancy and in a maturing scar. A good model system, therefore, for looking at regression of blood vessels includes assessment of their function, as in, for instance, intravital microscopy, surface videomicroscopy, or laser Doppler. The same principle of functionality of tissues to increase survival *in vivo* is also true for other tissues that are being engineered. It has been convincingly shown, for instance, that the survival of skeletal muscle increases tremendously while under continuous tension, whereas they rapidly regress or undergo apoptosis when introduced unanchored. For bone transplants to survive they need to be exposed to external pressures and strains as well.

The interest in vasculogenesis and bone marrow–derived stem cells that stimulate angiogenesis is rapidly growing. It will be an exciting field, but currently it has not reached a mature level. The key questions that need to be addressed are (1) How are the cells identified with respect to differentiation stage? (2) Do the cells survive and get incorporated in newly formed blood vessels in reasonable numbers? (3) If selected cells are truly pluripotential, how will differentiation into endothelial cells be stimulated? (4) How do stem cells home to a site of ischemia, inflammation, tissue growth, or tissue repair?

Finally, in order for stem cell therapy to be accepted as angiogenic strategy in cardiovascular disease, both improved perfusion and improved function of the heart need to be shown in blinded studies with proper cellular controls.

SELECTED REFERENCES

1. Post, M.J. *et al.* 2000. Adenovirus mediated gene therapy through intramyocardial injections: percutaneous intramyocardial versus surgical epicardial delivery. Cardiac Vasc. Regenerat. **1:** 106–113.
2. Sato, K. *et al.* 2000. Efficacy of intracoronary versus intravenous FGF-2 in a pig model of chronic myocardial ischemia. Ann. Thorac. Surg. **70:** 2113–2118.
3. Sato, K. *et al.* 2001. Efficacy of intracoronary or intravenous VEGF165 in a pig model of chronic myocardial ischemia. J. Am. Coll. Cardiol. **37:** 616–623.

The Tissue-Engineered Small-Diameter Artery

ROBERT T. TRANQUILLO

University of Minnesota, Minneapolis, Minnesota 55455, USA

KEYWORDS: tissue engineering; small-diameter vascular graft; smooth muscle cells; tissue scaffolds; elastin fibers; arterial mechanical properties; biocompatibility; endothelial cells; stem cells

The characteristics proposed for an "ideal" tissue-engineered small-diameter artery include the following:[1-4] It is biocompatible, that is, nonthrombogenic and nonimmunogenic, and resistant to infection as well (all of which are associated with a confluent nonactivated endothelium). Moreover, it results in an acceptable wound-healing response (without fibrosis). It possesses appropriate mechanical properties, including physiological compliance (viscoelasticity) and, critically, adequate strength, without any propensity for permanent creep that leads to aneurysm. It possesses physiological transport properties, that is, appropriate permeability to solutes and cells. Finally, it exhibits key physiological properties, such as vasoconstriction/relaxation responses. From a practical standpoint, suturability and simplicity of handling are necessary, and, from a commercial standpoint, it must be fabricated in a process that scales well with quantity and be a product that can be shipped and stored.

There are four main approaches currently being investigated, all of which satisfy an apparent preqrequisite to biocompatibilty of a small-diameter graft—that no permanent synthetic materials are used. One approach is acellular, based on implanting decellularized tissues treated to enhance biocompatibility, strength, and cell adhesion/invasion leading to cellularization with host cells.[5] The other three approaches involve implantation of constructs possessing some degree of cellularity. The most recent of these is based on the concept of *self-assembly*, wherein cells are cultured on tissue culture plastic in medium inducing high ECM synthesis.[6,7] This leads to sheets of *neotissue* that are subsequently processed into multilayer tubular form. The other two approaches rely on a polymeric scaffold. One is based on forming a tube of a synthetic biodegradable polymer and then seeding the cells (which would not survive the conditions of polymer synthesis), relying on active cell invasion or an applied force to achieve cellularity.[8-11] The other is based on a tube of a biopolymer, typically a reconstituted type I collagen gel,[12,13] formed with and compacted by tissue cells, where an appropriately applied mechanical constraint to the compaction yields circumferential alignment of fibrils and cells characteristic of the arterial media.[14-16] It is this last feature that is most attractive about a biopolymer-based tissue-

Address for correspondence: Robert T. Tranquillo, University of Minnesota, 421 Washington Ave. SE, Minneapolis, MN 55455. Voice: 612-625-6868; fax: 612-626-7246.

tranquillo@cems.umn.edu

Ann. N.Y. Acad. Sci. 961: 251–254 (2002). © 2002 New York Academy of Sciences.

engineered artery. This follows from two axioms, (i) that native artery function, particularly mechanical function, depends on *structure* (particularly *alignment* of the smooth muscle cells and collagen fibers in the medial layer) as much as it depends on composition, and (ii) that the tissue-engineered artery should serve as a functional remodeling template, so that while providing function during the remodeling, the artificial tissue also provides a template for the alignment of the remodeled tissue.

To some extent, all these approaches rely on the ability of cells (transplanted or host) to adhere to and migrate within the construct, and to remodel its composition and/or structure. This last point is key, as remodeling confers biocompatibility, in principle, by virtue of complete resorption of the initial scaffold. Of course, the initial scaffold must be replaced by functional cell-derived ECM on the same time scale. Remodeling also determines the ultimate mechanical, transport, and biological properties. Thus, attaining the ideal tissue-engineered artery is dependent on a large number of fabrication variables that affect remodeling, including polymer type and structural characteristics (density, stiffness, pore size, and, if relevant, fiber diameter), cell type and source (fibroblasts vs. SMC vs. endothelial cells, species, passage number, adult vs. neonatal differentiated cells, cells originating from stem cells), medium composition (serum and/or supplementing factors), nutrient availability (diffusional limitations determined by construct thickness and cell loading), mechanical stress state (static vs. cyclic, mechanical vs. hydrodynamic), autocrine factors (related to construct thickness and cell loading), and cell phenotype. In a fascinating, but poorly understood, feedback process, the cell phenotype, which dictates the nature of its future remodeling, is determined by the nature of its past remodeling.

The four approaches have met the various criteria of the ideal tissue-engineered artery to varying degrees. Generally, the first criterion of concern following biocompatibilty is adequate mechanical (burst) strength. While there is no standard for what the initial burst strength must be, burst pressures exceeding 2000 mm Hg have been reported at the *in vitro* stage for all approaches, except for the biopolymer-based tissue-engineered artery. There have been relatively few *in vivo* studies. The one published *in vivo* study using the acellular approach (chemically cross-linked submucosal collagen from small intestine) reported 100% patency in rabbits out to 13 weeks, with invasion and indications of organotypic organization of invading smooth muscle and endothelial cells.[5] The one published *in vivo* study using the self-assembly approach was limited by use of xenogeneic cells; the absence of an endothelium (to avoid hyperacute rejection) yielded low patency over the week studied.[6] A recent case involving replacement of the pulmonary artery in a juvenile used the cellularized synthetic degradable polymer approach with autologous cells.[10] There was no evidence of graft occlusion or aneurysmal changes after seven months.

Obviously, many challenges remain. The greatest will be meeting all criteria for the ideal tissue-engineered artery simultaneously. For example, high burst strength is often at the expense of a compliance mismatch,[6] which can lead to intimal hyperplasia at the suture line. Conversely, constructs that possess physiological compliance lack burst strength.[17] Notably, no approach has yet resulted in all the key features of the media, namely circumferential alignment of the smooth muscle cells and collagen fibers *and* elastin lamellae. In fact, mature elastin fibers have only been reported in the self-assembly approach, and in association with fibroblasts, not SMCs.[6] Elasticity is critical to abolish permanent creep and is naturally conferred by elastin lamellae. The developmental down-regulation of elastogenesis by SMCs continues to be a major hurdle. However, media-equivalents fabricated using neonatal rat SMCs and fibrin gel (instead of collagen gel), which exhibit enhanced collagen synthesis in the presence of TGF-β,[12,13,18] have recently been found to possess organized elastin fibers.[19] They have also been discovered in media-equivalents fabricated using adult rat SMCs and collagen gel, when subject to long-term cyclic distention.[20] Being able to induce reversion from the synthetic phenotype, which is desirable for remodeling during cell-based construct fabrication *in vitro*, back to the contractile phenotype at implantation, is another SMC-related issue. The immune response remains a challenge, with the high immunogenicity of the endothelial cell being the primary obstacle to using nonautologous cells without immunosuppression, and thus a barrier to widespread application and low-cost mass production by any approach. Other mass production challenges include reproducibility and associated quality control of cell-based constructs, economical and reliable bioreactors when stretch and flow conditioning are implemented, and the efficacy of cryopreservation, which appears the most attractive option for meeting the shipping and storage criterion. Finally, the ability to control the events that occur postimplantation (e.g., further construct remodeling and angiogenesis for blood supply) so that function is maintained or enhanced is a major challenge. The advantages associated with a defined cellularity at implantation must justify the difficulty and costs of meeting these challenges.

Successfully meeting these challenges presents many opportunities. It requires advances in cell and tissue biology; in particular, cell responses to combined matrix, growth factor, and mechanical signals require further understanding. The major achievement of a nonimmunogenic ("universal donor") endothelial cell requires advances in immunobiology. Adult stem cells and genetically modified cells offer the promise of realizing elastogenesis from induced SMCs. Advances in biomedical engineering are clearly required as well. Synthetic degradable fibrillar matrices would combine the advantages of the biopolymer approach (primarily, alignment from constrained compaction) with those of synthesis. Hybrid scaffolds with a biocompatible elastic component, such as elastin peptide-based polymers, provide more opportunities in biomaterials. Elucidating composition-structure-mechanical property relationships for cell-based constructs, a paradigm of materials science, is a prerequisite to their rationale design and fabrication. Appropriate fatigue testing and test standards for such constructs need to be defined. Biomedical engineers will also need to elucidate transport property relationships for the same reason. Finally, imaging methods that allow monitoring of any aspect of remodeling *in vitro* and, particularly, *in vivo* would be invaluable. Many of these challenges and opportunities are common to other areas of tissue engineering, necessitating continuation of the multifaceted, interdisciplinary research ongoing in a growing number of laboratories.

REFERENCES

1. MAYER, J.E., JR., T. SHIN'OKA & D. SHUM-TIM. 1997. Tissue engineering of cardiovascular structures. Curr. Opin. Cardiol. **12:** 528–532.
2. CONTE, M.S. 1998. The ideal small arterial substitute: a search for the Holy Grail? FASEB J. **12:** 43–45.

3. NIKLASON, L.E. 1999. Techview: medical technology. Replacement arteries made to order. Science **286:** 1493–1494.
4. NEREM, R.M. 2000. Tissue engineering a blood vessel substitute: the role of biomechanics. Yonsei Med. J. **41:** 735–739.
5. HUYNH, T., G. ABRAHAM, J. MURRAY, *et al.* 1999. Remodeling of an acellular collagen graft into a physiologically responsive neovessel. Nat. Biotechnol. **17:** 1083–1086.
6. L'HEUREUX, N. *et al.* 1998. A completely biological tissue-engineered human blood vessel. FASEB J. **12:** 47–56.
7. L'HEUREUX, N. *et al.* 2001. A human tissue-engineered vascular media: a new model for pharmacological studies of contractile responses. FASEB J. **15:** 515–524.
8. SHINOKA, T. *et al.* 1998. Creation of viable pulmonary artery autografts through tissue engineering. J. Thorac. Cardiovasc. Surg. **115:** 536–545; discussion 545–546.
9. NIKLASON, L.E. *et al.* 1999. Functional arteries grown in vitro. Science **284:** 489–493.
10. SHIN'OKA, T. *et al.* 2001. Transplantation of a tissue-engineered pulmonary artery. N. Engl. J. Med. **344:** 532–533.
11. NIKLASON, L.E. *et al.* 2001. Morphologic and mechanical characteristics of engineered bovine arteries. J.Vasc. Surg. **33:** 628–638.
12. NEIDERT, M.R *et al.* Enhanced fibrin remodeling in vitro for improved tissue-equivalents. Biomaterials. In press.
13. GRASSL, E.D. *et al.* 2002. Fibrin as an alternative biopolymer to type I collagen for fabrication of a media-equivalent. J. Biomed. Mater. Res. **60:** 607–612.
14. L'HEUREUX, N. *et al.* 1993. In vitro construction of a human blood vessel from cultured vascular cells: a morphologic study. J. Vasc. Surg. **17:** 499–509.
15. BAROCAS, V.H. *et al.* 1998 Engineered alignment in media-equivalents: magnetic prealignment and mandrel compaction. J. Biomech. Eng. **120:** 660–666.
16. SELIKTAR, D. *et al.* 2000. Dynamic mechanical conditioning of collagen-gel blood vessel constructs induces remodeling in vitro. Ann. Biomed. Eng. **28:** 351–362.
17. GIRTON, T.S. *et al.* 2000. Mechanisms of stiffening and strengthening in media-equivalents fabricated using glycation. J. Biomech. Eng. **122:** 216–223.
18. GRASSL, E.D., T.R. OEGEMA & R.T. TRANQUILLO. 2002. A fibrin-based arterial media-equivalent. Submitted.
19. LONG, J.L. & R.T. TRANQUILLO. 2002. Elastic fiber production in cardiovascular tissue-equivalents. Submitted.
20. ISENBERG, B.C. & R.T. TRANQUILLO. 2002. Effects of long-term cyclic distention on the development of collagen-based media-equivalents. Submitted.

Vascular Assembly in Engineered and Natural Tissues

Breakout Session Summary

Moderators

THOMAS C. SKALAK, *University of Virginia*

CHARLES D. LITTLE, *University of Kansas*

Panelists

LARRY V. McINTIRE, *Rice University*

KAREN K. HIRSCHI, *Baylor College of Medicine*

ROBERT T. TRANQUILLO, *University of Minnesota*

MARK POST, *Dartmouth Medical School*

JOHN RANIERI, *Sulzer Biologics*

BROAD STATEMENT

A convergence of problems in reparative medicine today requires renewed attention to improving our basic understanding, developing predictive modeling techniques, and creating rational therapeutic methods of guiding vascular assembly in diseased natural tissues as well as in engineered tissues. Pioneers and leaders in this field now recognize that a complex spatial and temporal interplay of multiple molecular and environmental signals determines vascular assembly and that many hurdles may be overcome though integrative, quantitative studies of molecular genetic determinants, cell behaviors *in vivo*, and functional cell aggregates. Current investigation and future goals are both driven by several key processes and objectives:

- Analysis of natural vascular assembly processes, including gene expression, spatial and temporal growth factor actions, stem cell lineages, and extracellular matrix–cell interactions in development and in adults;

- Analysis and harnessing of adaptive/remodeling processes in vascular systems;

- Quantitative experimental analysis and predictive computer modeling of vascular system assembly and maintenance;

- Design of synthetic constructs that guide/optimize vascular assembly;

- Physical linkage of micro- and macroscale vascular systems.

Address for correspondence: Thomas C. Skalak, Department of Biomedical Engineering, University of Virginia, Box 800759, Charlottesville, VA 22908. Voice: 804-924-0270; fax: 804-982-3870.

tskalak@Virginia.edu

Ann. N.Y. Acad. Sci. 961: 255–257 (2002). © 2002 New York Academy of Sciences.

VISION

Rational design of vascular systems assembly will take a prominent place in the practice of both preventative and restorative medicine. It is already clear that a complex spatial and temporal interplay of signals, including both genetic and environmental cues, are needed for normal small vessel growth, remodeling, and maintenance. Thus, the next phase of research and development in this field will almost certainly require innovative ways to study and design multicomponent, multisignal vascular systems, both in natural and engineered tissues. An emerging theme is that predetermination of final vascular or tissue structure in the microfabrication stage of engineered tissues is unlikely to succeed alone, but rather the harnessing of the natural, adaptive powers of the resident cells to guide functional vascular assembly and maintenance will provide the ability to supply large, complex tissues.

Biomedical engineers, vascular biologists, and developmental biologists are ideally suited to form teams to help address the behavior of these complex systems. The long list of components typically addressed today via the reductionist approach to studying therapeutic vascular assembly may be expressed either as a part of normal vascular development (vasculogenesis), later during vessel remodeling to a differentiated functional network with arterioles, capillaries, and venules, or during adult maintenance of a functional network. The implication of these observations is that several different design paths for engineering of *de novo* vascular networks in engineered constructs or remodeled networks *in vivo* can be envisioned.

OBJECTIVES

In order to achieve the goal of vascular assembly in engineered and natural tissues, it will be necessary to (1) advance fundamental understanding of vascular assembly: this includes stem cell biology, cell behaviors in development of functional networks, network growth and maintenance by environmental cues, artery wall remodeling, and venous/lymphatic assembly; (2) construct quantitative, predictive models of vascular assembly; and (3) fabricate engineered constructs (large and small blood vessels and networks) as well as *in vivo* remodeling processes.

OBSTACLES

Here we outline obstacles to achieving the above objectives. (1) Network development involves a complex temporal and spatial interplay of cellular, chemical, and mechanical stimuli. The tissue microenvironment affects local remodeling events. (2) Imaging and measurement tools for monitoring of vessel development, including gene expression, protein localization, chronic functional and cell phenotypic study, and mechanical forces *in vivo* are poorly developed or lack the needed spatial resolution, with the exception of optical methods used in the embryo or thin chambers. (3) Persistence of functional networks is little studied at present. (4) Control of stem cell differentiation is not quantitatively understood. (5) Self-assembly of tissue-engineered vessels must be scaled and modified to offer reproducible, economic production/storage/distribution capabilities. (6) We must identify embryonic processes that can be recapitulated in the adult or in synthetic vascular constructs. (7) Diffusive constraints limit cell-seeding densities in large constructs, necessitating novel approaches to vascularization.

CHALLENGES

In order to achieve the above objectives, it will be necessary to (1) understand the genetic circuits (i.e., genes and interplay of gene products) controlling vascular assembly and incorporate them into optimized designs and models; (2) move from a "single important molecule" approach to a multisignal, multicomponent approach; (3) develop modeling and simulation methods to predict cell motions, associations, and differentiation state, based on the secretion, diffusion, binding, and functional effects of a complex interplay of molecular signals in a field of interacting discrete cells that include endothelial, smooth muscle, and fibroblast precursors; (4) develop methods for delivery of cell-based therapies and multiple growth factors via microcarriers in a prescribed temporal and spatial sequence to perform guided arterialization in areas demanding blood supply; (5) assembly (tissue engineering) of arteries, veins, and lymphatics; (6) design methods for guiding the inosculation process (connection of engineered constructs to host vessels)—a micro- to macrointerfacing problem; (7) understand genetic regulation of on/off switches for various steps in vascular assembly and maintenance; (8) identify gene families that control vascular assembly in development and adult; (9) understand the biology of mesodermal precursor cells; (10) develop site- and cell-specific sampling methods and DNA microarray analysis to identify new candidates and gene circuitry of existing candidates; (11) develop micro- and nanofabrication methods allowing for creation of spatial gradients in ECM, signal molecules, and cell structures, and for high-spatial resolution actuation and sensing of mechanical stimuli; and (12) increase the respect in academia for integrative, interdisciplinary, collaborative scholarship that seeks to build new functional understanding from existing knowledge bases.

ACKNOWLEDGMENTS

The Vascular Assembly in Engineered and Natural Tissues Breakout Session was coordinated by Dr. Christine Kelley (NHLBI, NIH).

Storage and Translational Issues in Reparative Medicine

MEHMET TONER[a] AND JEFFERY KOCSIS[b]

[a]Center for Engineering in Medicine and Surgical Services, Massachusetts General Hospital, Harvard Medical School, Boston, Massachusetts, USA

[b]Neuroscience Research Center, Yale University School of Medicine, West Haven, Connecticut, USA

KEYWORDS: cryopreservation; freeze–thaw protocol; vitrification; desiccation; tissue-engineered products

INTRODUCTION

Living cells are increasingly in demand for a wide variety of exciting therapeutic strategies. Tissue engineering efforts have already led to the development of several commercial products to replace organ function permanently or temporarily; it is expected that many more will follow. Cell transplantation is being used for many clinical indications to replace or enhance certain organ functions. Gene therapy is based fundamentally on overexpression or down-regulation of one or more cellular functions to provide an enhancement of cell function for various therapeutic goals, such as wound healing and tissue repair. Recent advances in stem cell biology considerably ratchet up the excitement for these burgeoning approaches by providing a potentially unlimited source of cells for reparative medicine.

As several approaches to living cell-based therapies come near to bearing tangible fruit in the form of clinical applications, the fundamental and practical issues associated with the translation of these new technologies from the laboratory bench to the clinic take a frontline position. Some of the important translational technologies include cell isolation; cell and tissue culture and differentiation; scale-up of bioreactors; biomaterials and scaffolds; long-term storage strategies; and safety and regulatory policies. Although there have been significant advances both at the fundamental and practical levels for most of these translational technologies, the tools and understanding needed for the storage of living cells and complex tissue constructs lag significantly behind.[1]

Address for correspondence: Mehmet Toner, Shriners Burns Hospital, 51 Blossom Street, Boston, MA 02114. Voice: 617-371-4883; fax: 617-371-4950.
mtoner@sbi.org

Ann. N.Y. Acad. Sci. 961: 258–262 (2002). © 2002 New York Academy of Sciences.

CRYOPRESERVATION APPROACHES

Long-term storage of living cells or tissues is needed to provide a readily available supply of cells and engineered tissue constructs to end users, such as medical centers, hospitals, clinics, and physician offices. Cryopreservation, the established modality for long-term storage of living cellular systems, can be subdivided into two approaches based on the overall methodology, namely, freeze–thaw and vitrification. Albeit both approaches use cryogenic temperatures (typically below −80°C), there are important fundamental differences.[2]

In *freeze–thaw* protocols, the cells are suspended in 1 to 2 M cryoprotective agents (e.g., glycerol, dimethyl sulfoxide, or ethylene glycol) and then cooled slowly to storage temperature under controlled conditions. During storage, frozen cells are suspended in an ice matrix with a loss of more than 95% of the total cell water, are squeezed significantly by mechanical forces, and remain at cryogenic temperatures. These rather extreme physicochemical conditions are very deleterious to living systems unless the conditions are very accurately controlled and optimal protocols are determined for each cell type or engineered tissue, for example. Because all manipulations, prior to freezing and after thawing, as well as the culture conditions dramatically affect the recovery of cells after a freeze–thaw cycle, the determination of optimal cryopreservation protocols is complicated. The physicochemical conditions that determine cell fate during cryopreservation include intracellular ice formation, mechanical forces, and cell dehydration. There is a pressing need to better understand the mechanisms by which various physicochemical changes damage cells, how ice is propagated in cellular systems, and how cryoprotectants protect cells.[2,3] Equally as important as knowledge of the physicochemical damaging processes is the understanding of the cellular and molecular response to freezing.[4] The roles of cell repair processes, apoptotic and necrotic cell injury, and endogeneous protective mechanisms, such as heat shock proteins, are pivotal to the development of successful freezing protocols for living cellular systems.

In *vitrification* protocols the ice formation is completely prevented through the use of high concentrations (6–8 M) of cocktails of cryoprotectants and other solutes, such that cells are transformed directly from liquid to glassy phase as the cooling proceeds. During storage, vitrified cells are thus in a glassy (or vitrified) matrix. Typically, there are no solute shifts and increased electrolyte concentrations associated with ice growth in glassy matrix. Thus, it is believed that the glassy state is innocuous to cells. The use of vitrification has been especially promising for complex tissue structures or three-dimensional structures that have been traditionally very difficult to preserve by freeze–thaw protocols.[5,6] The major challenge for the vitrification-based approaches is then the loading and removal of complex mixtures of cryoprotectants at very high concentrations without causing damage. This is not a trivial process, and it requires controlled conditions during exposure to high concentrations of solute as well as the cooling and warming to and from cryogenic temperatures to prevent crystallization and devitrification, respectively.

A more recent approach for cryopreservation involves the use of small sugars to stabilize cells at very low concentrations. Typically, trehalose is used at about 0.1 to 0.2 M with or without the addition of conventional cryoprotectants such as DMSO.[7,8] Since mammalian cell membranes are practically impermeant to trehalose or other disaccharides, the plasma membrane is permeabilized using one of several approaches, including thermal poration, electroporation, microinjection, and genetically engineered pore formers. Results from experiments using trehalose show that fibroblasts and keratinocytes survive cryopreservation.[8] The major advantage of nontoxic sugars is the potential to infuse freeze–thawed cells directly into patients without the cumbersome steps involved for the removal of cryoprotectants such as DMSO.

STORAGE IN DRY STATE: DESICCATION

Another recent development is to understand the desiccation tolerance of mammalian cells and the development of novel strategies to store living cells at ambient temperatures. In *desiccated* (or *dry*) storage, the living cells are put into a stasis state at ambient temperatures by dehydrating the cells using either freeze-drying or convective drying techniques. This approach is based on the removal of water to achieve a glassy (i.e., amorphous or vitrified) state in and around cells at ambient temperatures. The glassy state is known to have an exceedingly high viscosity ($>10^{12}$ Pa) inhibiting most chemical, biological, and physical processes that yield cell deterioration. The potential benefit of dry storage to the living cell therapies is tremendous. Dried storage eliminates many of the problems associated with cryogenic storage, including the transport issues, thermal fluctuations during cryogenic storage, heavy weight, regulatory problems, and cryoprotectant removal, among others.

In nature, many organisms can survive in the dry state or in the glassy state for extended times, a phenomenon called *anhydrobiosis*.[9,10] Anhydrobiosis can occur in a variety of organisms, including plant seeds, bacteria, yeast, brine shrimp, fungi and their spores, and cysts of certain crustaceans. Extensive studies using these organisms have revealed that there are a series of complex molecular and physiological adaptations permitting these organisms to survive drying stress. Among these adaptations is the accumulation of internal sugars or mixtures of sugars, such as, trehalose, sucrose, and raffinose. Sugars are believed to play a major role in the stabilization of membranes, proteins, and other key cellular structures in the dry state. The mechanism of sugar protection is an active area of research that includes study of the role of the glassy state in long-term stabilization of living cells and the interaction of sugars with biological molecules and supramolecular structures to afford stabilization.

It is also important to note that it is necessary that sugars be present on both sides of the cell membrane in order to afford protection to intracellular structures.[10] Because of the large size of sugars, several approaches are being developed to load mammalian cells with sugars; many others will need to be developed in the future. Potential approaches for sugar loading include thermal poration, electroporation, genetically engineered membrane pores, and genetic engineering of cells to synthesize sugars internally. Recent studies with mammalian cells loaded with sugars have been very encouraging, in that some cellular functions are recovered after rehydration of desiccated cells.[11–14] Many parameters influence the stability of the dry state, including glassy state formation, light, oxygen, and moisture level. This research area is still in its infancy, and there is a pressing need to understand the behavior of living cells in the dry state. There are also many fundamental and technological issues that need to be resolved before the mammalian cells can be gotten into "suspended animation" at ambient conditions. Current work is focused on learning from the naturally occurring survival schemes and subsequent implementation of some of nature's strategies into mammalian cells through the use of genetic and metabolic engineering approaches.

OUTLOOK

It is necessary that future research encompass a number of interdisciplinary concepts in order to "successfully" store living cells, especially those that are very sensitive to any perturbation in environmental conditions. The key issue is the definition of what is meant by "successful." As the field of cell-based therapies matures, the requirements on the cryopreservation protocols become very challenging. Cryopreserved cells and tissues need to be reconstituted with minimal structural and functional compromise from storage, especially for applications where immediate functional capacity is necessary (e.g., hepatocyte transplantation in acute fulminant liver failure). There is a need to reconstitute the full biological potential of cells with minimal poststorage manipulation and without the help of highly skilled personnel. For scarce cells, the postthawing viability and yield are very important to ensure that all cells recovered from cryopreservation are viable and functional. This is also true for tissue-engineered products. Significant resources are used by the creation of tissue-engineered products, and the integrity and viability of this valuable product after thawing and rehydration must be maintained as nearly as possible to the fresh product. Tissue-engineered products also provide an additional level of challenge on account of their complex architecture and three-dimensional structure. Such demanding requirements make it imperative to develop interdisciplinary research programs to investigate physicochemical, engineering, molecular biology, and clinical aspects of how to deliver living cells on a readily available basis to the end users.

Some of the key scientific and technical issues involved in the achievement of the long-term storage of living cellular systems are (1) an understanding of the molecular mechanism of cell injury during cryopreservation protocols, (2) developing new strategies for stabilization of cells in storage, such as the use of molecular ice blockers, (3) the use of apoptotic cell death modifiers, (4) the investigation of the mechanism by which various organisms and animals survive extreme environmental conditions and cold temperatures, (5) the genetic engineering of key aspects of anhydrobiotic organisms into mammalian cells to induce stability in storage, and (6) developing a fundamental understanding of physicochemical processes leading to cell injury (e.g., ice–cell interaction, solute-induced damage, thermal effects of complex supramolecular structures, and scale-up issues). It is especially important to understand the role of glassy state in the stability of dried mammalian cells in storage.[8,15,16] The mobility of molecules near and below the glass transition temperature is of paramount importance to achievement of desiccation tolerance in cells. It is also important to develop new approaches to achieve cell stasis. The development, through the use of rational design principles, of novel cryoprotectants that are nontoxic and beneficial at very small concentrations is crucial so that cells can be infused into patients without any need to remove the cryoprotectant. The use of small amounts of sugars with excellent protective abilities in the frozen state seems to provide a very exciting avenue towards this avenue.

In summary, a number of translational issues need to be addressed for reparative medicine to have its full impact in medicine. Basic cell and tissue storage technologies require a multidisciplinary approach involving physical sciences, engineering, molecular and cellular biology, and clinical medicine.

REFERENCES

1. KARLSSON, J.O.M. & M. TONER. 2000. Cryopreservation: foundations and applications in tissue engineering. *In* Principles of Tissue Engineering. 2nd edit. R. Lanza, R. Langer & J. Vacanti, Eds.: 293–307. Academic Press. New York.
2. KARLSSON, J.O.M. & M. TONER. 1996. Long-term storage of tissues by cryopreservation. Biomaterials **17**: 243–256.
3. ACKER, J.P. & L.E. McGANN. 2001. Innocuous intracellular ice: experimental evidence for ice growth through membrane pores. Biophys. J. **81**: 1389–1397.
4. BAUST, J.M., M.J. VOGEL, R. VAN BUSKIRK & J.G. BAUST. 2001. A molecular basis of cryopreservation failure and its modulation to improve cell survival. Cell Transplant. **10**: 561–571.
5. SONG, Y.C., B.S. KHEIRABADI, F.G. LIGHTFOOT, *et al.* 2000. Vitreous cryopreservation maintains the function of vascular grafts. Nat. Biotechnol. **18**: 296–299.
6. SONG, Y.C., P-O. HAGEN, F.G. LIGHTFOOT, *et al.* 2000. In vivo evaluation of the effects of a new ice-free cryopreservation process on autologous vascular grafts. J. Invest. Surg. **13**: 279–288.
7. BEATTIE, G.M., J.H. CROWE, A.D. LOPEZ, *et al.* 1997. Trehalose: a cryoprotectant that enhances recovery and preserves function of human pancreatic islets after long-term storage. Diabetes **46**: 519–523.
8. CHEN, T., A. FOWLER & M. TONER. 2000. Literature review: supplemented phase diagram of the trehalose-water binary mixture. Cryobiology **40**: 277–282.
9. CROWE, J.H., J.F. CARPENTER & L.M. CROWE. 1998. The role of vitrification in anhydrobiosis. Annu. Rev. Physiol. **60**: 73–103.
10. POTTS, M. 1994. Desiccation tolerance of prokaryotes. Microbiol. Rev. **58**: 755–804.
11. CHEN, T., J.P. ACKER, A. EROGLU, *et al.* Beneficial effect of intracellular trehalose on the membrane integrity of dried mammalian cells. Cryobiology. In press.
12. GUO, N., I. PUHLEV, D.R. BROWN, *et al.* 2000. Trehalose expression confers desiccation tolerance on human cells. Letters to the Editor. Nat. Biotechnol. **18**: 168–171; GARCIA DE CASTRO, A., J. LAPINSKI & A. TUNNACLIFFE. Nat. Biotechnol. 2000. **18**: 473.
13. PUHLEV, I., N. GUO, D.R. BROWN & F. LEVINE. 2001. Desiccation tolerance in human cells. Cryobiology **42**: 207–217.
14. WOLKERS, W.F., N.J. WALKER, F. TABLIN & J.H. CROWE. 2001. Human platelets loaded with trehalose survive freeze-drying. Cryobiology **42**: 79–87.
15. MILLER, D.P., J.J. DE PABLO & H. CORTI. 1997. Thermophysical properties of trehalose and its concentrated aqueous solutions. Pharm. Res. **14**: 578–590.
16. SUN, W.Q. 1997. Glassy state and seed storage stability: the WLF kinetics of seed viability loss at T > Tg and the plasticization effect of water on storage stability. Ann. Bot. **79**: 291–297.

Neural Tissue Transplantation in Syringomyelia

Feasibility and Safety

DOUGLAS K. ANDERSON

Malcolm Randall VAMC and the University of Florida, Gainesville, Florida 32610, USA

KEYWORDS: syringomyelia; spinal cord injury; MRI

This translational project was initiated based on the philosophy that transplantation of fetal spinal cord (FSC) tissue is a promising strategy, with potential to serve as a key component in any therapeutic regimen used to treat spinal cord injury (SCI). However, before FSC grafts can be tested for their ability to promote functional improvement in the injured human spinal cord, it is necessary to determine the feasibility and safety of transplanting FSC tissue into the injured spinal cord of humans, which were the primary objectives of this study. Since safety and feasibility were the primary aims of this initial clinical pilot study, it was not advisable to use patients with stable deficits. It was, therefore, decided that a more suitable patient population for this initial study would be those patients whose deficits continue to worsen due to a progressive expansion of the lesion cavity (i.e., syringomyelia), thereby requiring surgical intervention in an attempt to stabilize the syrinx. Thus, a third aim of this pilot study was to determine the extent to which FSC tissue placed in previously drained syringes can fill these cavities, prevent their further expansion, and stabilize neurological function in this group of patients.

Prior to translating any findings from the laboratory to the clinic, necessary and sufficient preclinical data must be available to justify studies in humans. What constitutes "necessary and sufficient" preclinical data has not been defined and, thus, preclinical support varies widely from study to study. For this study, we had 15 years of preclinical experience with intraspinal grafting, including the repeated demonstration that FSC grafts reliably filled chronic lesion cavities in the spinal cords of both rats and cats; that MRI was predictive of graft survival; and, most important, FSC grafts did not compromise spared host neurological function due to overgrowth or after immunological rejection. In addition, since the donor tissue required virological and bacteriological testing, a series of experiments was undertaken to evaluate the feasibility of cyropreserving human FSC for 3 or 6 days prior to

Address for correspondence: Douglas K. Anderson, University of Florida, P.O. Box 100244, 100 Newell Drive, Gainesville, FL 32610. Voice: 352-392-6641; fax: 352-846-0250.
anderson@mbi.ufl.edu

Ann. N.Y. Acad. Sci. 961: 263–264 (2002). © 2002 New York Academy of Sciences.

transplantation in immunosuppressed SCI rats. Findings from these studies demonstrated that storage of 6- to 9-week gestational age human FSC tissue in a hibernation medium at 8°C was effective for maintaining adequate viability, as evidenced by an overall transplant survival in this series of 52%, with the highest survival occurring in the tissue stored for three days as solid pieces. Thus, in the present study we endeavored to transplant solid pieces of donor FSC tissue with the shortest storage time (i.e., 3–5 days).

A vital component of this study was the noninvasive assessment of graft survival and growth in human recipients. As stated previously, we had demonstrated that MRI can be used to predict graft survival in animals with SCI. Thus, determination of transplant survival by MRI was an important ingredient of this study. However, with our current technology, we have been unable to definitively differentiate graft from recipient tissue, which means that conclusive evidence of graft survival will require advanced MRI techniques, like labeling the implanted cells with MR contrast agents. To insure that these grafts do not worsen the recipient's neurological deficit (i.e., safety), a variety of assessments of neurological function were employed. These procedures ranged from objective to subjective, with the most objective being neurophysiological testing (i.e., evoked potentials, H-reflex evaluation) and the most subjective being a quality of life assessment. In addition, sensory and motor function was scored using the American Spinal Injury Association (ASIA) scale and the Functional Independence Measures (FIM), and pain was evaluated using two validated numerical rating scales. Results to date from this battery of outcome measures indicate that the intraspinal transplantation of human FSC tissue into patients with progressive syringomyelia is feasible and procedurally safe.

In summary, any translational study should strictly adhere to the first rule of medicine, "do no harm," by minimizing the risk/benefit ratio to the extent possible. Central to this credo is determining what the necessary and sufficient preclinical studies are and then rigorously applying the findings from these studies to the human situation. Conversely, initial clinical studies can also provide information for reshaping ongoing laboratory research or designing new studies, thereby providing a "bench-to-bedside-to-bench" continuum. Finally, translational studies should become the templates for more extensive clinical testing, including the development of new surgical approaches and/or additional quantitative outcome measures.

ACKNOWLEDGMENT

This work was carried out in collaboration with Edward D. Wirth, M.D., Ph.D., Department of Neuroscience and McKnight Brain Institute, University of Florida, Gainesville, Florida.

SELECTED REFERENCE

1. WIRTH, E.D. *et al.* 2001. Feasibility and safety of neural tissue transplantation in patients with syringomyelia. J. Neurotrauma **18:** 911–929.

Stabilization of Tissue-Engineered Products for Transportation and Extended Shelf-Life

KELVIN G.M. BROCKBANK

Organ Recovery Systems, Charleston, South Carolina 29403, USA

KEYWORDS: vitrification; cryopreservation; tissue engineering

Conventional approaches to cryopreservation that provide the cornerstone of isolated cell storage have not been successfully extrapolated to more complex natural or engineered multicellular tissues. Tissues are much more than simple aggregates of various cell types; they have a highly organized, often complex, structure that influences their response to freezing and thawing. The formation of extracellular ice, in particular, which is generally regarded as innocuous for cells in suspension, is known to be a hazard to structured tissues and organs. Cryopreservation is a complex process of coupled heat and mass transfer generally executed under nonequilibrium conditions. Advances in the field were modest until the cryoprotective properties of glycerol and dimethyl sulfoxide (DMSO) were discovered in the mid-1900s.[1,2] Many other cryoprotective agents (CPAs) have since been identified. Combinations of CPAs may result in additive or synergistic enhancement of cell survival by minimization of intracellular ice during freezing.[3]

Restriction of the amount and size of extracellular ice crystal formation during cryopreservation can be achieved by using high concentrations of CPAs that promote amorphous solidification, known as vitrification, rather than crystallization.[4] Vitrification is a relatively well-understood physical process, but its application to the preservation of biological systems has not been without problems because the high concentrations of CPAs necessary to facilitate vitrification are potentially toxic. To limit toxic effects it is necessary to use the least toxic CPAs at the lowest concentrations that will still permit glass formation (at cooling rates that are practical for bulky mammalian tissues).[4] Comparison of the effects of vitrification and conventional frozen cryopreservation upon venous contractility demonstrated that vitrification is superior to conventional cryopreservation methods.[4] Vitrification has more recently been used effectively for a variety of other tissues including myocardium, skin, and articular cartilage.

Address for correspondence: Kelvin G.M. Brockbank, Organ Recovery Systems, 701 East Bay Street, Suite 433, Charleston, SC 29403. Voice: 843-722-6756 x16; fax: 843-722-6657.
kbrockbank@organ-recovery.com

Ann. N.Y. Acad. Sci. 961: 265–267 (2002). © 2002 New York Academy of Sciences.

Molecular approaches to minimize damage due to cryopreservation in cells and tissues need more support. Molecular ice control techniques are in development using designer molecules that specifically interact with ice nuclei, resulting in either prevention of ice nucleus development or modification of ice crystal phenotype. These molecules promise to have benefits in both freezing and vitrification preservation protocols by rendering ice crystals either less damaging or by permitting reduction of CPA concentrations, respectively, in the next three to five years. In addition, a better understanding of molecular signaling pathways involved in cell injury due to cryopreservation is required. It is likely that both inhibition of apoptotic cell death and stabilization of cell membranes during cryopreservation may have significant benefits.

However, both conventional freezing and vitrification approaches to preservation have limitations. First, both of these technologies require low temperature storage and transportation conditions. Neither can be stored above their glass transition for long without significant risk of product damage due to ice formation and growth. Both technologies require competent technical support during the rewarming and CPA elution phase prior to product utilization. This is possible in a high technology surgical operating theater, but not in a doctor's outpatient office or in third-world environments. In contrast, theoretically, a dry product would have none of these issues because it should be stable at room temperature and rehydration should be feasible in a sterile packaging system.

Drying and vitrification have previously been combined for matrix preservation of cardiovascular and skin tissues but not for live cell preservation in tissues or engineered products. However, nature has developed a wide variety of organisms and animals that tolerate dehydration stress by a spectrum of physiological and genetic adaptation mechanisms. Among these adaptive processes the accumulation of large amounts of disaccharides, especially trehalose and sucrose, are especially noteworthy in almost all anhydrobiotic organisms including plant seeds, bacteria, insects, yeast, brine shrimp, fungi and their spores, cysts of certain crustaceans, and some soil-dwelling animals.[5-7] The protective effects of trehalose and sucrose may be classified under two general mechanisms: (1) "the water replacement hypothesis" or stabilization of biological membranes and proteins by direct interaction of sugars with polar residues through hydrogen bonding, and (2) stable glass formation (vitrification) by sugars in the dry state.

The stabilizing effect of these sugars has also been shown in a number of model systems, including liposomes, membranes, viral particles, and proteins during dry storage at ambient temperatures.[8-10] On the other hand, the use of these sugars in mammalian cells has been somewhat limited, mainly because mammalian cell membranes are impermeable to disaccharides or larger sugars. Recently, a novel genetically modified pore formerly has been used to reversibly permeabilize mammalian cells to sugars with significant postcryopreservation and to a lesser extent drying cell survival.[11] Such permeation technologies, which may also include use of pressure or electroporation, may provide some of the most likely opportunities for preservation of tissues in the 5–10-year vision, either by permitting cryopreservation with nontoxic cryoprotectants or drying. However, it should be noted that most organisms that reach a dried state during dormancy, and drought, do so by air drying (not freeze drying), which suggests this may be innocuous to cells under certain conditions.

Studies of anhydrobiotic organisms may suggest methods for conditioning mammalian cells for storage by either cryopreservation or drying in the tissue-engineered products of the future.

REFERENCES

1. POLGE, C., A.Y. SMITH & A.S. PARKES. 1949. Revival of spermatozoa after vitrification and de-hydration at low temperatures. Nature **164:** 666.
2. LOVELOCK, J.E. & M.W.H. BISHOP. 1959. Prevention of freezing damage to living cells by dimethylsulphoxide. Nature **183:** 1394–1395.
3. BROCKBANK, K.G.M. & K.M. SMITH. 1993. Synergistic interaction of low-molecular-weight polyvinylpyrrolidones with dimethylsulfoxide during cell cryopreservation. Transplant. Proc. **25:** 3185–3197.
4. SONG, Y.C. *et al.* 2000. Vitreous cryopreservation maintains the function of vascular grafts. Nature Biotechnol. **18:** 296–299.
5. CROWE, J.H. *et al.* 1998. The role of vitrification in anhydrobiosis. Annu. Rev. Physiol. **60:** 73–103.
6. POTTS, M. 1994. Desiccation tolerance of prokaryotes. Microbiol. Rev. **58:** 755–805.
7. CROWE, J.H. *et al.* 1988. Interactions of sugars with membranes. Biochim. Biophys. Acta **947:** 367–384.
8. CROWE, J.H. *et al.* 1993. Anhydrobiosis. Annu. Rev. Physiol. **54:** 579–599.
9. BIEGANSKI, R.M. *et al.* 1998. Stabilization of active recombinant retroviruses in an amorphous dry state with trehalose. Biotechnol. Prog. **14:** 615–620.
10. WOMERSLEY, C. *et al.* 1986. Inhibition of dehydration-induced fusion between liposomal membranes by carbohydrates as measured by fluorescence energy-transfer. Cryobiology **23:** 245–255.
11. EROGLU, A. *et al.* 2000. Intracellular trehalose improves the survival of cryopreserved mammalian cells. Nature Biotechnol. **18:** 163–167. [U.S. patent for preservation of biological materials allowed, 2001].

Engineering Human Tissues for *in Vivo* Applications

L. GERMAIN, F. GOULET, V. MOULIN, F. BERTHOD, AND F.A. AUGER

LOEX, Hôpital du Saint-Sacrement du CHA, 1050 Chemin Sainte-Foy, Québec, QC G1S 4L8, Canada and Department of Surgery, Laval University, QC, Canada

KEYWORDS: tissue engineering; wound repair

Tissue engineering is a rapidly developing field. This technology could offer a new alternative for wound repair and organ replacement. It is based on the ability of living cells, with or without biomaterials, to be assembled as three-dimensional tissues. The *in vivo* applications extend from specialized dressings that improve host tissue repair (e.g., ulcer) to permanent grafts that restore the function of the tissue (e.g., skin grafting for burn patients).

The presence of living cells in grafts has the advantages of potential self-renewal and regeneration after wounding by migration of the living cells. Depending on the application foreseen, either allogeneic or autologous grafts will be indicated. The challenges in the production and distribution of these two types of tissue-engineered products differ. In the case of allogeneic tissues, cell banks can be generated. Particular attention should be devoted to the absence of infectious agents in the cell banks in order to avoid one of the main drawbacks in organ transplantation—the transmission of diseases (e.g., AIDS, hepatitis C). Although some allogeneic tissues from mesenchymal cell types may be transplanted without rejection, the tissues containing epithelial and/or endothelial cells must be autologous. For autologous tissue grafts, special care has to be taken to avoid the introduction of infectious agents during the culture period, but there is no need for extensive viral testing of the cells and biopsy (e.g., for AIDS).

During cell expansion in culture, the cellular morphology assessment by phase contrast microscopy, cell viability and number must be evaluated. The characterization of cell types present in the cultures (using known differentiation markers) will be particularly important when producing cell banks that could be stored frozen in liquid nitrogen. Upon thawing, a high cell viability must be obtained to insure the high quality of the starting material in the production of engineered tissues. If pluripotent stem cells are chosen as the cell source, the extent of differentiation into the adequate cell type must be carefully monitored with several differentiation markers.

Address for correspondence: L. Germain, Hôpital du Saint-Sacrement du CHA, 1050 Chemin Sainte-Foy, Québec, QC G1S 4L8, Canada. Voice: 418-682-7663; fax: 418-682-8000.
lucie.germain@chg.ulaval.ca

Ann. N.Y. Acad. Sci. 961: 268–270 (2002). © 2002 New York Academy of Sciences.

Each tissue represents a different challenge in its production by tissue engineering. Therefore, the properties measured will vary. For example, because the blood vessel must sustain the blood pressure readily after grafting, the measurement of the mechanical resistance will be important.

The shipping and conservation of engineered tissues are also a challenge and would be facilitated if such tissues could be frozen. However, methods yielding high viability of cells within these tissues after thawing must be developed to insure the quality of the final graft. This parameter will be particularly crucial when tissues are destined to be permanent tissue replacement.

In conclusion, tissue engineering is undoubtedly a fascinating new addition to the therapeutic armamentarium of care providers for the twenty-first century. Once the production and transport of these products are well controlled, tissue engineering will have a major impact in the field of tissue and organ transplantation because of its immense potential and role in patient care.

ACKNOWLEDGMENTS

This work was supported by Medical Research Council of Canada/Canadian Institute for Health Research, Fonds de la Recherche en Santé du Québec (FRSQ), Heart and Stroke Foundation and Fondation des Pompiers du Québec pour les Grands Brûlés. L.G. is the holder of the Canadian Research Chair in Stem Cells and Tissue Engineering. F.G. and F.B. are recipients of scholarships from the FRSQ.

REFERENCES

1. AUGER, F., M. RÉMY-ZOLGHADRI, G. GRENIER & L. GERMAIN. 2000. The self-assembly approach for organ reconstruction by tissue engineering. e-biomed: a journal of regenerative medicine 1: 75–86.
2. BERTHOD, F., L. GERMAIN, R. GUIGNARD, et al. 1997. Differential expression of collagens XII and XIV in human skin and in reconstructed skin. J. Invest. Dermatol. 108: 737–742.
3. BERTHOD, F., L. GERMAIN, H. LI, et al. 2001. Collagen fibril network and elastic system remodeling in a reconstructed skin transplanted on nude mice. Matrix Biol. 20: 463–473.
4. BLACK, A.F., F. BERTHOD, N. L'HEUREUX, et al. 1998. In vitro reconstruction of a human capillary-like network in a tissue-engineered skin equivalent. FASEB J. 12: 1331–1340.
5. DUPLAN-PERRAT, F., O. DAMOUR, C. MONTROCHER, et al. 2000. Keratinocytes influence the maturation and organization of the elastin network in a skin equivalent. J. Invest. Dermatol. 114: 365–370.
6. GERMAIN, L. & F.A. AUGER. 1995. Tissue engineered biomaterials: biological and mechanical characteristics. In Encyclopedic Handbook of Biomaterials and Bioengineering, part B: applications. D. Wise, D.J. Trantolo, D.E. Altobelli, et al., Eds. 1: 699–734. Marcel Dekker. New York.
7. GERMAIN, L., M. MICHEL, J. FRADETTE, et al. 1997. Skin stem cell identification and culture: a potential tool for rapid epidermal sheet production and grafting. In Skin Substitute Production by Tissue Engineering: Clinical and Fundamental Applications. M. Rouabhia, Ed.: 177–210. Landes Bioscience. Austin, TX.
8. GERMAIN, L., M. RÉMY-ZOLGHADRI & F.A. AUGER. 2000. Tissue engineering of the vascular system: from capillaries to larger blood vessels. Med. Biol. Eng. Comput. 38: 232–240.

9. GOULET, F., L. GERMAIN, D. RANCOURT, *et al.* 1996. Tissue engineered ligament. *In* Ligaments and Ligamentoplasties. L'H. Yahia, Ed.: 367–377. Springler-Verlag. Berlin, Germany.
10. LAPLANTE, A.F., L. GERMAIN, F.A. AUGER & V. MOULIN. 2001. Mechanisms of wound reepithelialization: hints from a tissue-engineered reconstructed skin to long-standing questions. FASEB J. **13:** 2377–2389.
11. L'HEUREUX, N., L. GERMAIN, R. LABBÉ & F.A. AUGER. 1998. A completely biological tissue-engineered human blood vessel. FASEB J. **12:** 47–56.
12. L'HEUREUX, N., J.C. STOCLET, F.A. AUGER, *et al.* 2001. A human tissue-engineered vascular media: a new model for pharmacological studies of contractile responses. FASEB J. **15:** 515–524.
13. NEREM, R.M. & D. SELIKTAR. 2001. Vascular Tissue Engineering. Annu. Rev. Biomed. Eng. **3:** 225–243.

Cytoprotection by Stabilization of Cell Membranes

RAPHAEL C. LEE

Departments of Surgery, Medicine and Organismal Biology (Biomechanics),
The University of Chicago, Chicago, Illinois 60637, USA

KEYWORDS: cryoprotection; poloxamers; cell membranes

The major constraint that limits tissue storage is the accumulation of cellular damage or wounding by the storage conditions and consequential loss of tissue viability. The noun wound is used in the medical field to refer to a disruption of tissue integrity that follows trauma. Subsequent wound healing results from exposure to supraphysiological forces or is the consequence of action by reactive chemical agents.[1,2] Cell wounding results from exposure to supraphysiological forces or is the consequence of action by reactive chemical agents. Cell wounds result from alteration of cellular molecular structure or disruption of molecular assemblies such as membranes. Unlike tissue healing, the healing of cellular wounds occurs by neighboring biomolecular interactions. Like tissue wound healing, cellular wound healing involves accelerated processes that are constitutively expressed in routine physiologic repair of cellular structures.[3] This discussion relates to pharmaceutical strategies that are useful for augmenting the cellular healing response.

CELLULAR WOUNDS

Damage to supramolecular assemblies, like the lipid bilayer, is perhaps the most common mode of cell injury during storage.[2] Loss of lipid bilayer integrity occurs after exposure to supraphysiologic temperatures,[4] during freezing and thawing injuries,[5] in free-radical–mediated radiation injury,[6] in barometric trauma,[2,7] and in electrical shock[1,8,9] and mechanical shear or crush forces.[3,10] Ischemia-reperfusion injury, which is mediated by the effects of reactive oxygen species (ROS), is probably the most common cause and is a substantial factor in many common medical illnesses.[6]

Although in each instance the final result is a cell wound, the modes of cellular membrane injury are through different pathways. ROS produces wounding of the cell membrane through peroxidation of phospholipids and oxidative deamination of

Address for correspondence: Raphael C. Lee, Departments of Surgery, Medicine and Organismal Biology (Biomechanics), The University of Chicago, Chicago, Illinois 60637. Voice: 773-702-6302; fax: 773-702-1634.

rlee@surgery.bsd.uchicago.edu

Ann. N.Y. Acad. Sci. 961: 271–275 (2002). © 2002 New York Academy of Sciences.

proteins. This altered lipid conformation results in bleb formation, followed by formation of membrane defects. Membrane electroporation results from the pull of water into the membrane by the supraphysiologic electric fields. Heating increases the kinetic energy of membrane amphiphilic lipids until their momentum overcomes the forces of hydration that act to hold the lipids within the membrane lamella. Under freeze conditions, ice nucleation in the cytoplasm can lead to factors that are very destructive to the cell membrane, including the mechanical disruption of the membrane by the ice crystal growth and the damaging effects of increasing salt concentration as the ice spreads and excludes ions.[11] Abrupt barometric pressures can lead to acoustic wave disruption of the cell membrane.[7]

MOLECULAR STRUCTURE–FUNCTION RELATIONSHIP

Molecular structure is critical for the formation of large supramolecular assemblies like the lipid bilayer membrane.[12] Structural integrity of cell membrane is essential for making possible the transmembrane physiological ionic concentration gradients at a metabolic energy cost that is affordable. Despite the effectiveness of the membrane barrier, approximately 40–85% of the metabolic energy expended by cells is used to maintain normal transmembrane ion gradients. Cellular wounds involving disruption of the membrane structure quickly lead to cell necrosis.

DRUGS TO AUGMENT CELL WOUND HEALING

The possibility of sealing cell membrane wounds using synthetic surfactants is now well established.[1,13,14] However, the exact mechanisms of sealing remains unknown.[15] Effective agents include the surfactant class of poloxamers, representing a group of tri-block copolymers. Poloxamer 188 (P188) made by BASF Corporation (Pluronic F68) was initially shown to seal cells against loss of carboxyfluorescein dye after electroporation.[1] In the following years, it has been demonstrated that P188 can also seal membrane pores in skeletal muscle cells after heat shock[16] and enhance the functional recovery of fibroblasts that have been lethally heat-shocked[13] or exposed to high-dose ionizing radiation.[17] Very recently, P188 has been shown to protect embryonic hippocampal neurons against death due to neurotoxin-induced loss of membrane integrity.[14] Other surfactants, such as Poloxamine 1107, of similar composition but quad-block copolymers, have been shown to reduce testicular ischemia-reperfusion injury[6] and hemoglobin leakage from erythrocytes after ionizing radiation.[18] In all the aforementioned investigations the observed phenomena were attributed to sealing of permeabilized cell membranes by the surfactants. In addition, the effect of P188 infusions in reducing duration and severity of acute painful episodes of sickle cell disease is presently also explained by beneficial surfactant–erythrocyte membrane interactions.[19]

Poloxamers and poloxamines belong to a class of water-soluble multi-block copolymers that have important "surface-active" properties. Poloxamer 188 is a tri-block copolymer often abbreviated as POE-POP-POE with POE and POP representing poly(oxyethylene) and poly(oxypropylene), respectively. The POE chains are hydrophilic   hydrophobic   hydrophilic

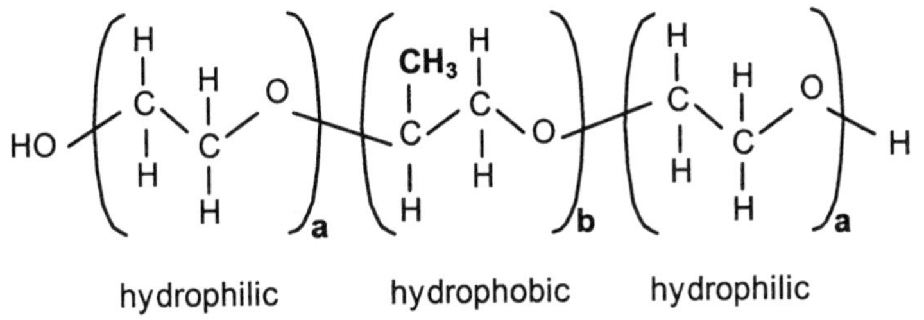

FIGURE 1. Chemical structure of poloxamers. The series of different poloxamers is constituted through varying numbers and ratios for **a** and **b**.

hydrophilic due to their short carbon unit between the oxygen bridges, whereas the POP center is hydrophobic due to the larger propylene unit (FIG. 1).

The poloxamer series covers a range of liquids, pastes, and solids, with molecular weights varying from 1100 to about 14,000. The ethylene oxide propylene oxide weight ratios range from about 1:9 to about 8:2. Poloxamer 188 (P188) has an average molecular weight of about 8400. It is prepared from a 1750 average molecular weight hydrophobe and its hydrophile comprises about 80% of the total molecular weight. The poloxamine series is slightly different from the poloxamer series in its chemical structure. The hydrophobic center consists of two tertiary amino groups carrying both two hydrophobic poly(oxypropylene) chains of equal length each followed by a hydrophilic poly(oxyethylene) chain. Thus, it still can be described as a tri-block copolymer, but is much bulkier than poloxamers (FIG. 2).

Most often, P188 is used at a sub-critical micelle concentration (sub-CMC) of 0.1 mM to1.0 mM for membrane repair *in vitro*.[1] Above their CMC surfactants self-aggregate to micelles causing the (active) surfactant monomer concentration to remain constant (= CMC) independently of the total surfactant concentration . The capability of these amphiphilic copolymers to repair cell membranes at 10^{-3} molar concentration levels distinguishes the sealing capability of copolymer surfactants from purely hydrophilic polymers such as poly(-ethyleneglycol) (PEG), which require molar concentrations.[10]

PEG has a long history of use for induction of membrane fusion applications.[20] It is hypothesized that PEG can force very close contact between vesicle membranes by lowering the activity of water adjacent to the membrane.[12] But even at the required high concentrations (e.g., 17.5%),[20] PEG-mediated vesicle fusion only occurs when the organization of lipid in the bilayers are substantially perturbed from their equilibrium values.

It is our working hypothesis that tri-block copolymers, like P188, with their hydrophobic center chains, act like cell wound targeted PEG molecules, thus requiring much lower concentrations to achieve fusion (sealing) of a permeabilized cell membrane. The membrane repair mechanism of these surfactants is becoming widely discussed. It is not yet understood whether the surfactants interact only with the disrupted parts of the membrane to seal the membrane wounds or whether their integration and interaction with the entire bilayer alters the membrane properties in a way to repair itself (e.g., decreased fluidity).[21,22]

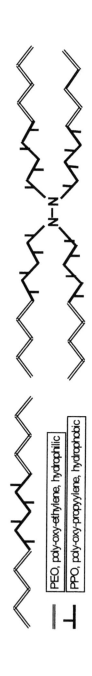

FIGURE 2. Schematic drawing to illustrate structural differences between poloxamers (*left*) and poloxamines (*right*). The PEO and PPO chain lengths vary among the members of the surfactant families.

REFERENCES

1. LEE, R.C. *et al.* 1992. Surfactant-induced sealing of electropermeabilized skeletal muscle membranes in vivo. Proc. Natl. Acad. Sci. USA **89:** 4524–4528.
2. MCNEIL, P.L. & R.A. STEINHARDT. 1997. Loss, restoration and maintenance of plasma membrane integrity. J. Cell Biol. **137:** 1–4.
3. MCNEIL, P.L. *et al.* 2000. Patching plasma membrane disruptions with cytoplasmic membrane. J. Cell Sci. **113:** 1891–1902.
4. BISCHOF, J.C. *et al.* 1995. Dynamics of cell membrane permeability changes at supraphysiological temperatures. Biophys. J. **68:** 2608–2614.
5. RUBINSKY, B. *et al.* 1992. The cryoprotective effect of antifreeze glycopeptides from Antarctic fishes. Cryobiology **29:** 69–79.
6. PALMER, J.S. *et al.* 1998. Surfactant administration reduces testicular ischemia-reperfusion injury. J. Urol. **159:** 2136–2139.
7. FISCHER, T.A.. *et al.* 1997. Cardiac myocyte membrane wounding in the abruptly pressure-overloaded rat heart under high wall stress. Hypertension **30:**1041–1046.
8. GAYLOR, D.G. *et al.* 1988. Significance of cell size and tissue structure in electrical trauma. J. Theor. Biol. **133:** 223.
9. TSONG, T.-Y. & Z.D. SU. 1999. Biological effects of electric shock and heat denaturation and oxidation of molecules, membranes and cellular functions. *In* Occupational Electrical Injury and Safety. Ann. N.Y. Acad. Sci. **888:** 211–232.
10. SHI, R. *et al.* 1999. Functional reconnection of severed mammalian spinal cord axons with polyethylene glycol. J. Neurotrauma **16:**727–738.
11. KARLSSON, J.O. *et al.* 1993. Nucleation and growth of ice crystals inside cultured hepatocytes during freezing in the presence of dimethyl sulfoxide. Biophysical J. **65:** 2524–2536.
12. ARNOLD, K. *et al.* 1990. Exclusion of poly(ethylene glycol) from liposome surfaces. Biochim. Biophys. Acta **1022:** 303–310.
13. MERCHANT, F.A. *et al.* 1998. Poloxamer 188 enhances functional recovery of lethally heat-shocked fibroblasts. J. Surg. Res. **74:** 1031–1040.
14. MARKS, J.D. *et al.* 2001. Nonionic surfactant prevents NMDA-induced death in cultured hippocampal neurons.FASEB J. 2001 April 15(6): 1107–1109.
15. LEE, R.C. *et al.* 1994. Promising therapy for cell membrane damage in electrical injury: a multidisciplinary approach to prevention, therapy and rehabilitation. Ann. N.Y. Acad. Sci. **720:** 239–245.
16. PADANILAM, J.T. *et al.* 1994. Effectiveness of poloxamer 188 in arresting calcein leakage from thermally damaged isolated skeletal muscle cells. Ann. N.Y. Acad. Sci. **720:** 111–123.
17. GREENEBAUM, B. *et al.* Poloxamer 188 prevents acute necrosis of adult skeletal muscle fibers after high-dose irradiation. Radiat. Res. Manuscript under review.
18. HANNIG, J. *et al.* 1999. Poloxamine 1107 sealing of radiopermeabilized erythrocyte membranes. Int. J. Rad. Biol. **75:** 379–385.
19. ADAMS-GRAVES, P. *et al.* 1997. RheothRx (Poloxamer 188) injection for the acute painful episode of sickle cell disease: a pilot study. Blood **90:** 2041–2046.
20. LEE, J.K. & B.R. LENTZ. Evolution of lipidic structures during model membrane fusion and the relation of this process to cell membrane fusion. Biochemistry **36:** 6251–6259.
21. SHARMA, V. *et al.* 1996. Poloxamer 188 decreases susceptibility of artificial lipid membranes to electroporation. Biophys. J. **71:** 3229–3241.
22. BAEKMARK, T.R. *et al.* 1997. The effects of ethylene oxide containing lipopoly-mers and tri-block copolymers on lipid bilayers of dipalmitoylphosphatidylcholine. Biophys. J. **73:** 1479–1491.
23. GABRIEL, B. & J. TEISSIE. 1994. Generation of reactive-oxygen species induced by electropermeabilization of Chinese hamster ovary cells and their consequence on cell viability. Eur. J. Biochem. **223:** 25–33.
24. ABLOVE, R.H. *et al.* 1996. Effect of high-energy phosphates and free radical scavengers on replant survival in an ischemic extremity model. Microsurgery **17:** 481–486.

Storage and Translational Issues in Reparative Medicine

Breakout Session Summary

Moderators

JEFFREY D. KOCSIS, *Yale University School of Medicine*

MEHMET TONER, *Harvard Medical School*

Panelists

DOUGLAS K. ANDERSON, *University of Florida College of Medicine*

KELVIN BROCKBANK, *Organ Recovery Systems, Inc.*

LUCIE GERMAIN, *Laval University*

RAPHAEL C. LEE, *University of Chicago*

BUDDY D. RATNER, *University of Washington*

Rapporteur

JEFFREY D. KOCSIS, *Yale University School of Medicine*

BROAD STATEMENT

Several issues must be addressed in translating any basic science research result into a clinical treatment. When the therapeutic agent is living tissue or cells, the translational issues become more complex. There are research questions relating to efficient and high-volume cell production, tissue typing or characterization, the health and consistency of cells, the long-term storage of living cells and tissues using cryopreservation and drying approaches, and the distribution of therapeutic products. There are also basic studies required to answer questions about efficacy and safety, regulation, and potential ethical issues related to the distribution of scarce resources. A third area of research is in clinical studies, starting with pilot studies and carrying through to phase 3 clinical trials.

VISION

Develop key technologies and approaches in order to seamlessly translate basic science ideas to clinical reality in reparative medicine. Especially focus on long-term storage issues to provide readily available supply of cells and engineered tissues to end-users including medical centers, hospitals, clinics, and doctors' offices.

Ann. N.Y. Acad. Sci. 961: 276–278 (2002). © 2002 New York Academy of Sciences.

OBJECTIVES

To facilitate this vision, the future research should aim to:

(1) Provide off-the-shelf readily available supply of cells and tissues for clinical applications;

(2) Preserve cells and tissues with minimal structural and functional compromise from storage and processing, especially for applications where immediate functional capacity is necessary, such as the use of hepatocytes for the treatment of acute fulminant liver failure;

(3) Reconstitute full biological potential with minimal post-storage manipulation, and develop novel storage technologies using nontoxic cryoprotectants to directly infuse cells into patients immediately upon thawing;

(4) Investigate new strategies for stabilization of cells and engineered tissues in storage including molecular ice blockers, cell death modifiers (e.g., apoptosis inhibitors, membrane stabilizers), and others;

(5) Understand basic cell and molecular biology of cell injury associated with various cryopreservation strategies including conventional slow freezing in the presence of 1–2-molar cryoprotectants and vitrification approaches utilizing 6–8-molar cocktail of cryoprotectants;

(6) Investigate mechanisms by which various organisms and animals survive extreme dehydration and cold temperatures and engineer key aspects of anhydrobiotic organisms into mammalian systems using genetic and metabolic engineering approaches;

(7) Overcome the existing technological barriers, including reliable large-scale controlled-rate freezers, and smart packages with built-in sensors to non-invasively monitor the stability of storage conditions; and

(8) Bring together engineers, physicists, biologists, and clinicians to create multidisciplinary groups to address complex biological and physicochemical processes associated with biopreservation.

OBSTACLES AND CHALLENGES

Among the major challenges that must be addressed in order to bring tissue engineered products to the clinic are: (1) fundamental knowledge about the mechanisms of damage inflicted on cells by cryopreservation-induced stresses needs to be obtained, and strategies to minimize cellular damage using genetic and molecular approaches need to be developed; and (2) alternative approaches to cryogenic storage need to be developed, such as dry storage at ambient temperature.

It must be kept in mind that (1) basic cell and tissue preservation technologies require a multidisciplinary approach involving the physical sciences, cell and molecular biology, clinical medicine, and engineering; and (2) preservation technologies lag behind other cell-processing technologies such as cell culture, isolation, purification, genetic engineering, and bioreactor design.

ACKNOWLEDGMENT

The Storage and Translational Issues in Reparative Medicine Breakout Session was coordinated by Dr. Frank Evans (NHLBI, NIH) and Dr. Bill Heetderks (NINDS, NIH).

Development of Artificial Blood Vessels

Seeding and Proliferation Characteristics of Endothelial and Smooth Muscle Cells on Biodegradable Membranes

PAUL I. MUSEY, SOBRASUA M. IBIM, AND NIRANJAN K. TALUKDER

Departments of Biological Sciences and Engineering, Clark Atlanta University, and Morris Brown College, Atlanta, Georgia 30314, USA

KEYWORDS: artificial blood vessels; polylactide-co-glycolide; biodegradable membranes

INTRODUCTION

A major objective of tissue engineering is the use of cultured cells to recreate tissues and organs as replacement parts that may be grafted into humans and animals.[1] These efforts have relied on synthetic materials to provide mechanical strength and serve as scaffolds for endothelialization with respect to vascular grafts. Although this approach has been successful to some extent, problems due to incomplete healing (growth of neointima) and intimal hyperplasia (overgrowth) of smooth muscle cells persist.[2–7] Intimal hyperplasia leads to vessel reclosure and is suspected to be the result of incomplete endothelialization over the anastomotic site, which, in turn, is due to compliance mismatch between prosthesis and adjacent artery. The current study is towards development of artificial vessel prostheses composed of cell-polymer composite membranes that will ultimately assume biological functions *in vivo*.

METHODS

Biodegradable membranes of polylactide-co-glycolide (PLAGA) 50:50 and 30:70 were fabricated by casting a 10% solution (w/v) of polymer in chloroform onto Teflon-coated petri dishes.[8,9] The dishes were allowed to dry at room temperature for 24 hr. The polymer discs (250 μm thick, and 3.2 cm in diameter) were removed and placed in cell culture plates for cell seeding.

Bovine aortic endothelial (EC) and smooth muscle cells (SMCs) were cultured in a humidified chamber with 5% CO_2, 95% air at 37°C using 6-well culture plates. Cells were seeded at approximately 5000 cells/cm^2 in Dulbecco's minimal essential

Address for correspondence: Paul I. Musey, Ph.D., Department of Biological Sciences, Clark Atlanta University, 223 James P. Brawley Drive, Atlanta, GA 30314. Voice: 404-880-6829; fax: 404-880-8065.

pmusey@cau.edu

Ann. N.Y. Acad. Sci. 961: 279–283 (2002). © 2002 New York Academy of Sciences.

media (DMEM), containing FBS and antibiotics. A standard volume of 2 ml of the media was used per well and changed every 2–3 days. Various times during incubation, replicate cultures were harvested and processed to assess cell growth, competence, and membrane degradation process.[8] When necessary, sodium bicarbonate (0.01g/10 ml) was incorporated into the polymer mixture before casting. Cell proliferation was assessed by manual count using a hemacytometer. Viability was assessed by dye exclusion and the cells were stained with Giemsa for light microscopy. Graded amounts of lactic and/or glycolic acids were added to media when necessary.

RESULTS: EFFECTS OF POLYMER DEGRADATION ON CELL GROWTH

FIGURE 1a represents the proliferation of endothelial (EC), smooth muscle (SM) and osteoblast (OB) cells on TCPS (control) and PLAGA membranes. The growth of EC and SM cells on biomembrane was approximately 75–80% of control TCPS up to day 8. Thereafter, cell proliferation and viability on PLAGA declined rapidly to undetectable levels while control cultures continued to proliferate. The rapid decline of cell viability on biomembrane was attributed to cytotoxicity of degradation products, lactic and glycolic acids, as the decline coincided with expected breakdown of the polymer to its composite monomers. The pH of spent media from the cultures was measured as they were renewed. FIGURE 1b represents changes in culture media pH. The recorded pH ranged from initial 7.5 to 3.8 at the various days of renewal with increasing rates of acidification up to day 34 as judged from ratios of units of pH change to time (days). Direct addition of either glycolic or lactic acids (0.1–10.0 mg/ml) to control cell cultures without PLAGA resulted in proportional cytotoxicities.

This degradation product effect was ameliorated by admixture of $NaHCO_3$ in the polymer fabrication. The media acidification was also successfully controlled by use of larger culture medium volume to buffer the acid released. The data again indicated that glycolic acid was more cytotoxic than lactic acid with SMCs tolerating lactic

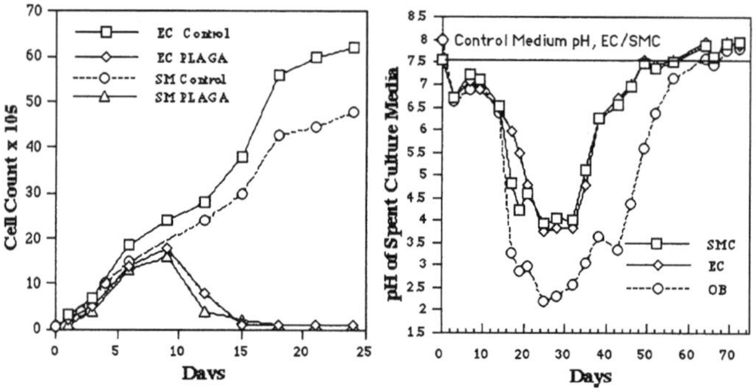

FIGURE 1. **(a)** Growth of endothelial and smooth muscle cells on TCPS and PLAGA membranes. **(b)** Effect of PLAGA degradation on pH of spent culture media.

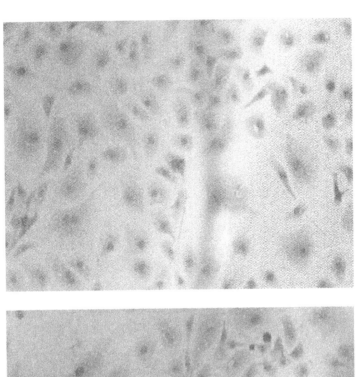

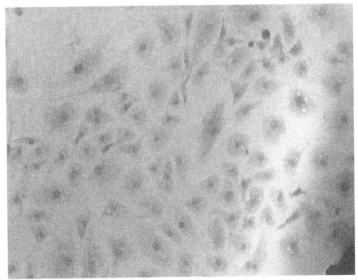

FIGURE 2. Attachment of seeded endothelial cells on PLAGA membrane (day 12). Panel 2 shows cell growth on PLAGA (*top*) and spread onto TCP culture plate (*bottom*).

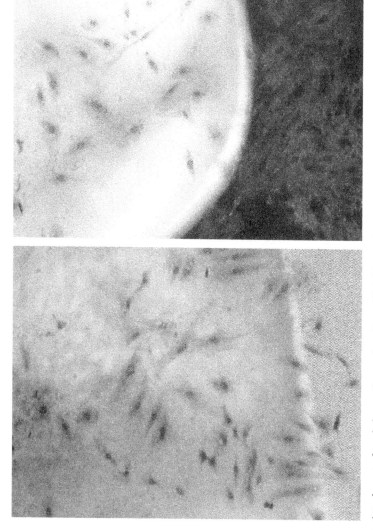

FIGURE 3. Attachment of seeded smooth muscle cells on PLAGA membrane (day 12). Panel 2 shows cell growth on PLAGA (*top*) and spread onto TCP culture plate (*bottom*).

acid environment better than ECs. FIGURES 2 and 3 represent light microscopic pictures of ECs and SMCs, respectively, on PLAGA.

CONCLUSION

Results indicate that metabolic breakdown products from PLAGA significantly lower the pH of the environment and that sustained cell viability on PLAGA membrane is partially dependent on the localized pH, especially in controlled limited *in vitro* cultures where there macrophages may not be present. Supplementation of control cultures with graded amounts of lactic and/or glycolic acids led to deterioration of cell viability as for PLAGA cultures. Bicarbonate and/or excess culture media may be used to control the acid toxicity, allowing the cells to remain viable throughout the degradation process. If the environmental pH is controlled, cell proliferation and viability on PLAGA is comparable to control cultures as evident from light microscopic pictures. Finally, sustained cell colonization of PLAGA copolymers without effective control of localized pH impairs their use as tissue scaffolds. Ongoing studies involve infiltration of porous PLAGA by fibroblast to form the platform of vascular tissue.

ACKNOWLEDGMENTS

This work was supported in part by NIH Grant NIH/NIGMS S06-GM08247. The technical assistance of Mr. Eno Ekong on this project is greatly appreciated.

REFERENCES

1. L'HEUREUX, N., F.A. AUGER, *et al.* 1998. A completely biological tissue-engineered human blood vessel. FASEB J. **12:** 47–56.
2. NOISHIKI, Y., Y. TOMIZAWA, *et al.* 1996. Autocrine angiogenic vascular prosthesis with bone marrow transplantation. Nature Med. **2:** 90–99.
3. PASIC, M., W. MULLERGLAUSER, *et al.* 1995 Seeding with omental cells prevents late neointimal hyperplasia in small-diameter Dacron grafts. Circulation **92:** 2605–2616.
4. WILLIAMS, S.K., D.G. ROSE, *et al.* 1994. Microvascular endothelial cell sodding of ePTFE vascular grafts: improved patency and stability of the cellular lining. J. Biomed. Mater. Res. **28:** 203–212.
5. ZILLA, P., M. DEUTSCH, *et al.* 1994. Clinical in vitro endothelialization of femoropopliteal bypass grafts: an actuarial follow-up over three years. J. Vasc. Surg. **19:** 540–548.
6. CLOWES, A.W., T.R. KIRKMAN & M.M. CLOWES. 1986. Mechanisms of arterial graft failure. II. Chronic endothelial and smooth muscle cell proliferation in healing polytetrafluoroethylene prostheses. J. Vasc. Surg. **3:** 877.
7. CLOWES, A.W., A.M. GOWN, *et al.* 1985. Mechanisms of arterial graft failure. Role of cellular proliferation in early healing of PTFE prostheses. Am. J. Pathol. **118:** 43.
8. LAURENCIN, C.T., C.D. MORRIS, *et al.* 1992. Osteoblast culture on bioerodible polymers: studies of initial cell adhesion and spread. Polymers Advan. Technol. **3:** 359–364.
9. IBIM, S.E.M., M. KWON, *et al.* 1997. Novel poly(phosphazenes)/poly (lactide-co-glycolide) blends: miscibility and degradation studies. Biomaterials **18:** 1565–1569.

In Vivo Remodeling of Surgically Constructed Vascular Anastomoses

Nonpenetrating, Arcuate-Legged Clips versus Standard Suture

WOLFF M. KIRSCH,[a] YONG HUA ZHU,[a] ROBERT STECKEL,[d]
WALDO CONCEPCION,[b] KERBY OBERG,[c] LENNART ANTON,[a] AND
NORMAN PECKHAM[e]

[a]*Neurosurgery Center for Research, Training, and Education,*
[b]*Transplantation Institute, [c]Department of Pathology and Human Anatomy,*
Loma Linda University, Loma Linda, California 92350, USA

[d]*Surgical Consulting, Norwalk, Connecticut 06854, USA*

[e]*Department of Pathology, Jerry L. Pettis VA Medical Center,*
Loma Linda, California 92354, USA

KEYWORDS: vascular anastomoses; titanium surgical clips

INTRODUCTION

A new mechanical system for facilitating vascular anastomosis (end-to-side, end-to-end) is described that enables the rapid and reproducible production of a nonpenetrated, compliant vascular reconstruction. The interrupted, nonpenetrated, flanged anastomotic line formed by clips has proven in both experimental and prospectively randomized trials to be both biologically and technically superior to junctions attainable by conventional suture.

METHODS

The VCS clip is an arcuate-legged, nonpenetrating titanium clip that autoregulates its final closing pressure on the basis of the width of interposed tissue between the clip tips.[1–3] Clip development was based on improving the quality of microsurgical cerebrovascular reconstructions, specifically the superficial temporal to middle cerebral artery bypass (EC-IC bypass).

After development of a proprietary method for stamping titanium clips with "work-hardened" properties and with GMP (Good Manufacturing Practice) approval, comprehensive Good Laboratory Practice Conditions studies were done. The pur-

---

Address for correspondence: Wolff M. Kirsch, M.D., Neurosurgery Center for Research, Training, and Education, Loma Linda University, Loma Linda, CA 92350.

Ann. N.Y. Acad. Sci. 961: 284–287 (2002). © 2002 New York Academy of Sciences.

pose of the study was to compare clip to conventional handsewn polypropylene suture for standard arterial, prosthetic, and venous reconstructions. Parameters evaluated and compared included relative intraoperative efficiency; physical properties of arterial, prosthetic, and venous reconstructions (e.g., burst pressures); patency at 1, 2, 4, 12 weeks and at 1 year; and the gross and microscopic appearance of the reconstructed vessels at 1, 2, 4, 12 weeks and at 1 year.

A total of 136 vascular procedures were performed on 38 adult male Suffolk cross sheep, 10 adult pigs, and 3 weanling pigs by three vascular surgeons. Constructs included end-to-end common carotid artery anastomosis (17 clip, 15 suture), linear femoral arteriotomies and closures 17 clip, 15 suture), linear jugular venotomies (7 clip, 8 suture), saphenous vein interpositional grafts to the common carotid arteries (4 clip, 4 suture), and common carotid interpositional synthetic prosthesis insertions to include Dacron and PTFE (18 clip, 18 suture), and 13 inferior vena cava end-to-end anastomosis (6 clip, 7 suture).

RESULTS

The overall patency rate for this study was 98% with one each of the Dacron and PTFE graft sutures and clips occluding due to infection. Significant differences were noted between clip and suture in terms of operative efficiency. Blood-tight reconstructions were achieved regularly and more rapidly with the clips, whereas hemor-

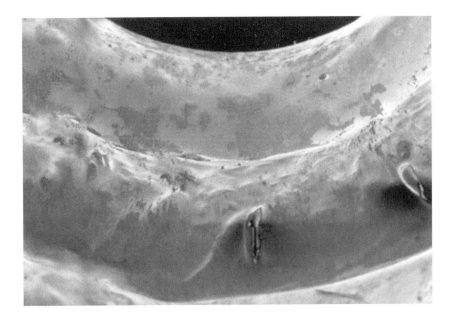

FIGURE 1. Scanning microscopic view of sutured anastomotic line (Dacron Interpositional ovine carotid graft) 1 year postoperation. Note the intimal hyperplasia (distal anastomotic line) and nonabsorbable suture in vessel lumen.

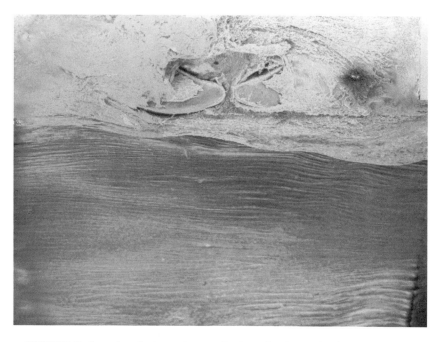

FIGURE 2. Scanning electron microscopic view of end-to-end ovine common carotid artery reconstruction (anastomotic line) 1 year post operation. Note large VCS clip tips in place with underlying compressed connection tissue matrix between clip tips and complete intimal resurfacing.

rhage occurred normally at sutured anastomosis. Speedier and stronger reconstructions were attained with clips (burst strength of clipped reconstructed carotid arteries averaged 15 mm Hg greater than sutured arteries at two weeks postoperative). Reconstructions studied at one year postoperative (4 clipped, 2 sutured) showed a significant difference in appearance between the clipped and sutured anastomotic lines, both by gross and microscopic examination (light and scanning electron microscopy). Clipped anastomotic lines were smoothly endothelialized without evidence of the ANH associated with sutures (FIGS. 1 and 2). Approval to market the device was granted in December 1993. The U.S. Surgical Corporation withheld general distribution of the device until two prospective, randomized clinical trials were done comparing clips to suture for vascular access construction.[3] These studies clearly established the safety and superiority of the VCS clip for vascular access construction, and since 1997 the VCS® clip has been marketed worldwide (FIG. 3).

CONCLUSIONS

(1) *In vivo* vascular anastomotic remodeling is improved with clips compared to suture.

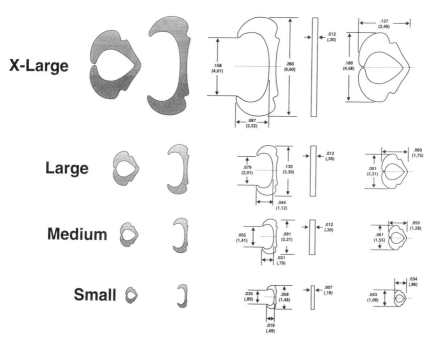

FIGURE 3. Sizes and dimensions of the four different VCS clips. Clips are fabricated from medical-grade titanium by a stamping process that work-hardens the material. The work-hardening effect increases the maximum yield strength of the clip.

(2) Reasons for clip benefits may be improved anastomotic compliance, no endothelial damage or intraluminal foreign body, and streamlined blood flow.

(3) Re-initiation of elastin fiber synthesis and assembly of normal elastin fibers in the arterial wall is possible in the presence of the intact endothelium.

REFERENCES

1. KIRSCH, W.M., Y.H. ZHU, R.A. HARDESTY, *et al.* 1992. A new method for microvascular anastomosis: report of experimental and clinical research. Am. Surg. **58**: 722–727.
2. KIRSCH, W.M., Y.H. ZHU, E. WAHLSTROM, *et al.* 1998. Vascular reconstructions with non-penetrating arcuate-legged clips. *In* Techniques in Vascular and Endovascular Surgery. J.S.T. Yao & W.H. Pearce, Eds.: **6**: 67–89. Appleton & Lange. Stamford, CT.
3. SCHILD, A.F. & J. RAINES. 1999. Preliminary prospective randomized experience with vascular clips in the creation of arteriovenous fistulae for hemodialysis. Am. J. Surg. **178**: 33–37.

Simple Method for Increasing Cell-Attachment Ability of Biodegradable Polyester

IN KAP KO AND HIROO IWATA

Institute of Frontier Medical Sciences, Kyoto University, Kyoto, 606-8507, Japan

KEYWORDS: cell attachment; biodegradable polyesters; PLA; tissue engineering

INTRODUCTION

Biodegradable polyesters such as poly(lactic acid) (PLA) and poly(glycolic acid) and their copolymers have been used widely as suitable materials in the field of tissue engineering.[1,2] Because of their poor cell attachment, various attempts have been made to increase that ability.[3] Some of these strategies involve tedious procedures or are not so effective. In this report, a simple method is introduced to effectively increase cell attachment ability of a PLA film. Engineered protein (Pronectin F),[4] which contains 13 arginine-glycine-aspartic acid (RGD) sequences in a molecule, well known as a cell-attachment site in many extracellular matrixes, was absorbed onto a surface of a PLA film to give cell attachment ability. Porcine endothelial cells were applied to the surface in serum-free medium (SFM) to examine cell attachment and proliferation.

MATERIALS AND METHODS

Pronectin F (Sanyo Chemical Industries, Japan) was physically adsorbed onto PLA films (Shimazu Co., Japan). Pronectin solution of various concentrations (0, 0.4, 2,10, 50 µg/ml) was added onto a circular film in each well of a 24-well culture plate, and the culture plate was put into a clean bench for 2 hours to allow Pronectin adsorption. The Pronectin solution was discarded and the film discs were washed with PBS three times to remove unadsorbed Pronectin. Amounts of adsorbed Pronectin on the PLA films were determined after hydrolysis by amino acid analysis using high-performance liquid chromatography [OPA(*o*-Phthaldialdehyde)-NaClO, HPLC] (Tosho Corp., Japan), according to the manufacturer's instructions.

To examine the effect of Pronectin adsorption on cell attachment, porcine aortic endothelial cells (PAECs) in serum free medium were applied on the films treated with Pronectin solutions of various concentrations, and attached cell numbers were

Address for correspondence: Dr. Hiroo Iwata, Institute of Frontier Medical Sciences, Kyoto University, 53 Kawara-cho, Shogoin, Sakyo-ku, Kyoto,606-8507, Japan. Voice: +81-75-751-4119; fax: +81-75-751-4144.

iwata@frontier.kyoto-u.ac.jp

Ann. N.Y. Acad. Sci. 961: 288–291 (2002). © 2002 New York Academy of Sciences.

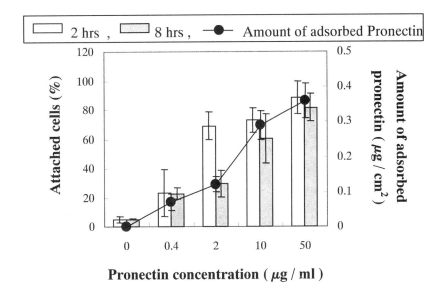

FIGURE 1. The effect of Pronectin concentration on initial cell attachment onto PLLA film for 2 or 8 hours. Cell attachment is expressed as a percentage of cell number attached to cell culture dish (TCPS). Amount of adsorbed Pronectin is expressed as a line curve.

determined by the MTT assay using a cell counting kit (WST-1). For the cell growth experiment, the films were rinsed with the serum-free medium to remove free or weakly binded cells after 2 hours' cell attachment and then incubated in a culture medium (D-MEM/F12) supplemented with 10% serum or in serum-free medium (SFM; Gibco BRL, Life Technologies, Inc., Japan) under 5% CO_2 at 37°C. After a certain time of culture, cell numbers on the films were determined to examine cell proliferation on film.

RESULTS AND DISCUSSION

FIGURE 1 shows the effects of Pronectin concentration on the cell attachment on PLA films. The amount of adsorbed Pronectin to PLA films increased with increasing Pronectin concentrations of the reaction solutions, which reached up 0.37 µg/cm^2 in 50 µg/ml Pronectin concentration. As the Pronectin adsorption concentration was higher, the percentage of cells attached (%) increased. When PAECs were seeded on the film adsorbed with 0.1 µg/cm^2 Pronectin, the percentage of cell attachment was 20% of that onto culture plate. On the films of more than 0.3 µg/cm^2 Pronectin adsorption, cells attached well regardless of attachment time. These results indicate that above 0.3 µg/cm^2 Pronectin adsorption is necessary for cells to attach well on PLA film.

Porcine aortic endothelial cells were cultured on the PLA film adsorbed of 0.37 µg/cm^2 Pronectin in media with/without serum. Cell growth patterns on film in SFM

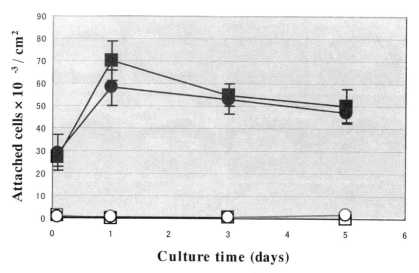

Culture time (days)

FIGURE 2. Comparison of cell growth on PLLA films in medium supplemented with 10% serum and in SFM: Untreated (*open circles*) and Pronectin-adsorbed film (*solid circles*) in 10% serum medium; untreated (*open squares*) and Pronectin-adsorbed film (*solid squares*) in SFM.

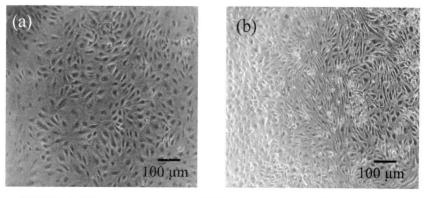

FIGURE 3. Morphologic pattern of PAECs on Pronectin-adsorbed PLLA films cultured for 5 days.

and a medium supplemented with 10% serum are shown in FIGURE 2. On untreated PLA film, a few cells attached for 2 hours. Also, no cell proliferation was observed regardless of difference in culture media supplemented with/without 10% serum. However, when cells attached well on Pronectin-adsorbed film, cells proliferated rapidly and formed monolayer during additional 1-day culture. Cells proliferated and reached maximum cell density after 1 day in culture. Although the number of cells decreased slightly between 1 and 3 days' cell culture, the cell monolayer was well maintained, as shown in FIGURE 3. Toward therapeutic applications, the use of serum-free culture medium is more desirable.[5] The use of serum presents many problems, such as high and fluctuating cost, limited availability, lot-to-lot variability, and potential for the presence of amphixenosis agents. In SFM, as shown in FIG. 3b, cells showed little spindle-shaped morphology compared with that in cells in 10% serum medium, and still maintained monolayer for the same culture period.

In this study, we found a very simple method for increasing cell-attachment ability to PLA with Pronectin. Also, PAECs attached and proliferated on Pronectin-adsorbed PLA film in serum-free medium (SFM).

REFERENCES

1. LANGER, R. & J. P. VACANTI. 1993. Tissue engineering. Science **260:** 920–926.
2. MOONEY, D.J., K. SANO, M. KAUFMANN, K. MAJAHOD, B. SCHLOO, J.P. VACANTI & R. LANGER. 1997. Long-term engraftment of hepatocyte transplanted on biodegradable polymer sponges. J. Biomed. Mater. Res. **37:** 413–420.
3. QUIRK, R.A., W.C. CHAN, M.C. DAVIES, S.J.B. TENDLER & K.M. SHAKESHEFF. 2001. Poly (L-lysine)-GRGDS as a biomimetic surface modifier for poly (lactic acid). Biomaterials **22:** 865–872.
4. BHADRIRAJU, K. & L.K. HANSEN. 2000. Hepatocyte adhesion, growth and differentiated function on RGD-containing proteins. Biomaterials **21:** 267–272.
5. GORFIEN, S., A. SPECTOR, D. DELUCA & S. WEISS. 1993. Growth and physical functions of vascular endothelial cells in a new serum-free medium (SFM). Exp. Cell. Res. **206:** 291–301.

Engineering of Capillary Patterns in Muscle by a Nonmitogenic Copper-Ribonucleoprotein Angiomorphogen [Angiotropin CuRNP Ribokine]

JOSEF H. WISSLER

ARCONS Applied Research Institute, D-61231 Bad Nauheim, Germany

KEYWORDS: capillary patterns; muscle; angiotropin CuRNP ribokine; angiogenesis; hemodynamics

INTRODUCTION

Integration of newly formed vascular patterns into homeostatic control mechanisms is a pivotal issue for reparative medicine, tissue engineering, implant, scaffold and transplant integration, and treatment of diseases. Muscular tissue function and regeneration[1,2] are highly dependent on both homeostasis of vascular patterns and regulation of hemodynamics. Thus, modes of application of an angiomorphogen in the form of angiotropin (AT) ribokine into muscle were investigated. Members of human and porcine AT cytokine (ribokine) family represent endogenous nonmitogenic endothelial cell-selective and vascularizing morphogens of leukocytic origin.[2–13] They are extracellular copper-ribonucleoprotein complexes (CuRNP) built up of copper ion, metalloregulated AT-related protein (ARP), and of extracellular eRNA bioaptamer units (ARNA, 72–78 nucleotides) of defined sequences. ARP have conserved domains with metal- (Cu, Zn, Ca, Mg) ion affinity (one canonical HxxxH and two Ca-binding EF-hand). It is homologous to RAGE-binding S100-EF-hand protein families, in particular to calgranulins (C, S100A12). ARNA are oxidant sensitive, highly modified and edited 5′ end-phosphorylated extracellular RNA showing copper as RNA-structuring metal ion. The metalloregulated copper ion complex (pCuRNP $_{N75-pARP}$) of porcine ARP and 75-base-ARNA bioaptamer (pCuRNA$_{N75}$) used in these studies is shown in FIGURE 1. The ribokine will be applied in the engineering of bioactive capillary patterns in muscle. Functional regulatory features of hemodynamics are a key issue in integration.

Address for correspondence: Josef H. Wissler, ARCONS Applied Research Institute, Postfach 1327, D-61231 Bad Nauheim, Germany. Fax: +49-6032-31725.

jhw@arcons-research.de

Ann. N.Y. Acad. Sci. 961: 292–297 (2002). © 2002 New York Academy of Sciences.

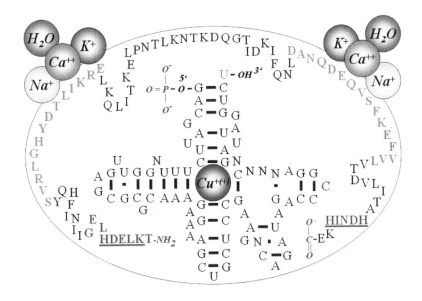

FIGURE 1. Schematic representation of the nonmitogenic angiotropin morphogen ribokine (pig AT, pCuRNP$_{N75\text{-pARP}}$) of average molecular mass in protonized form of 35075 Da used in this study. It consists of a metalloregulated angiotropin-related protein (ARP, *upper entity:* 91 amino acids, 10617 Da by Maldi/MS) and a Cu ion-dependent angiotropin-related extracellular, highly modified and edited 5′ end-phosphorylated eRNA bioaptamer (ARNA, *center entity:* 75 nucleotides). These are strongly complexed together by Cu and contain tightly bound Ca, Na and K (but not Zn, Mg and other metal) ions. In ARP, the domains with affinity for Cu ion of ARNA are at amino- and carboxy- ends (*underlined*). The two Ca ion-binding S100-EF-hand sites are shown in grey boldface. EFVVL is part of an exposed ARP epitope for a polyclonal antibody.[5] The cluster of Ca, Na, K ions and H$_2$O is schematic for multiple stoichiometric ratios.[4,13] In ARNA, N are modified bases; among these is isoguanine (151 Da). This is a biosynthetic product of redox/[OH*-]radical reactions and a structural link of hypoxia and nutritional factors (metal ions, reductones, vitamins)] to tissue blood vessel pattern remodeling at the morphogen level.[4–6,13] Intrinsic cellular or enzymatic (nuclease) activities of AT, its eRNA, and protein constituents may be modulated and controlled by Cu, Ca, Mg, or Zn ions. Zn ions are preferentially binding to canonical HxxxH only in copper- and eRNA-deficient ARP. It suggests AT, ARP, ARNA, and ions as components of a Fenton-redox reaction type, OH*-radical-dependent and metalloregulated biomolecular switch operative in vascular remodeling. A more detailed description is presented elsewhere.[4–6,13]

MATERIALS AND METHODS

Endogenous, not recombinant,[5] AT, a copper-ribonucleoprotein cytokine (CuRNP ribokine) (FIG.1) was used with more than 90% pCuRNP$_{N75\text{-pARP}}$, the remainder being homologous isoribokines (7% pCuRNP$_{N72\text{–}78\text{-pARP}}$).[2–6,13] It was purified from serum-free culture supernatants of isolated, lectin-activated porcine (mononuclear) leukocytes. The biotechnical and bioreactor details are described elsewhere.[3,4,13]

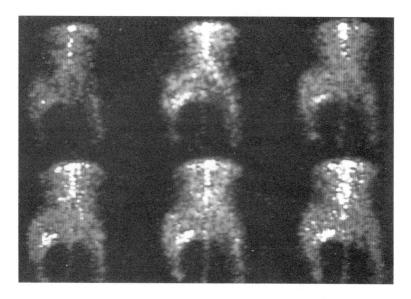

FIGURE 2. Neohypervascularization of skeletal muscle by angiotropin morphogen: Sequence of whole-body ^{99}Tc angioscintigrams (method described in Ref. 14) of a rabbit showing kinetics of entry and distribution of ^{99}Tc marker. Snapshots are in 3-s intervals from left to right. They begin 3 s after introduction of ^{99}Tc into the vasculature (*upper panel*). They show a rabbit two days after application of 500 fmol nonmitogenic endothelial cell-vascularizing angiotropin morphogen (left hind-leg muscle) and of morphogen-devoid buffer solution (right hind-leg muscle). In contrast to reference (right), Tc marker appears in the circulation of the left hind-leg muscle almost as fast and intense as in large axial vessels of the body. It suggests hemodynamics in remodeling vasculature. Similar results are obtained by 50 fmol angiotropin.

For preliminary experiments, the same AT was isolated from wound fluids (homogenates) of ischemic or infarcted inflamed heart muscle sites in dog and pig 24 h after occlusion of the left anterior descending branch of left coronary artery by transfemoral catherization.[2,12] AT, 50 and 500 femtomole, in phosphate-buffered physiologic saline (PBS, 100 ml) were focally applied intramuscularly (tissue application described in Ref. 11) into resting skeletal muscles of left hind-leg muscle of non-trained rabbits (New Zealand strain). No accessory application aids like polymer implants, amplifiers, delivery-retarding–, slow-release– or other systems were in use. [99]Technetium (Tc) angioscintigrams of rabbits that had undergone intravenous application of marker[14] were used to follow hemodynamics as a measure of morphogen-induced remodeling (formation, turnover, biofunction) of vascular patterns. Morphogen-devoid buffer treatment (right hind-leg muscle) was used as control. Calculation of peak counts per picture element (%) vs. time after ^{99}Tc injection was used[14] as a quantitative measure of remodeling.

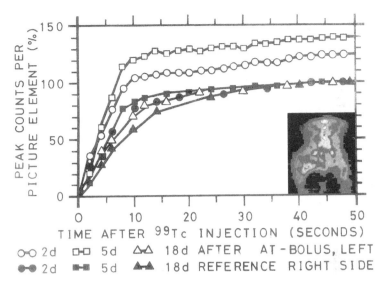

FIGURE 3. Neohypervascularization of skeletal muscle by angiotropin morphogen: Whole-body ^{99}Tc angioscintigram of rabbits (*inset*) and quantitative evaluation of ^{99}Tc marker in terms of peak counts per picture element (%) vs. time after ^{99}Tc injection (s). Different periods (2, 5, and 18 days) after angiotropin application (500 fmol, left hind-leg muscle) referenced to morphogen-devoid control (right hind-leg muscle) are evaluated. Capillary morphogenesis reactions in skeletal muscle tissue (rabbit) induced by the nonmitogenic endothelial cell-vascularizing morphogen lead to transient neohypervascularization with reversible, long-lasting increase in hemodynamics with a maximum of up to 140% between 2 and 10 days. Comparable effects are obtained by 50 fmol angiotropin.

RESULTS AND DISCUSSION

Biofunctions of newly formed capillary patterns were investigated in terms of hemodynamics and turnover as they affect integration into tissue homeostasis control mechanisms. In FIGURE 2, the kinetics of ^{99}Tc entry and tissue distribution are shown in angioscintigrams of rabbits as an indicator of direct blood flow after treatment. Comparison with controls (right hind-leg skeletal muscle) indicates asymmetric ^{99}Tc distribution (right vs. left). It shows that AT applied (left) in minute amounts (fmol) induces fast formation and remodeling of functionally efficient, additional circulation (hyper)patterns with active hemodynamics. The ^{99}Tc distribution appears almost as rapid and intense as in the large axial vessels of the body. Other techniques could show that dynamics of flow are a property of newly formed three-dimensional networks of capillaries in tissue[2–4,7–13] (e.g., by making capillary structures and functions accessible to visual inspection by microparticle distribution and to characterization by laser Doppler-flow velocimetry).

By a series of flow kinetics studies, quantitation in terms of peak counts per picture element vs. application time of AT to muscle (FIG. 3) demonstrates integration of newly formed hypervascular patterns and hemodynamic functions into tissue homeostatic control mechanisms. The results indicate that AT-induced capillary morphogenesis leads to transient neohypervascularization of muscle with reversible, long-lasting increase in hemodynamics (about 15 days, with a maximum between 2 and 10 days). Normal hemodynamic situations are reestablished after 20 days, suggesting no other deleterious effects, uncontrolled growth, and biofunctions in muscle resulting from AT application and action.

SUMMARY AND CONCLUSIONS

By a single application of nonmitogenic endothelial cell-selective and vascularizing angiotropin morphogen, a metalloregulated endogenous copper-complexed ribonucleoprotein cytokine (CuRNP ribokine), three-dimensional blood capillary (hyper)patterns with active flow formed in resting skeletal muscles of non-trained rabbits. [99]Tc angioscintigrams for measuring hemodynamics in remodeling capillary patterns showed these to be functionally active and to persist for about 15 days. Results suggest that the newly formed patterns are fully integrated and subject to homeostatic control mechanisms in muscular tissue. Thus, the final "physiological" pattern formation is suggested to be functionally integrated with respect to host tissue involving other endogenous factors formed by morphogen-activated cells.[2–4,11,13] Taken together with known properties, that is, its capacity for remodeling bioactive blood vessel patterns in tissue, for supporting tissue functions and preventing tissue necrosis,[2–13] angiotropin shows beneficial perspectives for further investigations in angiotherapy and engineering of biocompatible artificial organs. Potential applications include support of muscular and associated tissue functions, prevention of necrosis, angiotherapeutic aspects of bypass operations and restenosis following percutaneous transluminal coronary angioplasty, as well other disorders and risks subject to situations of changes in vascular patterns.

ACKNOWLEDGMENTS

The cooperation and advice of Prof. M. Gottwik, Klinikum Nuernberg; Prof. L.M.G. Heilmeyer, University of Bochum; Prof. H.P. Jennissen, University of Essen; Dr. E. Logemann, University of Freiburg; and H. Renner, Bad Kreuznach are gratefully acknowledged.

REFERENCES

1. MAURO, A., Ed. 1979. Muscle Regeneration. Raven Press. New York.
2. WISSLER, J.H. 1982. Inflammatory mediators and wound hormones: chemical signals for differentiation and morphogenesis in tissue regeneration and healing. In Proceedings 33th Mosbach Colloquium 1982: Biochemistry of Differentiation and Morphogenesis. L. Jaenicke, Ed.: 257–274. Springer Verlag. Heidelberg.
3. WISSLER, J.H. 1984. Large scale and biotechniques for the production and isolation of leucocytic effector substances of regenerative tissue morphogenesis by culturing cells in serum-free, synthetic fluids: design, preparation and use of a novel medium. In Developments in Nuclear Medicine, Vol. 7: Blood Cells in Nuclear Medicine, Part 2: Migratory Blood Cells. G.F. Fueger, Ed.: 41–102; 393–471. Martinus Nijhoff Publishers. Boston.

4. WISSLER, J.H., E. LOGEMANN, H.E. MEYER, *et al.* 1986. Structure and function of a monocytic blood vessel morphogen (angiotropin) for angiogenesis *in vivo* and *in vitro*: A copper-containing metallo-polyribonucleo-polypeptide as a novel and unique type of monokine. Protides Biol. Fluids **33:** 643–650 and **34:** 517–536.

5. WISSLER, J.H., W.M. AMSELGRUBER, M. SCHWEIGER & E. LOGEMANN. 1997. Angiotropin ribokine: natural and recombinant non-mitogenic leukocytic copper-RNP endothelial cell-vascularizing angio-morphogens and cellular and enzymatic activities of their S100-EF-hand-protein and RNA units. Mol. Biol. Cell **8** (Suppl.): 231a; **9** (Suppl.): 71a; FASEB J. **12:** A1463.

6. WISSLER, J.H. & E. LOGEMANN. 2000. Ribokines with modified and edited extracellular eRNA: endogenous metallo-regulated copper-ribonucleoprotein cytokines (CuRNP) and their components (S100-EF-hand protein, RNA, precursors) in cellular signal transduction. Biol. Chem. **381:** S246.

7. HÖCKEL, M., T. BECK & J.H. WISSLER. 1984. Neomorphogenesis of blood vessels in rabbit skin by a highly purified polypeptide mediator (monocyto-angiotropin) and associated tissue reactions. Int. J. Tissue React. **6:** 323–331.

8. HÖCKEL, M., J. SASSE & J.H. WISSLER. 1987. Purified monocyte-derived angiogenic substance (angiotropin) stimulates migration, phenotypic changes and "tube formation" but not proliferation of capillary endothelial cells *in vitro*. J. Cell. Physiol. **133:** 1–13.

9. HÖCKEL, M., W. JUNG, P. VAUPEL, *et al.* 1988. Purifed monocyte-derived angiogenic substance [angiotropin] induces controlled angiogenesis associated with regulated tissue proliferation in rabbit skin. J. Clin. Invest. **82:** 1075–1090.

10. HÖCKEL, M. & J.F. BURKE. 1989. Angiotropin treatment prevents flap necrosis and enhances dermal regeneration in rabbits. Arch. Surg. **124:** 693–698.

11. WISSLER, J.H., H. RENNER & E. LOGEMANN. 1989. Mechanisms in neovascularization of avascular ocular tissue by a monocytic angio-morphogen. *In* Modern Trends in Immunology and Immunopathology of the Eye. A.G. Secchi & I.A. Fregona, Eds.: 174–188. Masson. Milan.

12. GOTTWIK, M., H. RENNER & J.H. WISSLER. 1982. Biochemical neovascularization of muscles by leukocyte-derived polypeptide effectors: morphogenesis and turnover of blood vessel patterns with active hemodynamics *in vivo*. Hoppe-Seyler's Z. Physiol. Chem. **363:** 938–939.

13. WISSLER, J.H. 2001. Engineering of blood vessel patterns by angio-morphogens (angiotropins): non-mitogenic copper-ribonucleoprotein cytokines (CuRNP ribokines) with their metalloregulated constituents of RAGE-binding S100-EF-hand proteins and extracellular RNA bioaptamers in vascular remodeling of tissue and angiogenesis *in vitro*. Materialwiss. Werkstofftech. [Mater. Sci. Eng. Technol.] **32:** 984–1008.

14. SAUER, E. & H. SEBENING, EDS. 1980. Myocard- und Ventrikelszintigraphie. Grundlagen und Anwendung. Mannheimer Morgen Verlag. Mannheim.

Noninvasive Monitoring of a Retrievable Bioartificial Pancreas *in Vivo*

IOANNIS CONSTANTINIDIS,[a,b] CHERYL L. STABLER,[b,c] ROBERT LONG, JR.,[a,b] AND ATHANASSIOS SAMBANIS[b,d]

[a]*Department of Radiology, Emory University, Atlanta, Georgia 30322, USA*

[b]*Georgia Tech/Emory Center for the Engineering of Living Tissues, Atlanta, Georgia 30332, USA*

[c]*Department of Biomedical Engineering, Georgia Institute of Technology, Atlanta, Georgia 30332, USA*

[d]*School of Chemical Engineering, Georgia Institute of Technology, Atlanta, Georgia 30332, USA*

KEYWORSD: **bioartificial pancreas; diabetes**

INTRODUCTION

The bioartificial pancreas is potentially an efficacious treatment for diabetes that can provide physiologic blood-glucose regulation without immunosuppressive medication. It can be administered with relative ease and is readily available.[1,2] Although various designs have been considered, the design most commonly used to generate these constructs is based on the microencapsulation of insulin-secreting cells in a biocompatible matrix that provides mechanical support and at least partial immunoprotection.[1] A variety of cells have been used in these constructs including mammalian islets and transformed β-cell lines;[3,4] the matrix most frequently utilized is the alginate/poly-L-lysine/alginate (APA) bead.[5] At present, our only means of assessing the efficacy for an implanted bioartificial pancreas is to measure the blood glucose concentration of the host. Developing a noninvasive imaging technique that can monitor the viability and function of encapsulated cells as well as the integrity of the matrix is of critical importance. Nuclear magnetic resonance (NMR) has the ability to provide both biochemical and structural information, under either *in vivo* or *in vitro* conditions.

Retrieval of all APA beads following an i.p. implantation is difficult because of the dispersal of the beads throughout the peritoneal cavity. A possible solution to this problem is to contain the beads within a device that can be easily retrieved. This device, however, cannot impose additional diffusion barriers and thus can be detrimental to the metabolic and secretory activity of the encapsulated cells. By containment the acquisition of localized NMR spectroscopy is feasible. This study shows our ef-

Address for correspondence: Ioannis Constantinidis, Department of Medicine, Division of Endocrinology, University of Florida, Gainesville, FL 32610. Voice: 352-846-2227; fax: 352-846-2231.

consti@medicine.ufl.edu

Ann. N.Y. Acad. Sci. 961: 298–301 (2002). © 2002 New York Academy of Sciences.

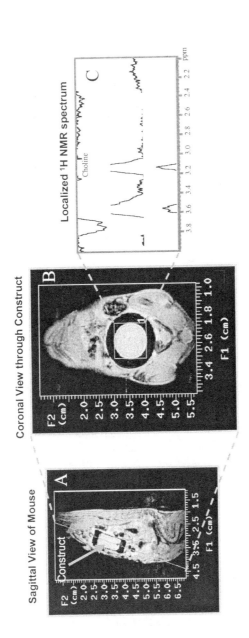

FIGURE 1. (A) A sagittal view through the CJ7/B6 mouse showing the location of the construct. The white line shown is the planing of the oblique slice through the construct shown in image **B**. **(B)** A coronal view through the mouse and the O-ring. The white square represents the volume of interest for the localized spectrum shown in image **C**. **(C)** A water-suppressed localized localized ^{1}H NMR spectrum acquired from within the O-ring.

forts in developing a NMR-based methodology to monitor a retrievable bioartificial pancreatic construct *in vivo*.

MATERIALS AND METHODS

βTC3 cells were obtained from the laboratory of Dr. Efrat (Albert Einstein College of Medicine) and cultured as previously described.[5] Encapsulation of βTC3 cells in APA beads was based on the initial protocol by Sun *et al.*[1] and modified to suit our needs.[5] The final APA beads measured ≈800 μm in diameter and contained initially 3.5×10^7 βTC3 cells/ml alginate.

In vivo ^{1}H experiments are performed with the aid of a quadrature "birdcage" coil using a 4.7 T horizontal bore VARIAN 200/33 *INOVA* spectrometer. Scout images were acquired to locate the construct, and localized ^{1}H NMR spectra were acquired using the double spin-echo PRESS sequence[6] with a TR = 3.00 s and TE = 90 ms. The voxel size was $4 \times 8 \times 8$ mm^3. Water suppression was achieved using CHESS pulses and the total acquisition time is 13 min.

RESULTS AND DISCUSSION

A prototype model that consists of a silicone O-ring (1.6 mm thickness and 1.27 cm diameter) with a polypropylene mesh of pore size 500 μm covering the top and bottom of the ring was considered for the containment of the APA beads. Approximately 0.2 ml of APA beads are injected into the ring and packed tightly within the construct. *In vitro* cultures confirmed that the metabolic and secretory activities of the encapsulated βTC3 cells contained within these constructs were statistically indistinguishable from free-floating beads.

In vitro experiments were performed to optimize the NMR acquisition parameters. *In vivo* experiments were initiated using CJ7/B6 (normal, nondiabetic) mice. One construct was surgically implanted per animal in the peritoneal cavity. Immediately after the implantation, the anesthetized animal was placed within the "birdcage" coil and positioned in the magnet. FIGURE 1 (panels A and B) shows two orthogonal T_2-weighted images of the mouse delineating the location of the O-ring; panel C shows a localized water-suppressed ^{1}H NMR spectrum from within the construct. The *in vivo* spectrum was acquired from a volume of 256 μl within 13 min. We have previously demonstrated that the choline resonance detected by ^{1}H NMR is a sensitive marker of cell proliferation and oxygen consumption by APA-encapsulated βTC3 cells.[7] Therefore, the ability to detect ^{1}H NMR spectra from encapsulated cells *in vivo* will permit us to monitor the proliferation and oxygenation of these cells. Such information is critical in assessing the functionality and possibly the longevity of an implanted construct.

ACKNOWLEDGMENTS

This work was supported by grants from the NSF (EEC-9731643) and NIH (DK47858, RR13003, and GM08433).

REFERENCES

1. SUN, A.M., W. PARISIUS, H. MACMORINE, *et al.* 1980. An artificial pancreas containing cultured islets of Langerhans. Artif. Organs **4:** 275–278.
2. LANZA, R.P. & W.L. CHICK. 1997. Transplantation of encapsulated cells and tissues. Surgery **121:** 1–9.
3. LANZA, R.P., D.M. ECKER, W.M. KUHTREIBER, *et al.* 1999. Transplantation of islets using microencapsulation: studies in diabetic rodents and dogs. J. Mol. Med. **77:** 206–210.
4. HIROTANI, S., R. EDA, T. KAWABATA, *et al.* 1999. Bioartificial endocrine pancreas (Bio-AEP) for treatment of diabetes: effect of implantation of Bio-AEP on the pancreas. Cell Transplant. **8:** 399–404.
5. STABLER, C., K. WILKS, A. SAMBANIS & I. CONSTANTINIDIS. 2001. The effects of alginate composition on encapsulated βTC3 cells. Biomaterials **22:** 1301–1310.
6. BOTTOMLEY, P.A., inventor. 1984. Selective volume method for performing localized NMR spectroscopy. U.S. patent no. 4480228.
7. LONG, R.C., JR., K.K. PAPAS, A. SAMBANIS & I. CONSTANTINIDIS. 2000. *In vitro* monitoring of total choline levels in a bioartificial pancreas: ^{1}H NMR spectroscopic studies of the effects of oxygen level. J. Magn. Reson. **146:** 49–57.

In Vivo Studies to Evaluate Tissue Engineering Techniques

STEPHEN F. BADYLAK

Purdue University, West Lafayette, Indiana 47907, USA

KEYWORD: *in vivo* tissue remodeling

Discussion of the use of *in vivo* methods as they relate to the field of tissue engineering can be divided into two general categories: (1) the use of *in vivo* remodeling systems as test beds for tissue-engineered constructs that have been developed in *in vitro* systems, and (2) the use of *in vivo* systems as an integral component of the actual remodeling process.

USE OF *IN VIVO* REMODELING SYSTEMS FOR THE EVALUATION OF TISSUE-ENGINEERED "DEVICES"

Sophisticated modeling techniques, *in vitro* cell culture, and state-of-the-art bioreactors are all valuable, important, and necessary tools for the development and evaluation of various tissue-engineering approaches. However, in the final analysis, the response of the complex machinery that we know as the mammal can be best mimicked and predicted only by testing in *in vivo* systems.

The choices of *in vivo* remodeling systems are many, including rodent animal models, larger animal models such as the dog, pig, and goat, and eventually subhuman primates and humans. The most appropriate *in vivo* test system for each tissue-engineering application will depend upon many variables including the type of tissue construct being tested (e.g., purified collagen as a dermal substitute vs. a cellularized synthetic scaffold for heart valve replacement), the questions being asked (e.g., local tissue remodeling response? vs. systemic immune response?), and cost considerations.

The evaluation of the host immune response to tissue-engineered body parts is an issue of immense importance. Tissue-engineered constructs are presently derived from a variety of synthetic and naturally occurring sources. Autologous cells, allogeneic and xenogeneic scaffolds, and synthetic materials with varying degrees of biocompatibility are presently being used to develop tissue-engineering approaches for the repair and restoration of numerous body systems. However, the hyperacute,

Address for correspondence: Stephen F. Badylak, Purdue University, 1296 Potter Building, Room 204, West Lafayette, IN 47907. Voice: 765-494-2994; fax:765-494-1193.
badylak@ecn.purdue.edu

Ann. N.Y. Acad. Sci. 961: 302–304 (2002). © 2002 New York Academy of Sciences.

acute, and chronic immune responses toward the end products cannot be reliably evaluated in any system short of a mammalian *in vivo* system. In fact, even with the most sophisticated and expensive *in vivo* systems (for example, the subhuman primate), there still remain questions regarding the eventual response of humans; thus, the necessity for phased human clinical trials. Consider for a moment the immune response to tissue-engineered devices. The most commonly used *in vivo* test systems for immune responses is the well-characterized mouse model. However, there are well-recognized differences in the histocompatibility antigens and other aspects of the immune system between the mouse and the human. Yet we continue to use the mouse as the model of choice for the evaluation of the immune response to various antigens because, short of human clinical trials, it is presently the best system that exists. Stated differently, *in vivo* systems to evaluate the remodeling of tissue-engineered constructs, including the immune response to constructs, are absolutely essential.

One of the most limiting aspects of existing approaches to the tissue engineering of complex tissues and organs is the ability to provide for a functional circulatory system when *in vitro* approaches are used. Even when organized three-dimensional cell and tissue development can be orchestrated in the laboratory, the ability to transfer a tissue-engineered replacement tissue or organ to the *in vivo* system (i.e., the patient/recipient) is limited by the nutrient supply. Unlike organ transplantation there is no preexisting arterial and venous "plumbing" that can be connected to the host circulatory system. Although elegant methods of *in vitro* tissue engineering can and have been developed, the practical limitations of transferring this technology to the patient often render the science non-usable. The use of *in vivo* systems to evaluate tissue-engineered devices thus becomes an essential precursor to human clinical applications.

THE USE OF *IN VIVO* REMODELING SYSTEMS AS PART OF THE TISSUE-ENGINEERING SOLUTION

The most recognized approach to tissue-engineered body parts is the method of creating specific tissues or organ structures in sophisticated cell culture laboratories and bioreactor systems, with subsequent transfer of the tissue-engineered constructs to the patient. An alternative approach to tissue engineering is the use of the *in vivo* mammalian system as the "bioreactor" for tissue or organ repair/replacement. The tissue-engineered scaffold or scaffold/cell combination is placed in the *in vivo* system for "on-site" remodeling into the final end product. The *in vivo* mammalian system serves as the ultimate bioreactor. With advances in genetic engineering, stem cell biology, biocompatible scaffolds, and increased recognition of cell-based approaches to tissue engineering, this approach may prove to be the optimal reduction to practice.

The above comments are not meant to diminish the importance and necessity of *in vitro* approaches or alternative approaches to the development of tissue-engineered solutions for the repair and restoration of body structures. However, in the end analysis, there appears to be an unavoidable, and in fact desirable, evaluation of our tissue-engineering approaches in *in vivo* systems. The development and understanding of *in vivo* models for tissue engineering will continue to be an important emphasis of efforts in reparative medicine.

REFERENCES

1. BADYLAK, S.F. *et al.* 2001. Marrow-derived cells populate scaffolds composed of xeno-geneic extracellular matrix. Exp. Hematol. **29:** 1310–1318.
2. CHEN, M.K. & S.F. BADYLAK. 2001. Small bowel tissue engineering using small intestinal submucosa as a scaffold. J. Surg. Res. **99:** 352–358.

Transition of Stem Cells to Therapeutically Functional Tissue-Specific Cells

MARKUS GROMPE

Oregon Health Science University, Portland, Oregon 97201, USA

KEYWORDS: stem cells; tissue specificity

A growing number of investigators are addressing the "plasticity" of stem cells and their potential for tissue repair and, in particular, the developmental potential of neural stem cells and adult hematopoietic stem cells. The usual approach has been the transplantation of stem cells, followed by detailed histological analysis in which donor markers and tissue-specific antigens are co-localized. For example, the co-localization of a donor Y-chromosome and a muscle-specific marker after bone marrow transplantation was interpreted to indicate the differentiation of hematopoietic stem cells into functional muscle fibers. Similarly, the expression of neuronal antigens in donor-derived CNS cells was perceived to indicate that bone marrow cells could differentiate into functional neurons. Despite the many interesting findings, however, only very few examples of functional correction of disease states by transplanted stem cells have been reported. One of these examples is the correction of an inherited liver disease, tyrosinemia type I (HT1), in a murine model by transplanted hematopoietic stem cells. This example was used to highlight some of the current problems in stem cell research and to make suggestions for further work. Several reports have claimed that 1–2% of hepatocytes are donor-derived 2–4 months after a simple bone marrow transplantation in several species (mouse, rat, human). This degree of repopulation would already be therapeutic for some disorders, such as hemophilia A. Indeed, HT1 mutant mice also have 30–50% donor-derived hepatocytes and are fully corrected functionally 5 months after BMT. We recently have performed a careful kinetic and quantitative analysis of this repopulation process. In contrast to previous reports, we found that at most 1/100,000 hepatocytes (500 cells per liver) were donor-derived 3 months after BMT. These very few cells expanded to repopulate the host liver only by the strong *in vivo* selection, which is characteristic of the HT1 model. We conclude that the transition of hematopoietic stem cells to hepatocytes is much too slow and much too inefficient to be of any clinical use without further improvement of the system. These findings raise several questions and also suggest some important goals for stem cell research aimed at tissue repair. These questions

Address for correspondence: Markus Grompe, Oregon Health Science University, 3181 Southwest Sam Jackson Park Road, Portland, OR 97201. Voice: 503-494-6888; fax: 503-494-6886.
grompem@ohsu.edu

Ann. N.Y. Acad. Sci. 961: 305–306 (2002). © 2002 New York Academy of Sciences.

include: (1) What criteria are valid to prove the transdifferentiation of stem cells? (2) Are immunohistochemical markers sufficient to assess the functionality of stem cell–derived differentiated cells? (3) Which markers should be used for which tissue? How many? (4) How can transplanted donor cells be given a selective growth advantage? (5) How does tissue remodeling occur during the repopulation process?

Current goals that will require achievement in order to move the field forward include: (1) generation of animal models that can be used to define tissue-specific stem cells and their function; (2) discovery of the factors that drive transdifferentiation of stem cells *in vivo* and *in vitro;* and (3) development of methods to facilitate/enable engraftment of transplanted cells.

SELECTED REFERENCES

1. GROMPE, M. 2001. Liver repopulation for the treatment of metabolic diseases. J. Inherit. Metab. Dis. **24:** 231–244.
2. GROMPE, M. 2001. Mouse liver goes human: a new tool in experimental hepatology. Hepatology **33:** 1005–1006.

In Vivo Remodeling

ARNOLD I. CAPLAN

Case Western Reserve University, Cleveland, Ohio 44106, USA

KEYWORDS: embryonic tissue development; tissue remodeling; mesenchymal stem cells; tissue engineering

Embryos and young animals are able to reform/regenerate tissues and organs following tissue excision or damage. Thus, two important cell systems are being cooperatively controlled: the first, the acute inflammatory response, has not developed or is muted in these young animals; and the second, the progenitor cell cascade, including active cell division and tissue-specific differentiation, is upregulated. In adults, the massive, acute inflammatory response is a defense and survival mechanism to insure that outside contaminants are eliminated from the wound site. The downstream result of this "quick-seal, quick-fix" mechanism is that scarring occurs at the site previously occupied by functional tissue. Thus, the "repaired" tissue may be compromised in its function while insuring that agents and organisms detrimental to the host are excluded.

A fundamental principle of reparative medicine which governs our efforts to regenerate differentiated tissue is to organize the reparative circumstances to *recapitulate selected aspects of embryonic developmental sequence,* including attempts to mimic the embryonic microenvironment in which tissue initiation, formation, and expansion take place.

Two prominent features of embryonic tissue development are (1) a large supply of uncommitted progenitor or stem cells and (2) an extracellular matrix scaffolding appropriate to the appropriate stage of embryonic development. Relevant to reparative medicine, adult mesenchymal stem cells have been isolated and culture-expanded and are thus available for delivery to sites of reparative tissue engineering. Likewise, we have identified hyaluronan and fibronectin as key scaffold components in embryonic mesenchyme and have developed technologies for using these macromolecules in a porous sponge formulation to deliver mesenchymal stem cells for the regeneration of subchondral bone and articular cartilage in the medial femoral compartment of the knee.

The primary challenge, at this time, is to effectively manipulate these complex, multicomponent systems in order to mimic the embryonic cascade of differentiated tissue formation in the adult organism at the adult size scale. For example, cartilage formation occurs in embryos at a scale of a few millimeters, while defects in the knee

Address for correspondence: Arnold I. Caplan, Case Western Reserve University, 2080 Adelbert Road, Cleveland, OH 44106. Voice: 216-368-3567; fax: 216-368-4077.

Ann. N.Y. Acad. Sci. 961: 307–308 (2002). © 2002 New York Academy of Sciences.

joint can be many centimeters in width and depth. The effective management of this system will require getting the proper signals to the appropriate cells in the embryonic sequence and within a three-dimensional and temporal framework to give not only appropriate differentiation, but also *integrated* function.

SELECTED REFERENCES

1. SOLCHAGA, L.A., V.M. GOLDBERG & A.I. CAPLAN. 2001. Cartilage regeneration using principles of tissue engineering. Clin. Orthop. Relat. R **391:** S161–S170 Suppl.
2. CAPLAN, A.I. & S.P. BRUDER. 2001. Mesenchymal stem cells: building blocks for molecular medicine in the 21st century. Trends Mol. Med. **7:** 259–264.

Regulated *in Vivo* Remodeling

HOWARD P. GREISLER

Loyola University Medical Center, Maywood, Illinois 60153, USA and
Hines Veterans Administration Hospital, Hines, Illinois 60141, USA

KEYWORD: tissue remodeling

While tissue engineering is often considered a new field holding previously unforeseen promise in the development of novel implantable therapeutic constructs, it is imperative that lessons learned during earlier generations of research not be ignored. Among these is the reality that *in vivo* remodeling of implants is *absolutely unavoidable*. Decades of work in vascular grafts focused on the identification of an inert synthetic material only to conclude that *all* materials elicit both favorable and unfavorable cellular responses. Similar efforts to optimize biological vascular grafts, including autografts, allografts, and xenografts, have been hampered by observations of post-implantation apoptosis and matrix degradation. Furthermore, alterations in cellular phenotypic characteristics resulting in often untoward functional activity may appropriately be considered an example of subcellular *in vivo* remodeling.

The quest in tissue engineering ought not to focus on whether or not to utilize *in vivo* remodeling approaches, but rather on how to best induce the most desirable remodeling.

The goals of regulated *in vivo* remodeling may include: (1) minimization of fibrosis and apoptosis; (2) recruitment of functional cells (e.g., hepatocytes, osteoblasts, endothelial cells, etc.); (3) recruitment of supporting matrix components; (4) induction of capillarization to nourish three-dimensional tissue constructs; (5) control of cellular phenotype; (6) minimization of untoward immunologic reaction to allogenic or xenogenic cells; and (7) circumventing the practical limitation and time constraints of *in vitro* utilization of autologous cells.

Many of the remodeling processes in response to implantation of both cellular and acellular implants are byproducts of a myriad of inflammatory and immunological pathways. Both chemical and biomechanical properties of implants induce the synthesis of a battery of intracellular messengers and intercellular cytokines and other effector molecules. These mediators may induce either untoward effect (i.e., fibrotic encapsulation, apoptosis, involution of pre-existing capillary networks, etc.) or desirable effects (i.e., angiogenic mechanisms, etc.).

Address for correspondence: Howard P. Greisler, Loyola University Medical Center, 2160 South First Avenue, Maywood, IL 60153. Voice: 708-216-8451; fax:708-216-6300.
hgreisl@Lumc.edu

Ann. N.Y. Acad. Sci. 961: 309–311 (2002). © 2002 New York Academy of Sciences.

Tissue engineering will ultimately be well served by advances in understanding of these mechanisms. Approaches may include: (1) Utilizing specific chemistries and/or biomechanics to induce preferential synthesis of desirable cytokines and growth factors. (2) Application of desirable naturally occurring genes and/or proteins within tissue-engineered constructs in an appropriate spatial and temporal distribution to optimize the *in vivo* biologic responses to the implants. (3) Generation and application of mutant genes and/or proteins within tissue-engineered constructs to effect more potent and/or more cell-specific responses after implantation.

It is likely that, in the near future, a combined approach, which may include *in vitro* production of incomplete cell/matrix constructs plus induced *in vivo* recruitment of the remaining autologous cellular components, may be most efficacious in promoting stable, biofunctional tissue-engineered implants without excessive immunologic response.

SELECTED REFERENCES

1. TASSIOPOULOS, A.K. & H.P. GREISLER. 2000. Angiogenic mechanisms of endothelialization of cardiovascular implants: a review of recent investigative strategies. J. Biomat. Sci.-Polym. E **11**: 1275–1284.
2. SHIREMAN, P.K. & H.P. GREISLER. 1998. Fibrin sealant in vascular surgery: a review. J. Long-Term Eff. Med. Implants **8**: 117–132.
3. AIJOKA, I., T. AKAIKE & Y. WANTABE. 1999. Hepatology **29**(2): 396–402.
4. BONADIO, J., E. SMILEY, P. PATIL & S. GOLDSTEIN. 1999. Localized, direct plasmid gene delivery *in vivo*: prolonged therapy results in reproducible tissue regeneration. Nature Med. **5**(7): 753–759.
5. GREISLER, H.P., E. ENDEAN, J.J. KLOSAK, *et al.* 1988. Polyglactin 910/polydioxanone biocomponent totally resorbable vascular prostheses. J.Vasc. Surg. **7**: 697–705.
6. GREISLER, H.P., C.W. TATTERSALL, J.J. KLOSAK, *et al.* 1991. Partially bioresorbable vascular grafts in dogs. Surgery **110**: 645–655.
7. GREISLER, H.P., S.C. HENDERSON & T.M. LAM. 1993. Basic fibroblast growth factor production *in vitro* by macrophages exposed to Dacron and polyglactin 910. J. Biomaterials Sci. Polymer Edu. **4**: 415–430.
8. GRAY, J.L., S.S. KANG, G.C. ZENNI, *et al.* 1994. FGF-1 affixation stimulates ePTFE endothelialization without intimal hyperplasia. J. Surg. Res. **57**: 596–612.
9. KANG, S.S., C. GOSSELIN, D. REN & H.P. GREISLER. 1995. Selective stimulation of endothelial cell proliferation with inhibition of smooth muscle cell proliferation by FGF-1 plus heparin delivered from fibrin glue suspensions. Surgery **118**: 280–287.
10. GREISLER, H.P., C. GOSSELIN, D. REN, *et al.* 1996. Biointeractive polymers and tissue engineered blood vessels. Biomaterials **17**: 329–336.
11. ZARGE, J.I., P. HUANG, V. HUSAK, *et al.* 1997. Fibrin glue containing FGF-1 and heparin with autologous EC reduces intimal hyperplasia in a canine carotid artery balloon injury model. J. Vasc. Surg. **25**: 840–849.
12. SHIREMAN, P.K., B. HAMPTON, W. BURGESS, *et al.* 1999. Modulation of vascular cell growth kinetics by local cytokine delivery from fibrin glue suspensions. J. Vasc. Surg. **29**: 859–873.
13. SHIREMAN, P.K., L. XUE, E. SZYLOBRYT, *et al.* 2000. The S 130K fibroblast growth factor-1 mutant S130K induces heparin-independent proliferation and is resistant to thrombin degradation in fibrin glue. J. Vasc. Surg. **31**: 382–390.
14. XUE, L., A.K. TASSIOPOULOS, S. WOLOSON, *et al.* 2001. Construction and biological characterization of an HB-GAM/FGF-1 chimera for vascular tissue engineering. J. Vasc. Surg. **33**(3): 554–560.
15. FRIEDMAN, T. 2000. Science **287**: 2163.
16. LAUFFENBURGER, D.A. & D.V. SCHAFFER. 1999. The matrix delivers. Nature Med. **5**(7): 733–734.

17. SAKIYAMA-ELBERT, S. & J. HUBBEL. 2000. J. Controlled Release: 389–402.
18. SCHENSE, J. & J. HUBBEL. 1999. Bioconjugate Chem. **10**(1): 75–81.
19. YE, X., V. RIVERA, P. ZOLTICK, *et al.* 1999. Science **283**: 88.

Consideration of Mechanical Factors

ROBERT E. GULDBERG

Georgia Institute of Technology, Atlanta, Georgia 30332, USA

KEYWORD: bone tissue engineering

The basic elements required for tissue engineering include an extracellular matrix scaffold, cells, and tissue-specific bioactive factors. Whether these elements are provided by the host or must be implanted in some combination within a construct depends critically on the *in vivo* biochemical, vascular, and mechanical environments. Using bone tissue engineering as an example, the initial repair response to a construct will vary substantially according to the local availability of osteoprogenitor cells, the vascularity of the wound bed, and the mechanical stability of the defect site. *In vivo* remodeling will subsequently reinforce, maintain, or degrade the tissue formed during repair. Thus, the initial and long-term biological responses to tissue-engineered constructs are strongly influenced by interactions between construct design parameters and the *in vivo* environment. The variability of these interactions among different species, patients, and anatomic sites represents a key challenge in tissue engineering.

Successful tissue regeneration must go beyond reproducing shape and structure to restore biological and mechanical function and long-term integration with surrounding native tissues. Initially, an implanted construct should be immune-acceptable and biocompatible and, in many cases, promote angiogenesis and the recruitment of progenitor cells. If these early response criteria are met, then an ordered repair sequence culminating in matrix synthesis and deposition may proceed. However, successful tissue formation during the repair phase does not alone guarantee long-term functional regeneration. The retention of newly-formed repair tissue and its integration with surrounding tissues depends on the subsequent remodeling response. For example, bone contains an intricate cellular communication network of osteocytes and is subject to local remodeling by bone-resorbing osteoclasts and bone-forming osteoblasts, serving under normal conditions to maintain skeletal structural integrity. In the absence of adequate mechanical stimuli, however, bone remodeling will also rapidly remove bone mass. This outcome is illustrated by a common model to test for bone-induction properties involving subcutaneous or intramuscular implantation of constructs into athymic rats and other animal species. Constructs that recruit local undifferentiated cells, induce differentiation along an

Address for correspondence: Robert E. Guldberg, Georgia Institute of Technology, 315 Ferst Drive, Atlanta, Georgia 30332. Voice: 404-894-6589; fax: 404-894-2291.
robert.guldberg@me.gatech.edu

Ann. N.Y. Acad. Sci. 961: 312–314 (2002). © 2002 New York Academy of Sciences.

osteogenic pathway, and subsequently mineralize are considered osteoinductive. The duration of these experiments is typically limited to approximately four weeks, however, because beyond that time a remodeling response begins to resorb the newly-formed bone.

We have recently studied the influence of the *in vivo* mechanical environment on the formation and retention of bone within tissue-engineered constructs implanted into hydraulic bone chambers.[1] The bone chamber model is a hollow threaded implant that closely resembles the titanium cages currently being used clinically to fuse spines. A unique aspect of this model is the ability to apply a controlled cyclic mechanical stimulus to tissue developing within the chamber. In bilaterally implanted chambers, we tested the hypothesis that tissue-engineered constructs under dynamic loading would promote bone repair *in vivo*. Tissue was allowed to form within the constructs without mechanical stimulation for four weeks followed by four weeks of loading on one side only. Relative to the baseline bone formation response at four weeks, mechanical loading increased the amount of mineralized matrix present at eight weeks by 125%. In contrast, the amount of mineralized matrix present in the no-load control chambers had decreased by 75% from four to eight weeks.

Mechanical factors can have both positive and negative effects on the biological response to construct implants. Repeated disruption of vascular invasion at the construct interface with native tissues can result in nonunion. For example, the optimal porosity or pore size for bone regeneration scaffolds remains an open question the answer to which may depend on mechanical and vascular factors at the defect site. The optimal pore size for bone ingrowth has often been reported to be in the range of 150–600 microns. However, Whang and coworkers[2] demonstrated substantial bone formation in non-load-bearing defects filled with polymer scaffolds possessing a median pore size less than 50 microns and porosity greater than 90%. The investigators proposed that high-porosity scaffolds with small pore sizes provided greater hematoma stabilization in the earliest phases of bone regeneration. The reason for uncertainty regarding optimal pore size may relate to variable interactions between the scaffold mechanical properties and the functional loading environment. The spatial requirement for vascularization and invasion of osteoprogenitor cells should be less than 20 microns based on the dimensions of blood vessels and bone cells. However, relative micromotion between the scaffold boundaries and surrounding tissue may disrupt vascular ingrowth into such small pores. Larger pores, although perhaps not optimal for hematoma stabilization or cell attachment, may allow vascular ingrowth to occur despite the presence of interface micromotion.

For load-bearing clinical applications, the implanted construct must therefore possess mechanical properties that provide adequate initial stability at the defect site and gradual transfer of physiological mechanical stimuli under functional loading to newly formed tissue.[3] Constructs possessing insufficient strength and fatigue resistance properties are at risk for plastic deformation or brittle failure under functional loads, leading to collapse of the internal porosity and subsidence of the implant. Martin and coworkers,[4] for example, found that a collagen sponge scaffold loaded with BMP effectively induced spine fusion in rabbits, but that the same scaffold failed as a delivery vehicle in a non-human primate spine fusion model due to mechanical collapse.

Although the influence of the local *in vivo* mechanical environment on repair and remodeling is well recognized in numerous tissue types, this potentially critical factor has not been well studied in the context of tissue-engineered constructs. An understanding of how mechanical signals affect construct integration and regeneration of function may provide microstructural design objectives for 3-D scaffold architectures and may have an impact on the selection of scaffold material, cell type or seeding density, and other construct parameters. Furthermore, it may be possible to exploit the adaptive potential of tissues by mechanically preconditioning constructs or cells prior to implantation. Many *in vitro and in vivo* test beds for reparative methods of mechanically functional tissues are non-load-bearing. Calvarial critical sized defects for bone repair and non-load-bearing articular cartilage defects for cartilage repair are two examples. Model systems that account for the three-dimensional, load-bearing functional requirements of tissues represent more rigorous test beds for these technologies and may ultimately prove to be more predictive of human clinical results.

REFERENCES

1. CASE, N.D. *et al.* 2000. Chondrocyte viability and cyclic mechanical loading promote ossification of tissue engineered cartilage constructs implanted into bone defects. Presented at the Annual Meeting of the Tissue Engineering Society, December, 2000.
2. WHANG, K. *et al.* 1999. Engineered bone regeneration with bioabsorable scaffolds with novel microarchitecture. Tissue Eng. **5:** 35–51.
3. DAVY, D.T. 1999. Biomechanical issues in bone transplantation. Orthop. Clin. N. Amer. **30:** 553–563.
4. MARTIN, G.J. *et al.* 1999. Posterolateral intertranverse process spinal arthrodesis with rhBMP-2 in a nonhuman primate: important lessons learned regarding dose, carrier, and safety. J. Spinal Disord. **12:** 179–186.

Is *in Vivo* Remodeling Necessary or Sufficient for Cellular Repair of the Heart?

DORIS A. TAYLOR

Duke University Medical Center, Durham, North Carolina 27710, USA

KEYWORDS: cardiomyocytes; myocardial repair; tissue engineering

INTRODUCTION

Because an adult heart lacks a pool of precursor, stem, or reserve cardiomyocytes, it is incapable of effective regeneration after injury or infarction. In fact, after an acute myocardial infarction (AMI), the injured area of the myocardium heals by forming a non-contracting fibrous scar. Treatment options for AMI and subsequent congestive heart failure (CHF) include medical management, heart transplantation, mechanical circulatory assist devices (LVADs), or other experimental attempts all of which suffer from specific limitations. In light of the limited efficacy and co-morbidity of current treatment options, additional alternative long-term therapeutic strategies are needed.

Tissue engineering has an important role in novel potential therapies for cardiac repair. These could range from cell-based therapeutics to synthetic or bioactive patches composed of biomaterials and/or cells and genes. Any of these products introduced into the injured myocardium must integrate both mechanically and electrically to be effective in the long term.

SUCCESSES

Cell-Based Myocardial Repair

For the treatment of cardiovascular disease states ranging from heart attack to end-stage heart disease, we are developing a cell transplantation technology (*cellular cardiomyoplasty*) to regenerate functioning muscle in previously infarcted, scarred, and dysfunctional myocardial tissue. To date, multiple cell types have been transplanted into injured myocardium; these cells include fetal cardiomyocytes, autologous skeletal myoblasts, smooth muscle cells, immortalized myoblasts, syngeneic skeletal myoblasts, fibroblasts, adult cardiac-derived cells (myofibroblasts), embryonic stem cells, and bone marrow–derived stromal cells. Many of these cell

Address for corrrespondence: Doris A. Taylor, Box 3345, Duke University Medical Center, Durham, NC 27710. Voice: 919-684-4484; fax: 919-684-8907.

dataylor@duke.edu

Ann. N.Y. Acad. Sci. 961: 315–318 (2002). © 2002 New York Academy of Sciences.

types have been transplansplanted successfully into injured myocardium. However, to date only skeletal myoblasts show promise as a clinical tool for myocardial repair.

Autologous Skeletal Myoblasts

By using patient-derived (autologous) skeletal myoblasts, we are able to overcome the major limitations associated with other cell-based therapies. Most notably, by using autologous cells we overcome the shortage of donor tissue, the need for immunosuppression, and the ethical dilemma associated with the use of allogeneic or embryonic cells. By using primary cells rather than immortalized or multipotent stem cells, we decrease the likelihood of tumor formation after cell transplantation. By using myogenic (versus non-myogenic) cells, we can potentially regenerate contractile muscle in previously infarcted heart. Finally, by using skeletal myoblasts rather than cardiocytes, we are able to obtain higher levels of cell engraftment and survival in infarcted regions of the heart. This increase in graft survival is due in part to the fact that the normal capillary density in skeletal muscle is similar to that of infarcted heart. Furthermore, skeletal muscle demonstrates relative resistance to the ischemic conditions found in the relatively avascular scar that results from an acute myocardial infarction.

There have been numerous reports of successful engraftment of autologous skeletal myoblasts into injured myocardium in multiple animal models of cardiac injury. Studies have shown that autologous skeletal myoblasts can differentiate and develop into striated cells within the damaged myocardium. To date, promising effects of skeletal myoblast cellular cardiomyoplasty have included:

- Adequate survival and engraftment of myoblasts in infarcted, necrotic, or toxin-injured heart;
- Survival and engraftment of myoblasts after single or multiple injections of varying numbers of myoblasts;
- Improved myocardial functional performance irrespective of the method used to assess function *in vitro* (dP/dt, force transduction) or *in vivo* (sonomicrometry, echocardiography) ; and
- Augmentation of myocardial performance in all animal species studied: rat, rabbit, dog, pig, sheep.

These results suggest that regeneration of functional muscle after myoblast implantation could offer a valuable treatment option for injured and failing heart. Clinical trials based on these data have begun both in Europe and in the United States.

CHALLENGES

Several challenges remain if we are to successfully repair infarcted or failing myocardium. The first challenge is defining the appropriate temporal window for cell implantation. Can myoblasts be delivered immediately post infarct or in end-stage myocardial dysfunction? The most significant challenges include electrically integrating any transplanted cells or tissues, ensuring long-term survival of these engineered grafts, guaranteeing nutrient delivery and blood supply in an infarcted region of tissue, and finally protecting surrounding myocardium from mechanical remodeling and decompensation secondary to these grafts. A major biologic challenge that remains is understanding the requirements for cells or patches to repair myocardial injury. Is diastolic repair sufficient? Is contractile benefit required?

Finally, an intriguing question is what cell type is best for myocardial repair? Although cardiocytes might seem the ideal target cell, biology predicts that cardiocytes will not survive in infarcted regions of heart. Empiric data support this. Stem cells have received much attention of late as an ideal target for cardiac repair. By their very nature, stem cells respond to their microenvironment and develop a corresponding phenotype. This would suggest that in normal myocardium, stem cells could become cardiocytes. But, it further suggests that, as shown by Chiu and others recently, when injected into infarcted heart, stem cells will develop characteristics of scar. Understanding this phenomenon and beginning to control it is an active area of tissue engineering.

OPPORTUNITIES

Defining the appropriate environment to guide cell development for myocardial repair remains an exciting opportunity. The changes in phenotype that we see secondary to myoblast engraftment in injured heart cannot simply be explained on the basis of electrical, mechanical or "chemical" environmental cues. Likely, it is some superimposition of all three. Defining and engineering an appropriate milieu to mediate cell phenotype, considering 2- and 3-dimensional structures and their impact on cell phenotype, and defining the role of electrical and mechanical cues are all exciting opportunities for this field.

Another tremendous opportunity is the potential to create bioactive patches for myocardial repair. Limiting myocardial dilatation by physical constraints has recently been demonstrated to have a positive impact on function. This technology in fact forms the basis of several start-up companies. However, combining physical meshes or patches with cells and genes is a widely underexplored field that has tremendous opportunity in the next few years.

CONCLUSION

Cell-based therapies for myocardial repair are here. Understanding their mechanism of action, their longevity, and how to improve their electromechanical integration is the next frontier. Creating bioactive patches that can respond to changes in contractility and that can relax in synchrony with the surrounding heart is also a frontier to be explored.

SELECTED REFERENCES

1. AMERICAN HEART ASSOCIATION. 1998. Heart and Stroke 1998: Statistical Update Cardiovascular Diseases. American Heart Association. Houston, TX. [www.amhrt.org/Scientific/HSstats98/03cardio.html].

2. CARPENTIER, A. *et al.* 1993. Dynamic cardiomyoplasty at seven years. J. Thorac. Cardiovasc. Surg. **106**:. 42–52.
3. SILVESTRY, S.C. *et al.* 1995. A model for assessing in vivo myocardial function and cardiac gene therapy in rabbits [abstract]. J. Mol. Cell. Cardiol. **27**: A57.
4. TAYLOR, D.A. *et al.* 1996. Skeletal myoblast therapy in cardiovascular disease. *In* Gene Transfer in Cardiovascular Biology: Experimental Approaches and Therapeutic Implications. K.L. Marsh, Ed.: 355–375. Kluwer Academics. Norwell, MA.
5. TAYLOR, D.A. *et al.* 1997. Delivery of primary autologous skeletal myoblasts into rabbit heart by coronary infusion: a potential approach to myocardial repair. Proc. Assoc. Am. Phys. **109**: 245–253.
6. ATKINS, B.Z. *et al.* 1998. Cryoinduced transmural myocardial injury decreases regional contractile fucntion in rabbits. J. Mol. Cell. Cardiol.
7. ATKINS, B.Z. *et al.* 1999. Reversing post-MI dysfunction: improved myocardial performance after autologous skeletal myoblast transfer to infarcted rabbit heart. Circulation Suppl. **100**: I–838.
8. ATKINS, B.Z. *et al.* 1999. Intracardiac transplantation of skeletal myoblasts yields two populations of striated cells in situ. Ann. Thorac. Surg. **67**: 124–129.
9. TAYLOR, D.A. *et al.* 1998. Regenerating functional myocardium: improved performance after skeletal myoblast transplantation. Nat. Med. **4**: 929–933.
10. LEWIS, C.W. *et al.* 1998. A load-independent in vivo model for evaluating therapeutic interventions in injured myocardium. Am. J. Physiol. **275**:. H1834–1844.
11. HUTCHESON, K.A. *et al.* 2000. Comparing the benefits of cellular cardiomyoplasty with skeletal myoblasts or dermal fibroblasts on myocardial performance. Cell Transpl. **9**: 359–368.
12. CHERWEK, D. *et al.* 1999. Relieving exercise intolerance secondary to heart failure: myoblast-mediated angiogenesis via VEGF delivery to ischemic skeletal muscle. Circulation Suppl. **100**: I-657.
13. ATKINS, B.Z. *et al.* 1999. Cellular cardiomyoplasty improves diastolic properties of injured heart. J. Surg. Res. **85**: 234–242.
14. MARELLI, D. *et al.* 1992. Cell transplantation for myocardial repair: an experimental approach. Cell Transpl. **1**: 383–390.
15. TAYLOR, D.A. & W.E. KRAUS. 1994. Primary rabbit skeletal myoblasts as a tool for gene therapy. In Vitro Cell Dev. Biol. **30A** [Part II: Hot Topics].
16. CHIU, R.C.-J., A. ZIBAITIS & R.L. KAO, 1995. Cellular cardiomyoplasty: myocardial regeneration with satellite cell implantation. Ann. Thorac. Surg. **60**: 12–18.
17. CHACHQUES, J. *et al.* 2000. Cellular therapy reverses myocardial dysfunction. *In* program of the American Association for Thoracic Surgery 80th Annual Meeting.
18. HUEMAN, M. *et al.* 1999. VEGF-mediated angiogenesis improves in vivo regional compliance in chronically injured rabbit myocardium. Circulation Suppl. **100**: I-838.
19. MENASCHE, P. *et al.* 2001. Myoblast transplantation for heart failure. Lancet **357**: 279–280.
20. ELLIS, M., S. RUSSELL & D. TAYLOR. 2001. Translating cell transfer for cardiovascualr disease to the bedside: a pre-clinical review and discussion of potential early trials. Cardio. Vasc. Regen. **3**: in press.
21. LABRECQUE, C., R. ROY & J.P. TREMBLAY. 1992. Immune reactions after myoblast transplantation in mouse muscles. Transplant. Proc. **24**: 2889–2892.
22. ROY, R. *et al.* 1993. Antibody formation after myoblast transplantation in Duchenne-dystrophic patients, donor HLA compatible. Transplant. Proc. **25**: 995–997.
23. TREMBLAY, J.P. *et al.* 1993. Results of a triple blind clinical study of myoblast transplantations without immunosuppressive treatment in young boys with Duchenne muscular dystrophy. Cell Transplant. **2**: 99–112.
24. HUARD, J. *et al.* 1994. Human myoblast transplantation in immunodeficient and immunosuppressed mice: evidence of rejection. Muscle & Nerve **17**: 224–234.

In Vivo Remodeling

Breakout Session Summary

Moderators
STEPHEN F. BADYLAK, *Purdue University*
MARKUS GROMPE, *Oregon Health and Science University*

Panelists
ARNOLD I. CAPLAN, *Case Western Reserve University*
HOWARD P. GREISLER, *Loyola University Medical Center*
ROBERT E. GULDBERG, *Georgia Institute of Technology*
DORIS A. TAYLOR, *Duke University Medical Center*

Rapporteur
STEPHEN F. BADYLAK

BROAD STATEMENT

The implantation of tissue-engineered devices or cells into living hosts is accompanied by unavoidable remodeling of both the implant and the host tissue. Tissue remodeling is a dynamic process that is related to normal developmental biology and is present from the earliest moments of fetal development. In the adult, this process exists as part of the tissue repair response and can result in several possible outcomes including tissue regeneration, scar tissue formation, or complete tissue loss.

VISION

An in-depth understanding of the signals and mechanisms that control this process at the molecular level is necessary to maximize the constructive outcomes and therapeutic applications that can potentially result from *in vivo* remodeling. The patterns of *in vivo* remodeling differ among species, within species, and even as a function of age or body location within the same individual. The remodeling events are fundamentally protective responses against tissue injury or tissue exposure to harmful substances, and thus are attempts to assure survival of both the individual and the species. Efforts to understand the molecular signals that control the remodeling process will have multiple positive payoffs, including an improved understanding of developmental biology, inflammation and immunity, scar tissue formation, tissue and organ regeneration, and wound healing.

Ann. N.Y. Acad. Sci. 961: 319–322 (2002). © 2002 New York Academy of Sciences.

OBJECTIVES

In vivo remodeling is the inevitable fate of all efforts at tissue engineering and reparative medicine. Regardless of whether the therapeutic approach is gene- or cell-based, scaffold-based, or based upon the use of selected growth factors or a combination approach, it will be subjected to the biologic variability that is part of *in vivo* remodeling. All events that precede the *in vivo* step, including the selection of specific cell types, the *in vitro* cell or tissue bioreactor conditions, and the utilization of desired scaffold materials, can generally be tightly controlled. However, the *in vivo* remodeling that occurs after application to the mammalian system has much less predictable results based upon our current knowledge base and current methodologies. Goals of research in this area should include:

(1) Characterization at the molecular level of the differences between wound healing events that occur at different ages; especially the differences between fetal vs. adult wound healing. The rationale for this goal is that the genetic makeup of an individual is constant throughout life, yet tissue response to injury can differ markedly between the fetus and the adult. Fetal wounds can heal with complete regeneration of tissues and organs, whereas adults tend to heal by scar tissue formation or, in selected instances, wound healing can fail completely. What are the specific differences in this wound healing response? What molecular signals control these differences based upon age of the individual? Can these signals be modified by therapeutic intervention? Without a thorough understanding of these issues, it is likely that only incremental advances will be made in this field.

(2) Understanding at the molecular level the differences between wound healing in normal and diseased individuals. The rationale for this goal is that reparative medicine and tissue-engineering efforts are targeted at the repair and restoration of damaged or missing tissues and organs. By definition, therefore, these efforts will be targeted at otherwise healthy individuals with congenital deformities, healthy individuals with traumatic injuries, and diseased individuals with degenerative, inflammatory, or neoplastic conditions. It is important to understand the differences that may exist in the *in vivo* remodeling events for these different conditions. Even if the therapeutic methods that are considered optimal are utilized for all of these different states, it is unlikely that the short- and long-term expectations for success will be the same.

(3) An understanding at the molecular level of the events that control and regulate the formation of scar tissue; especially in the nervous system. The rationale for this goal is that scar tissue formation is a common result of *in vivo* remodeling following tissue injury in the adult. Scar tissue is the bane of surgeons and is associated with loss of tissue or organ function, undesirable adhesions, and less than optimal cosmetic results. Scar tissue is considered to be one of the major barriers to regeneration of central and peripheral nerve tissue. The formation of scar tissue has significantly more detrimental sequelae in the central nervous system than in other body systems such as the integumentary system or musculoskeletal system. An understanding of the controlling signals for scar tissue formation is essential before therapeutic methods can be developed to minimize or eliminate it.

OBSTACLES

Several significant obstacles exist to a more thorough understanding of the events that regulate *in vivo* remodeling in mammalian systems. Identification and enumeration of these obstacles will permit the development of a rational approach to overcoming these limitations and provide for advancement in the field. These obstacles provide targets for future resource deployment. Although the following list is by no means comprehensive, it represents the panel's collective opinion as to the most dominant current rate-limiting issues:

(1) Methods to quantitatively and qualitatively study gene expression in host and donor cells at the single-cell level are poorly developed. Until such methods can be developed, it will be very difficult to make significant progress in our understanding of normal developmental biology events, the mammalian response to tissue injury, transplanted cells, or foreign materials (e.g., scaffolds), or the regulators of cellular proliferation, hypertrophy, differentiation or spatial three-dimensional organization.

(2) Inter-individual biologic variability in higher mammals makes controlled study of the events that regulate *in vivo* remodeling very difficult.The reasons for the differences in *in vivo* remodeling events following the same tissue injury in separate individuals is very poorly understood. Are these differences in the rate of remodeling or the extent of remodeling or the effects of nutrition or age inevitable? Can these differences be controlled? The presence of this biologic variability makes controlled study design difficult and expensive. This variability also raises significant questions with regard to choice of animal models for particular types of tissue injury.

(3) There is a paucity of methods for tracking the fate of implanted tissue engineered genes, cells, scaffolds and composite devices in real time. The ability to monitor in real time the fate of cells or scaffolds, gene expression patterns, and the degradation of implanted tissue-engineered devices is essential to the iterative research process and thus, to progress in the field.

CHALLENGES

A concerted multidisciplinary effort to identify and understand the molecular events that control wound healing, especially the differences in wound healing between the fetus and the adult, is imperative. Such an effort will likely require the combined and coordinated efforts of molecular biologists, developmental biologists, and tissue engineers. Such an effort should be inclusive of individuals not only with specialized expertise in selected basic science fields, but also those individuals with knowledge of the translational efforts necessary to quickly mobilize this knowledge to the tissue-engineering laboratory and the bedside. An understanding, at the molecular level, of those controlling events that result in constructive remodeling of tissues vs. those that result in destructive remodeling or lack of remodeling is necessary.

It is likely that the traditional dividing lines that tend to exist between our understanding of the processes of inflammation, immunity, scar tissue formation, developmental biology, and wound healing will require rethinking and/or elimination. *In vivo* remodeling incorporates all of these events simultaneously. Understanding the similarities and differences in these events will allow the design of rational cell, gene, scaffold or combination approaches as therapeutic methods.

Development of improved methods to track the fate of implanted tissue-engineered devices is needed. These methods should include strategies for monitoring the ultimate fate of implanted cells, degradation of implanted scaffolds, and quantitative expression of specific genes. It is likely that autologous, allogeneic, and xenogeneic genetic material, cells, and scaffold materials will have potential benefit in future strategies for reparative medicine. Furthermore, both synthetic and naturally occurring materials are likely to have roles in optimal therapeutic modalities. The biologic fate of these products is of critical importance in the development of rational treatment methods. Without the tools to track the *in vivo* fate of these products and the function of the target tissue or organ, progress will be severely limited.

And, finally, it is recommended that a registry of human patients that receive tissue-engineered products be developed. This database should include such information as gender, ethnic background, selected genotypic information, age, and state of health/disease. This information can be used to later establish predictors of success and failure for tissue-engineered products.

ACKNOWLEDGMENTS

The *in Vivo* Remodeling Breakout Session was coordinated by Dr. Arlene Chiu (NINDS, NIH) and Dr. Nancy Freeman (NIDCD, NIH).

Immunological Concerns with Bioengineering Approaches

DAVID M. HARLAN,[a] CHRISTOPHER L. KARP,[b] POLLY MATZINGER,[c]
DAVID H. MUNN,[d] RICHARD M. RANSOHOFF,[e] AND DENNIS W. METZGER[f]

[a]National Institute of Diabetes, Digestive and Kidney Diseases, NIH,
Bethesda, Maryland, USA

[b]Children's Hospital Research Foundation, University of Cincinnati,
Cincinnati, Ohio, USA

[c]National Institute of Arthritis and Immune Diseases, NIH, Bethesda, Maryland, USA

[d]Medical College of Georgia, Augusta, Georgia, USA

[e]Lerner Research Institute, Cleveland Clinic Foundation, Cleveland, Ohio, USA

[f]Center for Immunology and Microbial Disease, Albany Medical College,
Albany, New York, USA

KEYWORDS: bioengineered tissue; immune response

INTRODUCTION

Scientists have made remarkable progress toward generation of cells and tissues for therapeutic benefit. Nevertheless, before further advances can be made difficulties associated with the recipient's response against the bioengineered tissue will need to be overcome. While very little has been published on the immune reaction toward bioengineered tissues, the materials will most likely induce immune reactions—for good as well as ill. Both inflammatory and immune reactions can be anticipated, even with autologous tissue. Past experience with allografts and xenografts has demonstrated that innate and adaptive immunity (defined below) represent formidable barriers. Inflammation likewise influences transplant engraftment, but may also be pivotal for proper tissue remodeling. For instance, tendon segments engineered from autologous mesenchymal stem cells, harvested from rabbit bone marrow, and cultured in a way to promote fibroblast differentiation can correct a tendon defect upon transplantation to a surgically constructed patellar tendon window. Surprisingly, however, approximately 25% of the implants develop into bone at the graft site (D.L. Butler, personal communication), although differentiation into bone has not been observed in vitro.[1] While the exact factors causing this bone development in vivo are not known, it stands to reason that local immune and/or inflammatory fac-

Address for correspondence: David M. Harlan, National Institute of Diabetes, Digestive, and Kidney Diseases, Building 10, Room 11S210, 10 Center Drive, Bethesda, MD 20892. Voice: 301-295-2654; fax: 301-295-6484.
davidmh@intra.niddk.nih.gov

Ann. N.Y. Acad. Sci. 961: 323–330 (2002). © 2002 New York Academy of Sciences.

tors stimulated by the surgical placement of the animal's own cells are likely to play a role in such osteoblast formation. Similarly, scientists and clinicians should not assume that bioengineered tissues will be inert. It is likely that the host immune system will influence the implanted bioengineered tissue, and vice versa.

IMMUNE SYSTEM OVERVIEW

All known multicellular organisms have immune systems that serve a defensive function against potential threats. Immunologists typically subdivide the immune system into two broad subcategories: the innate immune system and the adaptive immune system.[2] The innate immune system is the more primitive of the two, and comprises a variety of cellular and humoral regulatory and effector mechanisms, including the complement system, phagocytic cells, NK cells, natural antibodies, γδ T cells, and a variety of antimicrobial peptides. The innate immune system uses a relatively small number of proteins for immune recognition. Encoded in the germline, these receptors recognize molecular patterns found on pathogens, but not present in higher eukaryotic organisms. In nomenclature devised by Janeway and Mezhitov, the structures recognized have been referred to as pathogen-associated molecular patterns (PAMPs).[3] Many are known, including:

- cell membrane or cell wall constituents of bacteria-like lipopolysaccharide (LPS), lipoteichoic acid (LTA), and peptidoglycan;
- diverse carbohydrate signatures (e.g., the mannans of yeast cell walls);
- unmethylated CpG motifs found largely in bacterial, but not mammalian DNA;
- double-stranded RNA (a signature of RNA virus replication);
- and N-formylated peptides (the N-terminal residue of all bacterial [and mitochondrial] proteins is formylmethionine).

The more recently evolved immune system subcategory is the adaptive immune system comprising cells with antigen-specific cell surface receptors that recognize specific antigens as opposed to non-specific molecular patterns. The cells composing the adaptive immune system are the B lymphocytes for specific antibody production and the αβ T lymphocytes for cellular immune responses.

The important questions thus become: How do the innate and adaptive immune systems interact, and how is the immune system regulated such that it responds when appropriate (invasion from some pathogen), or lies dormant when appropriate (so as to avoid autoimmune reactions)? Traditionally, immunologists have held the fundamental belief that the immune system discriminates between self and non-self and that, if the system were perfect, it would attack everything that is non-self and be totally tolerant of anything that is self. Over the years, many immunological findings have not fit with that hypothesis. For example, if each individual's immune system learns "self" at an early age, then why are new antigens that appear at puberty not considered "foreign" and destroyed? How can normal individuals harbor both T and B cells capable of reacting to self-antigens like DNA, keratin, and myelin basic protein, yet not have destructive autoimmunity? Why are liver transplants rejected less vigorously than hearts? Why is a newly lactating breast not rejected when it begins to make new proteins? Why is the immune system so bad at dealing with tumors, even when transformed cells demonstrably express new, "non-self" antigens? These questions led Matzinger and her colleagues to propose an alternative "danger model" hypothesis to explain immune system activation.[4,5] The model starts with the idea that the immune system defines "danger" as anything that causes tissue stress or destruction. According to this model, antigen-presenting cells are activated by alarm signals from stressed or damaged tissues. Without this activation, no primary immune response can occur. As the putative danger signals are incompletely understood, and if the danger model is close to reality, then it stands to reason that cellular manipulations integral to the creation of a bioengineered product could result in an immune activation cascade.

Innate immunity has been viewed as "a poor cousin" of the adaptive immune system, serving to control infection until the adaptive immune system has had time to mobilize. This is clearly an oversimplification, and incomplete, if not incorrect. The enormous plasticity of the adaptive immune system lacks an essential feature. That important characteristic is that the somatically mutated, clonally distributed adaptive immune system antigen receptors, while theoretically able to recognize any molecular structure, are unable to distinguish self from non-self, or a dangerous, previously unseen antigen from an innocuous, previously unseen one. These discriminative functions are largely provided by the phylogenetically more primitive, "hardwired" modes of pathogen recognition specified by the innate immune system.

The receptors that recognize PAMPs have been named pattern-recognition receptors (PRRs). Such receptors are located both on cell surfaces and in plasma. While many such PRRs had previously been characterized, a revolution in the understanding of innate immune activation recently occurred with recognition that vertebrate homologues of the *Drosophila* Toll protein, the TLR (Toll-like receptor) family, mediate signal transduction in response to a variety of microbial PAMPs including, paradigmatically, lipopolysaccharide (LPS) [reviewed in Aderem and Ulevitch[6]]. Defined downstream signaling involves (at least) both NF-κB and MAP kinase pathways. Genes activated as a result of such signaling include those encoding a plethora of inflammatory mediators and regulatory molecules (e.g., cytokines, chemokines, adhesion molecules, acute phase proteins, and co-stimulatory molecules required for efficient activation of T cells by antigen-presenting cells). Interestingly, in line with the danger model of Matzinger, endogenous molecules associated with cellular stress, damage, or destruction, such as the breakdown products of hyaluron, heat-shock proteins, and interferons (IFNs) can also be recognized. In some cases (such as heat-shock protein and fibronectin) the receptors recognizing these endogenous products are the same as those that bind to PAMPs, such as TLR4 and TLR2. In others (such as the breakdown products of hyaluron), the receptors are not yet known.

Another well-studied aspect of normal immune system function warrants mention. Many tissues (e.g., the central nervous system) are relatively devoid of immune system cells unless the brain generates signals to call immune system cells into action. During brain homeostasis, immune system cells tend to follow a well-regulated circulatory pattern from the blood stream to the lymphatic system, and back again until a perturbation or disturbance of homeostasis in some tissue of the body sounds the trumpet to generate a more robust immune cell infiltration of parenchymal tissues.[10,11] Several chemical signals can generate the siren call for immune cells.

These signals include certain complement components, certain cytokines that act to upregulate cell adhesion molecules, and even some PAMPs. Most predominantly, however, the class of chemoattractant peptides called chemokines[12] has been found to play an important role. Chemokines selectively signal subpopulations of leukocytes via G-protein–coupled receptors (GPCR). Studies of central nervous system (CNS) inflammation have implicated the monocyte chemoattractant called CCL2/MCP-1 as playing an early and important role. That is, CCL2 levels increase in CNS tissues after local trauma and before hematogenous cells are found within that tissue. Further, using a variety of different stimuli, parenchymal cells were found competent to express chemokines within a few hours of injury.[13–15] In mice lacking CCL2, recruitment of hematogenous cells to the injured nervous system was impaired. Thus, tissues themselves appear capable of generating signals to the immune system (both innate and acquired), indicating danger or distress.

How might this knowledge, incomplete as it is, be manipulated toward the goal of developing therapies to foster a salutary interaction between the immune system and a bioengineered tissue? The innate immune system discussed above contains many cells, called "antigen-presenting cells" (e.g., macrophages and dendritic cells), which present foreign antigens derived from graft-associated proteins to T lymphocytes. Antigen-presenting cells are thus in a unique position to regulate the process of rejection or tolerance. In the "danger model" epistemology, it may well be that the antigen-presenting cell serves a useful integrative function in determining the context of the antigen presentation. That is, antigen-presenting cells are always presenting antigens, both self and foreign. The context of that presentation, however, must determine how the T lymphocytes will interpret the antigens being presented.

Traditionally, macrophages and dendritic cells have been viewed as "accessory" cells, functioning solely to assist T-cell activation. Recently, however, it has become evident that antigen-presenting cells can also play an important negative regulatory role, by inhibiting T-cell responses and promoting tolerance. This function has been reported both for macrophages[16] and for certain subsets of dendritic cells.[17–20] The exact mechanism of the tolerance so induced is incompletely understood, but an important feature of the tolerance induced by immunoregulatory dendritic cells is that it appears capable of propagating. That is, starting from a small number of dendritic cells, tolerance can be induced throughout the host immune system via a phenomenon sometimes called "infectious tolerance."[21] Thus, perhaps not all host T cells need be exposed individually to the tolerogenic antigen-presenting cells. Tolerance can even be induced when the same antigen is simultaneously presented by other, non-suppressive dendritic cells,[22] and tolerogenic dendritic cells appear capable of reversing even established immunity.[23] Taken together, these properties suggest that the initial presentation of antigen may create a secondary population of immunoregulatory T cells, which then spread antigen-specific tolerance to other T cells.[24]

The specific molecular mechanism by which dendritic cells induce tolerance is almost certainly multifactoral. The cells may present antigen without the requisite co-stimulation (e.g., due to low expression of B7-family molecules and/or interference with the CD40/CD154 receptor interactions [see Ref. 25 for a review]). In fact, long-term allograft acceptance has been reported now in a variety of non-human primate allotransplant models through this mechanism.[26–33] Others have suggested mechanisms dependent upon immunomodulatory cytokines or contact factors. As an additional potential mechanism, certain macrophages and dendritic cells have been shown to suppress T-cell activation *in vitro* via expression of the enzyme indoleamine 2,3-dioxygenase, or IDO.[34] IDO catalyzes the breakdown of tryptophan along the kynurenine pathway (reviewed in Ref. 35), and is induced by pro-inflammatory signals such as IFNγ. *In vitro*, expression of IDO by macrophages rapidly depletes tryptophan from the culture medium, which prevents T-cell proliferation and acquisition of effector function. Mellor and Munn have proposed that a similar sequence of events occurs *in vivo*, with tryptophan catabolism by IDO-expressing antigen-presenting cells acting to lower local interstitial tryptophan concentrations below the threshold required for T-cell activation. In support of that hypothesis, experiments using a pharmacologic inhibitor of IDO (1-methyl-tryptophan) administered to mice have shown that functional IDO activity is required in order to prevent maternal $CD8^+$ T cells from rejecting an antigenically "foreign" fetus during gestation,[36] and to prevent excessive activation of maternal complement, which would otherwise be deposited in a T-cell-driven hemorrhagic vasculitis in the placenta.[37] In humans, IDO is expressed widely in immunoregulatory antigen-presenting cells throughout the immune system. Relatively large numbers of IDO-expressing antigen-presenting cells are also found in infiltrating lymph nodes that drain certain malignancies, and in HIV-infected lymphoid tissue. Transfection of the IDO gene that is under control of a strong constitutive promoter into a variety of murine cell lines results in suppression of allogeneic T-cell responses *in vitro*. Thus, IDO is a physiologic regulator of T-cell responses and may prove useful in the engineering of immunosuppressive cell lines.

Coincident with research exploring the immune system response to bioengineered organs, other groups are exploring promising approaches to the use of acellular bioengineered materials for tissue remodeling. Extracellular matrix (ECM) has been used successfully in many species as an avascular, resorbable bioscaffold for repair of cardiovascular tissue,[38] abdominal wall defects, urinary bladder defects[39] and ligament and tendon damage.[40] Even when used as xenografts, ECMs induce site-specific tissue remodeling without evidence of immunological rejection.

Porcine small intestinal submucosa (SIS) is a naturally derived ECM that is being utilized for tissue remodeling and repair in numerous species, including humans. To determine whether SIS induces an immune response and, if so, the nature of that response, mice have been implanted subcutaneously with SIS and the graft sites analyzed histologically for rejection and for cytokine expression. SIS implantation was found to cause an acute inflammatory response, followed by tissue remodeling. Seven days after implantation, IL-4 and IL-10, but not IFN-γ, were expressed at the graft site.[41] In addition, the implanted mice produced anti-SIS antibodies that were restricted to the IgG1 isotype. Reimplantation of SIS into mice led to a secondary anti-SIS antibody response, yet it was still restricted to IgG1. The observed immune responses were T-cell-dependent since T-cell "knock-out" (KO) mice expressed neither IL-4 at the implant site nor serum anti-SIS antibodies. Nevertheless, all SIS grafts were accepted in the absence of T cells. These initial results suggest that while porcine ECM is immunogenic, the resulting immune response is restricted to the Th2 pathway, consistent with acceptance rather than rejection. Additional experiments were performed using mice treated with IL-12 at the time of implantation, STAT6 KO, or IL-10 KO mice to determine whether induced Th1 anti-SIS immune responses would lead to tissue rejection. While all these mice mounted SIS-specific Th1 responses, in no case was rejection observed. Thus, acceptance and remodeling of xenogeneic ECM is not dependent upon lack of Th1-like immunity.

SIS is known to contain the galactose α1,3 galactose (αgal) epitope that normally represents a major barrier to xenotransplantation in humans due to the presence of natural anti-αgal antibody.[41,42] In order to understand the potential consequences of αgal expression on transplanted SIS, mice were utilized that contained a genetic disruption in the α1,3 galactosyltransferase gene. These mice respond to immunization with the galactose α1,3 galactose (αgal) epitope by producing anti-αgal antibody. Nevertheless, while these mice mounted a more robust inflammatory response to SIS, they ultimately accepted the graft. These results demonstrate that naturally occurring anti-αgal antibodies do not influence the ability of xenogeneic ECMs to serve as bioscaffolds for tissue remodeling.

To date, nearly all bioengineered tissues used therapeutically have been either avascular tissues like heart valves,[43,44] or relatively simple tissues like corneal grafts.[45] It is obvious that for years, surgeons have been able to transplant complex organs such as the kidney, liver, heart, lung, and pancreas with ever increasing therapeutic benefit, but that organ supply prevents many from benefiting. Scientists hope to be able some day to grow such organs in the laboratory, but it is evident that success will be achieved with simpler tissues first. The recently reported success achieved with islet cell transplantation in patients with type 1 diabetes mellitus,[46] coupled with reports suggesting that islets might be propagated *in vitro*,[47] or be induced to differentiate from stem cells,[48] has suggested that it may be possible soon to generate a slightly more complex mini-organ like the pancreatic islet for more widespread therapeutic benefit.

REFERENCES

1. AWAD, H.A. *et al.* 2000. In vitro characterization of mesenchymal stem cell-seeded collagen scaffolds for tendon repair: effects of initial seeding density on contraction kinetics. J. Biomed. Mater. Res. **51:** 233–240.
2. HARLAN, D.M. & A.D. KIRK. 1999. The future of organ and tissue transplantation: can T-cell costimulatory pathway modifiers revolutionize the prevention of graft rejection? JAMA **282:** 1076–1082.
3. MEDZHITOV, R. & C. JANEWAY, JR. 2000. Innate immune recognition: mechanisms and pathways. Immunol. Rev. **173:** 89–97.
4. MATZINGER, P. 1998. An innate sense of danger. Semin. Immunol. **10:** 399–415.
5. MATZINGER, P. 1994. Tolerance, danger, and the extended family. Annu. Rev. Immunol. **12:** 991–1045.
6. ADEREM, A. & R.J. ULEVITCH. 2000. Toll-like receptors in the induction of the innate immune response. Nature **406:** 782–787.
7. OHASHI, K. *et al.* 2000. Cutting edge: heat shock protein 60 is a putative endogenous ligand of the toll-like receptor-4 complex. J. Immunol. **164:** 558–561.
8. OKAMURA, Y. *et al.* 2001. The extra domain A of fibronectin activates Toll-like receptor 4. J. Biol. Chem. **276:** 10229–10233.
9. VABULAS, R.M. *et al.* 2001. Endocytosed heat shock protein 60s use TLR2 and TLR4 to activate the toll/interleukin-1 receptor signaling pathway in innate immune cells. J. Biol. Chem. **276:** 31332–31339.
10. TEDLA, N. *et al.* 1998. Regulation of T lymphocyte trafficking into lymph nodes during an immune response by the chemokines macrophage inflammatory protein (MIP)-1 alpha and MIP-1 beta. J. Immunol. **161:** 5663–5672.
11. TSOKOS, G.C. & S.N. LIOSSIS 1998. Lymphocytes, cytokines, inflammation, and immune trafficking. Curr. Opin. Rheumatol. **10:** 417–425.

12. YOSHIE, O. *et al.* 2001. Chemokines in immunity. Adv. Immunol. **78:** 57–110.
13. GLABINSKI, A.R. *et al.* 1996. Chemokine monocyte chemoattractant protein-1 is expressed by astrocytes after mechanical injury to the brain. J. Immunol. **156:** 4363–4368.
14. McTIGUE, D.M. *et al.* 1998. Selective chemokine mRNA accumulation in the rat spinal cord after contusion injury. J. Neurosci. Res. **53:** 368–376.
15. SCHREIBER, R.C. *et al.* 2001. Monocyte chemoattractant protein (MCP)-1 is rapidly expressed by sympathetic ganglion neurons following axonal injury. Neuroreport **12:** 601–606.
16. ATTWOOD, J.T. & D.H. MUNN. 1999. Macrophage suppression of T cell activation: a potential mechanism of peripheral tolerance. Int. Rev. Immunol. **18:** 515–525.
17. BANCHEREAU, J. & R.M. STEINMAN. 1998. Dendritic cells and the control of immunity. Nature **392:** 245–252.
18. FAIRCHILD, P.J. & H. WALDMANN. 2000. Dendritic cells and prospects for transplantation tolerance. Curr. Opin. Immunol. **12:** 528–535.
19. FEARON, D.T. & R.M. LOCKSLEY. 1996. The instructive role of innate immunity in the acquired immune response. Science **272:** 50–53.
20. LECHLER, R. *et al.* 2001. Dendritic cells in transplantation: friend or foe? Immunity **14:** 357–368.
21. QIN, S. *et al.* 1993. "Infectious" transplantation tolerance. Science **259:** 974–977.
22. GROHMANN, U. *et al.* 2001. CD40 ligation ablates the tolerogenic potential of lymphoid dendritic cells. J. Immunol. **166:** 277–283.
23. DHODAPKAR, M.V. *et al.* 2001. Antigen-specific inhibition of effector T cell function in humans after injection of immature dendritic cells. J. Exp. Med. **193:** 233–238.
24. RONCAROLO, M.G. *et al.* 2001. Differentiation of T regulatory cells by immature dendritic cells. J. Exp. Med. **193:** F5–F9.
25. HARLAN, D.M. & A.D. KIRK. 2000 The future of organ and tissue transplantation: can T-cell costimulatory pathway modifiers revolutionize the prevention of graft rejection? JAMA **282:** 1076–1082.
26. BLUESTONE, J.A. 1995. New perspectives of CD28-B7-mediated T cell costimulation. Immunity **2:** 555–559.
27. HARLAN, D.M. & A.D. KIRK. 1998. Anti-CD154 and the prevention of graft rejection. Graft **1:** 60–70.
28. KENYON, N.S. *et al.* 1999. Long-term survival and function of intrahepatic islet allografts in baboons treated with humanized anti-CD154. Diabetes **48:** 1473–1481.
29. KENYON, N.S. *et al.* 1999. Long-term survival and function of intrahepatic islet allografts in rhesus monkeys treated with humanized anti-CD154. Proc. Natl. Acad. Sci. USA **96:** 8132–8137.
30. KIRK, A.D. *et al.* 1997 CTLA4-Ig and anti-CD40 ligand prevent renal allograft rejection in primates. Proc. Natl. Acad. Sci. USA **94:** 8789–8794.
31. KIRK, A.D. *et al.* 1999. Humanized anti-CD154 monoclonal antibody treatment prevents acute renal allograft rejection in non-human primates. Nat. Med. **5:** 686–693.
32. KIRK, A.D. *et al.* 2001. Induction therapy with monoclonal antibodies specific for CD80 and CD86 delays the onset of acute renal allograft rejection in non-human primates. Transplantation **72:** 377–384.
33. LEVISETTI, M.G. *et al.* 1997. Immunosuppressive effects of human CTLA4-Ig in a non-human primate model of allogeneic pancreatic islet transplantation. J. Immunol. **159:** 5187–5191.
34. MUNN, D.H. *et al.* 1999. Inhibition of T cell proliferation by macrophage tryptophan catabolism. J. Exp. Med. **189:** 1363–1372.
35. MELLOR, A.L. & D.H. MUNN. 1999. Tryptophan catabolism and T-cell tolerance: immunosuppression by starvation? Immunol. Today **20:** 469–473.
36. MUNN, D.H. *et al.* 1998. Prevention of allogeneic fetal rejection by tryptophan catabolism. Science **281:** 1191–1193.
37. MELLOR, A.L. *et al.* 2001. Prevention of T cell-driven complement activation and inflammation by tryptophan catabolism during pregnancy. Nat. Immunol. **2:** 64–68.
38. LINDBERG, K. & S.F. BADYLAK. 2001. Porcine small intestinal submucosa (SIS): a bioscaffold supporting in vitro primary human epidermal cell differentiation and synthesis of basement membrane proteins. Burns **27:** 254–266.

39. BADYLAK, S.F. *et al.* 1998. Small intestinal submucosa: a rapidly resorbed bioscaffold for augmentation cystoplasty in a dog model. Tissue Eng. **4:** 379–387.
40. SUCKOW, M.A. *et al.* 1999. Enhanced bone regeneration using porcine small intestinal submucosa. J. Invest. Surg. **12:** 277–287.
41. ALLMAN, A.J. *et al.* 2001. Xenogeneic extracellular matrix grafts elicit a TH2-restricted immune response. Transplantation **71:** 1631–1640.
42. MCPHERSON, T.B. *et al.* 2000. Galalpha(1,3)Gal epitope in porcine small intestinal submucosa. Tissue Eng. **6:** 233–239.
43. CARPENTIER, S.M. *et al.* 2001. Biochemical properties of heat-treated valvular bioprostheses. Ann. Thorac. Surg. **7(5 Suppl):** S410–S412.
44. EDMUNDS, L.H., JR. *et al.* 1997. Directions for improvement of substitute heart valves: National Heart, Lung, and Blood Institute's Working Group report on heart valves. J. Biomed. Mater. Res. **38:** 263–266.
45. TSAI, R.J. *et al.* 2000. Reconstruction of damaged corneas by transplantation of autologous limbal epithelial cells. N. Engl. J. Med. **343:** 86–93.
46. SHAPIRO, A.M. *et al.* 2000. Islet transplantation in seven patients with type 1 diabetes mellitus using a glucocorticoid-free immunosuppressive regimen. N. Engl. J. Med. **343:** 230–238.
47. BONNER-WEIR, S. *et al.* 2000. *In vitro* cultivation of human islets from expanded ductal tissue. Proc. Natl. Acad. Sci. USA **97:** 7999–8004.
48. LUMELSKY, N. *et al.* 2001. Differentiation of embryonic stem cells to insulin-secreting structures similar to pancreatic islets. Science **292:** 1389–1394.

Islet Cell Allotransplantation as a Model System for a Bioengineering Approach to Reparative Medicine

Immunological Concerns

DAVID M. HARLAN

National Institute of Diabetes, Digestive, and Kidney Diseases,
Bethesda, Maryland 20892, USA

KEYWORDS: islet cell allotransplantation; type 1 diabetes

Cellular transplantation therapy for insulin-deficient diabetes mellitus represents an interesting model system for a bioengineering approach to reparative medicine. Type 1 diabetes mellitus (T1DM) is a common and costly illness, with an estimated 1 million Americans diagnosed, and the disease accounts for at least 8% of annual health care expenditures. Recently, and following the lead established by investigators in Edmonton Canada,[1,2] several institutions have succeeded in at least temporarily restoring patients with long-standing T1DM to a state of insulin-independence and normal blood sugars. Remarkably, the health-restoring therapy requires the transplantation of only a small tissue volume, typically less than 5.0 milliliters of a tissue called "islets of Langerhans." and the transplant procedure is relatively non-invasive as the cells are infused via a percutaneously placed intraportal catheter. Two problems, however, stand in the way of this becoming a standard therapeutic procedure: an inadequate islet supply and immune-mediated destruction once the health-restoring islets are transplanted.

One goal for investigators interested in developing clinical islet cell transplantation is therefore to generate a renewable islet supply. The pancreatic islets of Langerhans are small cell clusters that represent a minority (less than 5%) of the organ's mass. Currently, pancreatic islets for transplant can only be obtained from brain-dead human donors, and in the U.S. only about 6,000 such donors are identified each year. In an effort to overcome this limitation, and recognizing that for 50 years individuals with T1DM were treated with insulin isolated from pigs or cows, several groups are isolating pig islets and testing those cells in non-human primate xenotransplant experiments. Still others, recognizing that islets are small cell clusters varying in size from about 75 microns to more than 500 microns in diameter, are working to "grow" these cells *in vitro*. While work to grow bioengineered tissues

Address for correspondence: David M. Harlan, National Institute of Diabetes, Digestive, and Kidney Diseases, NIH, Building 10, Room 11S210, 10 Center Drive, Bethesda, MD 20892. Voice: 301-295-2654; fax: 301-295-6484.

davidmh@intra.niddk.nih.gov

Ann. N.Y. Acad. Sci. 961: 331–334 (2002). © 2002 New York Academy of Sciences.

and/or organs is making great progress, it stands to reason that success in fostering the generation of relatively small and simple tissues like islets will pave the way for more complex organs/tissues. Considerable excitement surrounded a recent report suggesting that mouse embryonic stem cells could be induced to develop *in vitro* into cell clusters remarkably similar to pancreatic islets.[3]

Another hurdle that must be cleared, however, is the immune barrier that otherwise destroys islets once they are transplanted. It has been known for years that individuals who must have pancreatic excision for medical reasons could have their own islets isolated from the recently excised pancreas, and that those islets could be reinfused back into their liver (via the portal vein) and there monitor blood sugar and secrete insulin so as to maintain glucose homeostasis. Unfortunately, the same technique did not work reliably when allogeneic islets were transplanted.[4] Particularly troublesome is the fact that T1DM is an autoimmune illness that develops in most cases because the individual's immune system has specifically targeted the endogenous pancreatic islet's insulin-producing cells (beta cells).[5] Thus, when islets are transplanted into an individual with T1DM, two general types of immune response must be considered and prevented. One, the immune response generated against the allogeneic (or in the case of animal islets, xenogeneic) tissue, and a recrudescence of the autoimmune response that caused the disease to begin with. While the literature is unclear, evidence for both kinds of immune response can be found.[6,7]

Currently, isolated islets are transplanted in the clinical setting via a catheter percutaneously placed into the recipient's portal vein to the liver.[8] Some data suggest that the liver may play a somewhat unique role in promoting allograft acceptance, in that one group has been able to generate skin allograft acceptance in rodents solely by injecting donor tissues into the recipient's portal vein.[9] Further, such islet placement results in islets secreting insulin into the portal circulation, which mimics the normal secretion pattern. Nevertheless, there are several caveats to this approach:

(1) In the native pancreas, although β cells account for less than 5% of the cell mass, they consume about 20% of the arterial blood flow.[10] The oxygen tension within the portal system is lower than that of arterial blood and this could negatively influence islet function.[11]

(2) As portal blood has passed through the intestinal capillary bed, it contains various toxins and high fat concentrations. These materials could have a negative impact on islet function.

With these hurdles and questions in mind, we have initiated studies to evaluate islet allograft therapy in a non-human primate model. Our initial aim was to establish a working non-human primate model using the Edmonton islet transplant protocol and to explore new sites for transplanting islets.

METHODS

Rhesus macaques underwent total pancreatectomy to induce diabetes and were transplanted with islets isolated from a different rhesus donor. Islets were infused into either the portal vein or the arterial celiac tree. In an effort to prevent islet rejection, primates were given daclizumab, FK506, and rapamycin. Islet function was assessed by daily glucose measurements and by periodic arginine-stimulated c-peptide levels.

RESULTS

We transplanted pancreatic islets into 9 rhesus monkeys: 6 infusions were given into the portal vein (one monkey was transplanted twice) and four infusions were given into the celiac artery. In 5 of the 6 portal vein infusion experiments, the animals achieved at least temporary (5 days or more) normoglycemia without requiring exogenous insulin. We found it difficult to maintain therapeutic immunosuppressive agent blood levels and succeeded in only two of the six animals. For those two animals, long-term insulin independence was achieved. In contrast, none of the animals given islets into their liver arterial supply achieved insulin independence. The primate model is expensive, associated with high morbidity, and is limited by the complicated drug regimen. The model does, however, answer critical questions that could not otherwise be addressed. We have evidence that the islets revascularize shortly after transplant, limited evidence for some islet cell mitosis, and have data suggesting that continuous immunosuppression is required to prevent rejection.

CONCLUSIONS

We have established a preclinical islet transplantation protocol at the National Institutes of Health. The Edmonton Protocol is feasible in primates, and our initial results indicate that the traditional portal vein infusion site is superior to arterial infusion. The model is expensive ($50,000 per primate per year) and associated with morbidity secondary to total pancreatectomy and drug toxicity.

ACKNOWLEDGMENTS

The following have contributed to the work presented here: B. Hirshberg, S. Montgomery, J. Gaglia, N. Patterson, A. Kirk, and D.M. Harlan of the NIDDK/Navy Transplantation and Autoimmunity Branch; J. Lee and K. Hines of the Clinical Center Department of Tansfusion Medicine; M. Wysoki and E.J. Read of the Radiology Department, Yale Medical Center, New Haven; and R. Chang of the Radiology Department Clinical Center, NIH, Bethesda, MD.

REFERENCES

1. SHAPIRO, A.M. et al. 2000. Islet transplantation in seven patients with type 1 diabetes mellitus using a glucocorticoid-free immunosuppressive regimen. N. Engl. J. Med. **343:** 230–238.
2. RYAN, E.A. et al. 2001. Clinical outcomes and insulin secretion after islet transplantation with the Edmonton protocol. Diabetes **50:** 710–719.
3. LUMELSKY, N. et al. 2001. Differentiation of embryonic stem cells to insulin-secreting structures similar to pancreatic islets. Science **292:** 1389–1394.

4. INTERNATIONAL ISLET TRANSPLANT REGISTRY. 1999. Report. University of Giessen, Germany.
5. ATKINSON, M.A. & M.K. MACLAREN. 1994. The pathogenesis of insulin-dependent diabetes mellitus. N. Engl. J. Med. **331:** 1428–1436.
6. OLACK, B.J. *et al.* 1997. Sensitization of HLA antigens in islet recipients with failing transplants. Transplant. Proc. **29:** 2268–2269.
7. ROEP, B.O. *et al.* 1999. Auto- and alloimmune reactivity to human islet allografts transplanted into type 1 diabetic patients. Diabetes **48:** 484–490.
8. BERNEY, T. & C. RICORDI. 1999. Islet transplantation. Cell Transplant. **8:** 461–464.
9. MORITA, H. *et al.* 1998. A strategy for organ allografts without using immunosuppressants or irradiation. Proc. Natl. Acad. Sci. USA **95:** 6947–6952.
10. JANSSON, L. 1994. The regulation of pancreatic islet blood flow. Diabetes Metab. Rev. **10:** 407–416.
11. CARLSSON, P.O., P. LISS, A. ANDERSSON & L. JANSSON. 1998. Measurements of oxygen tension in native and transplanted rat pancreatic islets. Diabetes **47:** 1027–1032.

Immune Responses to Tissue-Engineered Extracellular Matrix Used as a Bioscaffold

DENNIS W. METZGER

Albany Medical College, Albany, New York 12208, USA

KEYWORDS: extracellular matrix biomaterials; immune response

Transplantation of xenogeneic and allogeneic organs is hampered by rejection mechanisms that involve antibody-mediated vascular damage, inflammatory cytokines, and activated macrophages, natural killer cells, and T cells. Nevertheless, extracellular matrix (ECM) biomaterials have been used successfully in many species as avascular, resorbable bioscaffolds for repair of cardiovascular tissue, abdominal wall defects, urinary bladder defects, and ligament and tendon damage. When used as xenografts, ECMs appear to induce site-specific tissue remodeling without any evidence of immunological rejection.

Porcine small intestinal submucosa (SIS) is an acellular, naturally derived ECM that is being utilized for tissue remodeling and repair in numerous species, including humans. To determine whether SIS induces an immune response and, if so, the nature of that response, mice were implanted subcutaneously with SIS and the graft sites were analyzed histologically for rejection and for cytokine expression. It was found that SIS implantation caused an acute inflammatory response, followed by tissue remodeling. Seven days after implantation, IL-4 and IL-10, but not IFN-γ, were expressed at the graft site. In addition, the implanted mice produced anti-SIS antibodies that were restricted to the IgG1 isotype. Re-implantation of SIS into mice led to a secondary anti-SIS antibody response that was still restricted to IgG1. The observed immune responses were T cell–dependent since T cell KO mice expressed neither IL-4 at the implant site nor serum anti-SIS antibodies. However, SIS grafts were still accepted in the absence of T cells. These initial results suggested that porcine ECM is, in fact, immunogenic, but the resulting immune response is restricted to the Th2 pathway, consistent with acceptance rather than rejection.

Additional experiments were performed to determine whether induced Th1 anti-SIS immune responses would lead to tissue rejection. For this purpose, a number of models were employed—mice treated with IL-12 at the time of implantation, mice deliberately immunized with SIS extract in Freund's adjuvant, STAT6 knockout mice, and IL-10 knockout mice—all of which mounted SIS-specific Th1 responses

Address for correspondence: Dennis W. Metzger, Albany Medical College, 47 New Scotland Avenue, Albany, NY 12208. Voice: 518-262-6750; fax: 518-262-6053.
metzged@mail.amc.edu

Ann. N.Y. Acad. Sci. 961: 335–336 (2002). © 2002 New York Academy of Sciences.

upon implantation. However, in no case was rejection observed. Thus, acceptance and remodeling of xenogeneic ECM is not dependent upon lack of Th1-like immunity.

It is known that SIS contains the galactose $\alpha1,3$ galactose (αgal) epitope, which is normally a major barrier to xenotransplantation in humans, due to the presence of natural anti-αgal antibody. In order to understand the potential consequences of αgal expression on SIS, mice with a genetic disruption in the $\alpha1,3$ galactosyltransferase gene, which results in production of anti-αgal antibody due to lack of αgal epitope expression, were used. It was found that these mice mounted a more robust inflammatory response to SIS, but ultimately accepted the graft. These results demonstrate that naturally occurring anti-αgal antibodies do not influence the ability of xenogeneic ECMs to serve as bioscaffolds for tissue remodeling.

Our recent experiments have investigated the components within SIS that are responsible for its apparent anti-rejection effects. Western blotting showed that porcine IgA is a major component of SIS. Furthermore, we have found that IgA-immunodeficient mice overexpress a variety of macrophage inflammatory molecules, including IL-1, TNF-α, and nitric oxide. Thus, IgA within SIS may act to inhibit macrophage cytokine production and downregulate inflammation, a possibility that is currently being investigated.

ACKNOWLEDGMENTS

This work was supported by NIH grant HL60359. The work presented is the result of collaboration between me and Stephen F. Badylak of the Department of Biomedical Engineering, Purdue University, West Lafayette, Indiana.

REFERENCE

1. ALLMAN, A.J. *et al.* 2001. Xenogeneic extracellular matrix grafts elicit a Th2-restricted immune response. Transplantation **71:** 1631–1640.

Immunological Barriers (Opportunities?) to the Use of Bioengineered Tissue

Taking Note of the Innate Immune System

CHRISTOPHER L. KARP,[a] RICHARD WENSTRUP,[a] AND DAVID L. BUTLER[b]

[a]Children's Hospital Research Foundation and [b]Department of Biomedical Engineering, University of Cincinnati, Cincinnati, Ohio USA

The promise of sustained replacement of damaged or defective body parts has considerable allure. Initial approaches in the lab and clinic have focused on allo- and xenotransplantation, approaches with formidable immunological barriers (and, in the case of xenotransplantation, potential zoonotic infectious barriers[1]) to success. With xenotransplantation, there is rapid and broad activation of both innate and adaptive immune systems.[2] This is also true with allotransplantation,[3] although more attention has traditionally been paid to adaptive immunity in studies of allotransplantation. It may reasonably be thought that implantation of bioengineered autologous cells or tissue will avoid all such immunological barriers, and indeed all notice by the implanted host's immune system. Actually, very little has been published on the immunology of bioengineered tissue. And there is good reason to believe that bioengineered autologous tissue may well be noticed by the innate immune system—for good as well as ill.

The innate immune system comprises a variety of cellular and humoral regulatory and effector mechanisms, including the complement system, phagocytic cells, NK cells, natural antibodies, γ/δ T cells, and a variety of antimicrobial peptides. Innate immunity has, for years, been thought of as a sort of poor cousin to the adaptive immune system—controlling infection until the real (i.e., adaptive) immune system has had time to mobilize. This is clearly false. The enormous plasticity of the adaptive immune system lacks an essential feature: the somatically mutated, clonally distributed receptors of adaptive immunity, while theoretically able to recognize any molecular structure, are unable to distinguish self from non-self, danger from a lack of danger. It has become clear recently that these discriminative functions are largely provided by the phylogenetically prior, hardwired modes of pathogen recognition specified by the innate immune system.

The innate immune system uses a relatively small number of proteins for immune recognition. Encoded in the germline, these receptors recognize molecular patterns found on pathogens but not present in higher eukaryotic organisms. In nomenclature

Address for correspondence: Christopher Karp, M.D., Director, Molecular Immunology Section, Children's Hospital Research Foundation, University of Cincinnati, TCHRF 1566, 3333 Burnet Avenue, Cincinnati, OH 45229-3039. Voice: 513-636-7608; fax: 513-636-5355.

chris.karp@chmcc.org

Ann. N.Y. Acad. Sci. 961: 337–340 (2002). © 2002 New York Academy of Sciences.

devised by Janeway and Medzhitov, the structures recognized have been referred to as pathogen-associated molecular patterns (PAMPs).[4] Many are known, including lipopolysaccharide (LPS), lipoteichoic acid (LTA), and peptidoglycan (all cell membrane or cell wall constituents of bacteria); diverse carbohydrate signatures (e.g., the mannans of yeast cell walls); the unmethylated CpG motif found largely in bacterial not mammalian DNA; double-stranded RNA (a signature of RNA virus replication); and N-formylated peptides (the N-terminal residue of all bacterial [and mitochondrial] proteins being a formylated methionine). The receptors that recognize these (and other) PAMPs have been named pattern-recognition receptors (PRRs). Such receptors are located both on cell surfaces and in plasma.

While many such PRRs had previously been characterized, a revolution in the understanding of innate immune activation has recently occurred with recognition of the fact that vertebrate homologues of the *Drosophila* Toll protein, the TLR (Toll-like receptor) family, mediate signal transduction in response to a variety of microbial PAMPs including, paradigmatically, lipopolysaccharide (LPS) [reviewed in Ref. 5]. Defined downstream signaling involves (at least) both NF-κB and MAP kinase pathways. Genes activated as a result of such signaling include those encoding a plethora of inflammatory mediators and regulatory molecules (e.g., cytokines, chemokines, adhesion molecules, acute-phase proteins, and co-stimulatory molecules required for efficient activation of T cells by antigen-presenting cells). Interestingly, both heat shock proteins and fibronectin have recently been implicated as endogenous ligands for TLR molecules.[6–8]

Why be concerned about such pattern recognition in the context of autologous bioengineered tissue? After all, implantation of genetically altered autologous fibroblasts that produce factor VIII has recently been shown to have promise as a form of gene therapy for hemophilia in humans.[9] We have recently used a rabbit model of tendon injury to investigate the feasibility of using *in vitro*-cultured autologous tendon implants.[10–13] In this system, mesenchymal stem cells from autologous bone marrow are cultured under conditions that lead to fibroblast differentiation, and the architectural constitution of a tendon segment. When placed into a surgically constructed patellar tendon window, such engineered grafts close the defect. However, approximately 25% of the time, frank bone develops in the graft after *in vivo* placement—something not seen *in vitro*.[14] Why this occurs is as yet unclear. However, while much is known about downstream transcriptional control of osteoblast differentiation,[15] it is also clear that the cytokine milieu plays an important, if complex, upstream role in osteoblast differentiation. Cytokines can inhibit (TGF-β[16]) or promote (IL-6[17]) such differentiation. TNF-α can play both roles; it is a direct inhibitor of osteoblast differentiation,[18] but also stimulates fibroblast production of IL-6.[19] Notably, these latter cytokines are paradigmatic proinflammatory cytokines produced upon activation of the innate immune system.

Leaving aside the question of possible contamination of the system with LPS (fibroblasts express TLRs[20,21]), surgical injury itself leads to upregulation of the production of heat shock proteins and fibronectin. As noted above, these appear to be ligands for TLRs. In general, the resulting inflammatory responses are essential for tissue remodeling. However, in the context of the architecture of bioengineered tissue, such responses may well be detrimental. Other facets of the innate immune system that are activated in the context of surgical injury include: (*a*) the complement cascade (yielding potent signals for leukocyte recruitment, and the activation of a variety of cell types—including fibroblasts[22] by terminal complement proteins); and (*b*) the recruitment of macrophages (with phagocytic capacities important for the clearing of cellular debris; with prodigious and adaptable biosynthetic capacities, including the production of a variety of growth factors; and with a central role in the orchestration of tissue inflammatory responses).

It is important to point out that such immune responses are not necessarily bad. They may, in fact, be essential for the successful engraftment of bioengineered tissue under normal circumstances. However, to enlist the beneficial effects and limit the pernicious effects of immune recognition, such recognition and immune activation need to be carefully understood. These are issues that are clearly understudied in the context of bioengineered tissue. It is possible, for example, that an immunoregulatory cytokine such as IL-10 (a prototypical "anti-danger" signal that downregulates a variety of innate immune responses) can be harnessed for the purpose of maintaining the function and architecture of bioengineered tissue. There is much to learn.

REFERENCES

1. BACH, F.H., J.A. FISHMAN, N. DANIELS, *et al.* 1998. Uncertainty in xenotransplantation: individual benefit versus collective risk. Nat. Med. **4:** 141–144.
2. CASCALHO, M. & J.L. PLATT. 2001. The immunological barrier to xenotransplantation. Immunity **14:** 437–446.
3. BALDWIN, W.M., 3RD, C.P. LARSEN & R.L. FAIRCHILD. 2001. Innate immune responses to transplants: a significant variable with cadaver donors. Immunity **14:** 369–376.
4. MEDZHITOV, R. & C. JANEWAY, JR. 2000. Innate immune recognition: mechanisms and pathways. Immunol. Rev. **173:** 89–97.
5. ADEREM, A. & R.J. ULEVITCH. 2000. Toll-like receptors in the induction of the innate immune response. Nature **406:** 782–787.
6. OKAMURA, Y., M. WATARI, E.S. JERUD, *et al.* 2001. The extra domain A of fibronectin activates Toll-like receptor 4. J. Biol. Chem. **276:** 10229–10233.
7. OHASHI, K., V. BURKART, S. FLOHE & H. KOLB. 2000. Cutting edge: heat shock protein 60 is a putative endogenous ligand of the toll-like receptor-4 complex. J. Immunol. **164:** 558–561.
8. VABULAS, R.M., P. AHMAD-NEJAD, C. DA COSTA, *et al.* 2001. Endocytosed heat shock protein 60s use TLR2 and TLR4 to activate the toll/interleukin-1 receptor signaling pathway in innate immune cells. J. Biol. Chem. **11:** 11.
9. ROTH, D.A., N.E. TAWA, JR., J.M. O'BRIEN, *et al.* 2001. Nonviral transfer of the gene encoding coagulation factor VIII in patients with severe hemophilia A. N. Engl. J. Med. **344:** 1735–1742.
10. AWAD, H.A., D.L. BUTLER, M.T. HARRIS, *et al.* 2000. In vitro characterization of mesenchymal stem cell-seeded collagen scaffolds for tendon repair: effects of initial seeding density on contraction kinetics. J. Biomed. Mater. Res. **51:** 233–240.
11. BUTLER, D.L. & H.A. AWAD. 1999. Perspectives on cell and collagen composites for tendon repair. Clin. Orthop. :S324–332.
12. AWAD, H.A., D.L. BUTLER, G.P. BOIVIN, *et al.* 1999. Autologous mesenchymal stem cell-mediated repair of tendon. Tissue Eng. **5:** 267–277.
13. YOUNG R.G., D.L. BUTLER, W. WEBER, *et al.* 1998. The use of mesenchymal stem cells in a collagen matrix in achilles tendon repair. J. Orthop. Res. **16:** 406–413.
14. AWAD, H.A., D.L. BUTLER, G.P. BOIVIN, *et al.* Effects of implant seeding density on MSC-mediated repair of patellar tendon injuries. In review.
15. YAMAGUCHI, A., T. KOMORI & T. SUDA. 2000. Regulation of osteoblast differentiation mediated by bone morphogenetic proteins, hedgehogs, and Cbfa1. Endocr. Rev. **21:** 393–411.

16. ALLISTON, T., L. CHOY, P. DUCY, *et al.* 2001. TGF-beta-induced repression of CBFA1 by Smad3 decreases cbfa1 and osteocalcin expression and inhibits osteoblast differentiation. EMBO J. **20:** 2254–2272.
17. BELLIDO, T., V.Z. BORBA, P. ROBERSON & S.C. MANOLAGAS. 1997. Activation of the Janus kinase/STAT (signal transducer and activator of transcription) signal transduction pathway by interleukin-6-type cytokines promotes osteoblast differentiation. Endocrinology **138:** 3666–3676.
18. GILBERT, L., X. HE, P. FARMER, *et al.* 2000. Inhibition of osteoblast differentiation by tumor necrosis factor-alpha. Endocrinology **141:** 3956–3964.
19. ELIAS, J.A. & V. LENTZ. 1990. IL-1 and tumor necrosis factor synergistically stimulate fibroblast IL- 6 production and stabilize IL-6 messenger RNA. J. Immunol. **145:** 161–166.
20. TABETA, K., K. YAMAZAKI, S. AKASHI, *et al.* 2000. Toll-like receptors confer responsiveness to lipopolysaccharide from *Porphyromonas gingivalis* in human gingival fibroblasts. Infect. Immun. **68:** 3731–3735.
21. WANG, P.L., Y. AZUMA, M. SHINOHARA & K. OHURA. 2000. Toll-like receptor 4-mediated signal pathway induced by *Porphyromonas gingivalis* lipopolysaccharide in human gingival fibroblasts. Biochem. Biophys. Res. Commun. **273:** 1161–1167.
22. VON KEMPIS, J., I. TORBOHM, M. SCHONERMARK, *et al.* 1989. Effect of the late complement components C5b-9 and of platelet-derived growth factor on the prostaglandin release of human synovial fibroblast-like cells. Int. Arch. Allergy Appl. Immunol. **90:** 248–255.

An Innate Sense of Danger

POLLY MATZINGER

*National Institute of Allergy and Infectious Diseases, National Institutes of Health,
Bethesda, Maryland 20894, USA*

KEYWORDS: self and non-self; immune system

For three quarters of a century, immunologists have based their theories and experiments on the fundamental belief that the immune system discriminates between self and non-self and that, if the system were perfect, it would attack everything that is non-self and be totally tolerant of anything that is self. I have abandoned this belief. Over the years there have accumulated too many immunological findings that don't fit with and too many questions that are not answered by this paradigm.

For example, if each individual's immune system learns "self" at an early age, then why are new antigens that appear at puberty not considered "foreign" and destroyed. How can normal individuals contain both T and B cells capable of reacting to self antigens like DNA, keratin, and myelin basic protein, yet not have destructive autoimmunity. Why are liver transplants rejected less vigorously than hearts? Why is a newly lactating breast not rejected when it begins to make new proteins? Why is the immune system so bad at dealing with tumors, even when they demonstrably express new, "non-self" antigens? Why do we need adjuvant? Why do we not normally respond to all the foreign antigens in food, to our commensal intestinal bacteria, or to our fetuses or to the sperm that begot them?

The answers to these questions are not easily found when we approach the immune system from a self–non-self viewpoint, although they fall easily into place when approached from the perspective that the immune system is more concerned with danger and potential destruction than with the distinction between self and non-self.

The danger model is based on the idea that the driving force for the immune system is the need to recognize danger. The model starts with the idea that the immune system defines "danger" as anything that causes tissue stress or destruction. Under this model, antigen-presenting cells are activated by alarm signals from stressed or damaged tissues. Without this activation, no primary immune response can occur. Some of the recent evidence in its favor has been shown and its implications for all of the questions above discussed.

Address for correspondence: Polly Matzinger, National Institute of Allergy and Infectious Diseases, National Institutes of Health, Building 4, Room 111, Bethesda, MD 20894. Voice: 301-496-6440; 301-496-4286.

pcm@helix.nih.gov

Ann. N.Y. Acad. Sci. 961: 341–342 (2002). © 2002 New York Academy of Sciences.

REFERENCES

1. MATZINGER, P. 1994. Tolerance, danger, and the extended family. Annu. Rev. Immunol. **12:** 991–1045.
2. RIDGE, J.P., E. FUCHS & P. MATZINGER. 1996. Neonatal tolerance revisited: turning on newborn T cells with dendritic cells. Science **271:** 1723–1726.
3. MATZINGER, P. 1998. An innate sense of danger. Semin. Immunol. **10:** 399–415.
4. GALLUCCI, S., M. LOLKEMA & P. MATZINGER. 1999. Natural adjuvants: endogenous activators of dendritic cells. Nat. Med. **5:** 1249–1255.
5. ANDERSON, C.C., J.M. CARROLL, S. GALLUCCI, *et al.* 2001. On the adjuvanticity of grafting J. Immunol. **166:** 3663–3671.

Tolerogenic Antigen-Presenting Cells

DAVID H. MUNN

Medical College of Georgia, Augusta, Georgia 30912, USA

KEYWORDS: **graft rejection; antigen-presenting cells**

The ultimate goal of transplant immunology is to create a state of stable tolerance toward graft-associated antigens, without causing global immunosuppression in the host. T cells provoke antigen-specific graft rejection by recognizing foreign antigens derived from graft-associated proteins. However, in order for T cells to initiate rejection, the antigens must first be presented to them by other cells—the "professional" antigen-presenting cells, macrophages, and dendritic cells. Antigen-presenting cells are thus in a unique position to regulate the process of rejection or tolerance.[1]

Until recently, macrophages and dendritic cells have been viewed as "accessory" cells, with the sole function of assisting T-cell activation. Now, however, it has become evident that antigen-presenting cells can inhibit T-cell responses and promote tolerance, thus serving as a negative-regulatory element.[2] Under physiologic conditions, inhibitory antigen-presenting cells are thought to help maintain tolerance toward self and innocuous environmental antigens. But these endogenous immunoregulatory pathways are also of interest for transplant immunology, because they are more specific and less toxic than the global immunosuppressive agents currently in use.

A number of reports have shown that macrophages can suppress T-cell activation *in vitro* and *in vivo* (reviewed in Ref. 3). However, the study of immunosuppressive macrophages has been complicated by the fact that there are many types of macrophages (e.g., isolated from different tissues, or derived *in vitro* under different culture conditions), and a variety of putative mechanisms for suppression (nitric oxide, prostaglandins, reactive oxygen species, etc.). It is not clear which of these are of biologic relevance *in vivo*. Moreover, macrophages reside in the tissues and do not transport graft antigens to the draining lymph nodes, which is where T-cell immune responses must be initiated. Thus, the real role of macrophages in regulating primary immune responses remains unclear.

In contrast, dendritic cells are highly migratory and readily transport tissue antigens to lymph nodes, where they are responsible for initiating primary T-cell responses.[4] Recently, it has been recognized that certain subsets of dendritic cells can induce tolerance to the antigens that they present.[1,2,5,6] In mice, tolerogenic den-

Address for correspondence: David H. Munn, Medical College of Georgia, Room IMMAG, Mail Stop CA-2010, Augusta, GA 30912. Voice: 706-721-7141; fax: 706-721-8732.
dmunn@mail.mcg.edu

Ann. N.Y. Acad. Sci. 961: 343–345 (2002). © 2002 New York Academy of Sciences.

dritic cells have been associated with a population bearing the CD8α^+ marker.[7] In humans, immature myeloid dendritic cells have been shown to be tolerogenic.[8] This tolerogenic activity is lost as myeloid dendritic cells mature in response to stimuli such as CD40 ligation or tumor necrosis factor-α, but can be maintained by cytokines such as interleukin-10 and transforming growth factor-β. Certain types of dendritic cells can also polarize CD4$^+$ T cells toward Th2-type responses, which may in turn be functionally immunosuppressive for Th1 reactions. [9]

One important feature of the tolerance induced by immunoregulatory dendritic cells is that it appears to be able to "propagate" or spread—starting from a small initial number of dendritic cells, it can induce tolerance throughout the host immune system. This is potentially of great significance for transplantation, because it means that all of the host T cells do not need to be exposed individually to the tolerogenic antigen-presenting cells. Tolerance can even be induced when the same antigen is being simultaneously presented by other, nonsuppressive dendritic cells,[10] and tolerogenic dendritic cells can reverse established immunity.[8] Taken together, these properties suggest that the initial presentation of antigen may create a secondary population of immunoregulatory T cells, which then spread antigen-specific tolerance to other T cells.[11]

The specific molecular mechanism by which dendritic cells induce tolerance remains unclear, and is probably multifactoral. It has been variously suggested that they may present antigen without the requisite co-stimulation (e.g., due to low expression of B7-family molecules), or they may actively produce immunomodulatory cytokines or contact factors. We and others have shown that, as an additional potential mechanism, certain Mϕs and dendritic cells can suppress T-cell activation *in vitro* via expression of the enzyme indoleamine 2,3-dioxygenase, or IDO.[12] IDO catalyzes the breakdown of tryptophan along the kynurenine pathway (reviewed in Ref. 13), and is induced by proinflammatory signals such as IFNγ. *In vitro*, expression of IDO by Mϕs rapidly depletes tryptophan from the culture medium, which prevents T-cell proliferation and acquisition of effector function. We have proposed that a similar sequence of events occurs *in vivo*, with tryptophan catabolism by the IDO-expressing APC acting to lower the local interstitial concentration of tryptophan below the threshold level required for T-cell activation. Although this hypothesis has received indirect support from our recent elucidation of a tryptophan-specific cell cycle regulatory checkpoint governing T-cell activation, the true mechanism by which IDO regulates T cells *in vivo* remains unknown.

Experiments using a pharmacologic inhibitor of IDO (1-methyl-tryptophan) administered to mice have shown that functional IDO activity is required in order to prevent maternal CD8$^+$ T cells from rejecting the antigenically "foreign" fetus during gestation,[14] and to prevent excessive activation of maternal complement, which would otherwise be deposited in a T-cell-driven hemorrhagic vasculitis in the placenta.[15] In the human, IDO is expressed widely in immunoregulatory antigen-presenting cells throughout the immune system. IDO-expressing APCs are also found heavily infiltrating lymph nodes draining certain malignancies, and in HIV-infected lymphoid tissue. Transfection of the IDO gene under a strong constitutive promoter into a variety of murine cell lines results in suppression of allogeneic T-cell responses *in vitro*. Thus, IDO is a both physiologic regulator of T-cell responses, and can be used to engineer immunosuppressive cell lines.

In conclusion, regulatory antigen-presenting cells have the potential to dramatically affect the host's decision to tolerate or reject a foreign graft. A strategy of engineering grafts to either contain a population of tolerogenic antigen-presenting cells, or to directly employ the molecular mechanisms used by such cells, may hold promise for inducing antigen-specific tolerance.

ACKNOWLEDGMENTS

The work presented is the result of collaboration with Andrew Mellor, Medical College of Georgia.

REFERENCES

1. FEARON, D.T. & R.M. LOCKSLEY. 1996. The instructive role of innate immunity in the acquired immune response. Science **272:** 50–54.
2. LECHLER, R. *et al.* 2001. Dendritic cells in transplantation: friend or foe? Immunity **14:** 357–368.
3. ATTWOOD, J.T. & D.H. MUNN. 1999. Macrophage suppression of T cell activation: a potential mechanism of peripheral tolerance. Int. Rev. Immunol. **18:** 515–525.
4. GRABBE, S. *et al.* 2000. Dendritic cells: multi-lineal and multi-functional. Immunol. Today **21:** 431–433.
5. BANCHEREAU, J. & R.M. STEINMAN. 1998. Dendritic cells and the control of immunity. Nature **392:** 245–252.
6. FAIRCHILD, P.J. & H. WALDMANN. Dendritic cells and prospects for transplantation tolerance. Curr. Opinion Immunol. **12:** 528–535.
7. FAZEKAS DE ST. GROTH, B. 1998. The evolution of self-tolerance: new cell arises to meet the challenge of self-reactivity. Immunol. Today **19:** 448–454.
8. DHODAPKAR, M.V. *et al.* 2001. Antigen-specific inhibition of effector T cell function in humans after injection of immature dendritic cells. J. Exp. Med. **193:** 233–238.
9. PULENDRAN, B. *et al.* 2001. Modulating the immune response with dendritic cells and their growth factors. Trends Immunol. **22:** 41–47.
10. GROHMANN, U. *et al.* 2001. CD40 ligation ablates the tolerogenic potential of lymphoid dendritic cells. J. Immunol. **166:** 277–283.
11. RONCAROLO, M-G. *et al.* 2001. Differentiation of T regulatory cells by immature dendritic cells. J. Exp. Med. **193:** F5–F9.
12. MUNN, D.H. *et al.* 1999. Inhibition of T cell proliferation by macrophage tryptophan catabolism. J. Exp. Med. **189:** 1363–1372.
13. MELLOR, A.L. & D.H. MUNN. 1999. Tryptophan catabolism and T-cell tolerance: immunosuppression by starvation? Immunol. Today **20:** 469–473.
14. MUNN, D.H *et al.* 1998. Prevention of allogeneic fetal rejection by tryptophan catabolism. Science **281:** 1191–1193.
15. MELLOR, A.L. *et al.* 2001. Prevention of T cell-driven complement activation and inflammation by tryptophan catabolism during pregnancy. Nat. Immunol. **2:** 64–68.

Chemokines in Neurological Trauma Models

RICHARD M. RANSOHOFF

The Lerner Research Institute, Cleveland, Ohio 44195, USA

KEYWORDS: axotomy; chemokines; neurological trauma

Introduction of synthetic materials or engineered cells into the nervous system (either central [CNS] or peripheral [PNS]) will engender a tissue reaction. For this reason, successful tissue engineering will require us to decipher mechanisms of inflammatory tissue reactions in the nervous system. Inflammation in the mammalian nervous system follows a diversity of insults, but in many cases the inflammatory reaction is restricted in cellular composition to mononuclear phagocytes. Signals that govern selective recruitment of mononuclear phagocytes to the nervous system have been enigmatic.[1] Classical studies by Perry *et al.* demonstrated that adhesion molecules are readily induced on CNS microvessels by injection of lipopolysaccharide (LPS), so that failure of neutrophil infiltration could not be accounted for by absence of an activated, adhesive microvascular substrate.[2] Studies in which chemokines were delivered either by recombinant adenoviruses or CNS-directed transgenes showed that the infiltration of CNS tissues by neutrophils required nothing more than the presence of a neutrophil-directed chemokine.[3,4] Chemokines are chemoattractant peptides which signal selectively to subpopulations of leukocytes *via* G-protein–coupled receptors (GPCRs). In concert with leukocyte and endothelial adhesion molecules, the chemokines govern the ceaseless process of leukocyte trafficking, both inflammatory and physiological. Chemokine expression profiles have been assessed in a variety of models of neural trauma. The results indicate that chemokine expression in post-traumatic inflammation is generally restricted to the monocyte chemoattractant CCL2/MCP-1, and occurs before hematogenous cell entry into neural tissues. Therefore CCL2 is an excellent candidate for a mediator of leukocyte recruitment in these settings. Evidence that chemokines selectively recruit target leukocytes focused attention on the possibility that restricted chemokine expression could account in part for the unusual monocyte-rich inflammation that follows nervous-system trauma. This line of thinking was supported also by the finding that parenchymal neural cells were competent to express chemokines under physiological circumstances.[5,6]

Address for correspondence: Richard Ransohoff, The Lerner Research Institute, 9500 Euclid Avenue, Room NC30, Cleveland, OH 44195. Voice: 216-444-8939; fax: 216-444-7927.
ransohr@ccf.org

Ann. N.Y. Acad. Sci. 961: 346–349 (2002). © 2002 New York Academy of Sciences.

SCIATIC NERVE AXOTOMY

In collaboration with Dr. John Griffin (Department of Neurology, Johns Hopkins University School of Medicine), we evaluated chemokine expression during Wallerian degeneration of peripheral nerves in mice and found selective early expression of CCL2; *in situ* hybridization localized CCL2 message in Schwann cells, preceding recruitment of macrophages. Both macrophage recruitment and clearance of debris were delayed in CCL2-null mice after sciatic axotomy (our unpublished observations).

AXOTOMY OF SUPERIOR CERVICAL GANGLION

Normal rat SCGs contain only resident ED2+ macrophages; by 48 hours post axotomy, abundant ED1+ macrophages enter ganglia. In collaboration with Drs. Richard Zigmond and Rebecca Schreiber (Department of Neurosciences, Case Western Reserve University School of Medicine), we showed that CCL2 mRNA increased dramatically by 6 hours post axotomy, with the message localized to neurons near the axotomy site.[7,8]

PENETRATING MECHANICAL TRAUMA TO THE CEREBRAL CORTEX

Within the first twenty-four hours after mechanical trauma to the central nervous system (CNS), reactive astrogliosis develops and injury sites are infiltrated by activated mononuclear phagocytes derived from blood-borne monocytes and endogenous microglia. In collaboration with Drs. V. Balasingam and V.W. Yong (Montreal Neurological Institute and McGill University), we analyzed the time course and cellular source of MCP-1 in mouse brain after penetrating mechanical injury, with particular focus on early time points before histological detection of infiltrating mononuclear phagocytes. Steady-state levels of CCL2 mRNA and protein increased within three hours after nitrocellulose membrane stab or implant injury to the adult mouse brain (but not after stab to the neonatal mouse brain). *In situ* hybridization combined with immunohistochemistry for the glial fibrillary acidic protein (GFAP) astrocyte marker showed that astrocytes were the cellular source of CCL2 at these early time points after mechanical brain injury.[9]

CRYOPROBE LESION TO THE CEREBRAL CORTEX

Cryolesion to the cortex is considered a highly reproducible protocol for generating neuronal injury without opening the skull or dura. In collaboration with Dr. Sean Murphy (then at the Department of Pharmacology, University of Iowa School of Medicine), we evaluated chemokine expression in the rodent CNS after cortical cryolesion. In the ipsilateral cortex, CCL2 gene expression was increased significantly at 6 hours, 20-fold at the 12- and 24-hour time points, and declined to control levels by 48 hours. Unexpectedly, there was significant CCL2 expression contralaterally, maximally about 40% of that seen in lesioned cortex with a superimposable time course.[10] This result was compatible with several possible explanations: that damaging of fibers within lesioned cortex could initiate chemokine expression contralateral to the lesion, as a result of signaling from neurons from which the fibers originated. Alternatively, injury to fibers in one region of cortex could signal to neurons in the contralateral projection field, resulting in chemokine expression.

Further experiments in collaboration with Dr. Roy Weller (Department of Neuropathology, University of Southampton School of Medicine, UK), showed that cortical cryolesions induced abundant CCL2 in the cerebrum one day after lesion placement. At this time point (day 9), there was no detectable CCL2 expression in spinal cord, demonstrating that universal chemokine production throughout the CNS was not provoked by a cortical cryolesion.[11]

SPINAL CORD PERCUSSION INJURY

Contusion injury of the rat spinal cord produces a reproducible clinical and histological evolution. Inflammation in this model is prominent and its role in recovery remains uncertain. We evaluated chemokine expression in the spinal cords of rats after percussion injury, in collaboration with Dr. Bradford Stokes (Ohio State University). After a single-level laminectomy at T8, rats were immobilized in a spinal frame and the dorsal surface of the spinal cord was rapidly and precisely displaced with a sterile impactor. Histological changes included immediate and transient neutrophil infiltration; microglial reaction and macrophage accumulation, beginning at three days and pronounced from seven to 28 days post injury. CCL2 gene expression was more than 20-fold increased by 6 hours, and >50-fold elevated by 12 hours post injury.[12,13]

Our results to date indicate that chemokine expression in the context of CNS and PNS inflammation is highly regulated and closely associated with the recruitment of specific populations of leukocytes into the target tissue. Thus, CCL2 expression correlates well with accumulation of macrophages in such diverse settings as sciatic nerve axotomy or spinal cord percussion injury. Our challenge now is to define the functions of the inflammatory reaction in these model systems, and in pathological challenges to the nervous system.

REFERENCES

1. HUANG, D., Y. HAN, M. RANI, et al. 2000. Chemokines and chemokine receptors in inflammation of the nervous system: manifold roles and exquisite regulation. Immunol. Rev. **177:** 52–67 [with cover illustration].
2. ANDERSSON, P.-B. & V. PERRY. 1991. The CNS acute inflammatory response to excitotoxic cell death. Immunol. Lett. **30:** 177–182.
3. BELL, M.D. & V.H. PERRY. 1995. Adhesion molecule expression on murine cerebral endothelium following the injection of a proinflammagen or during acute neuronal degeneration. J. Neurocytol. **24:** 695–710.

4. BELL, M.D., D.D. TAUB, S.J. KUNKEL, et al. 1996. Recombinant human adenovirus with rat MIP-2 gene insertion causes prolonged PMN recruitment to the murine brain. Eur. J. Neurosci. **8:** 1803–1811.
5. TANI, M., M.E. FUENTES, J.W. PETERSON, et al. 1996. Neutrophil infiltration, glial reaction and neurological disease in transgenic mice expressing the chemokine N51/KC in oligodendrocytes. J. Clin. Invest. **98:** 529–539.
6. GLABINSKI, A., M. TANI, R. STRIETER, et al. 1997. Synchronous synthesis of α- and β-chemokines by cells of diverse lineage in the central nervous system of mice with relapses of experimental autoimmune encephalomyelitis. Am. J. Pathol. **150:** 617–630.
7. SCHREIBER, R.C., A.M. SHADIACK, T.A. BENNETT, et al. 1995. Changes in the macrophage population of the rat superior cervical ganglion after postganglionic nerve injury. J. Neurobiol. **27:** 141–153.
8. SCHREIBER, R., K. KRIVACIC, B. KIRBY, et al. 2001. Monocyte chemoattractant protein (MCP)-1 is rapidly expressed by sympathetic ganglion neurons following axonal injury. NeuroReport **12:** 601–606.
9. GLABINSKI, A.R., M. TANI, V. BALASINGAM, et al. 1996. Chemokine monocyte chemoattractant protein-1 (MCP-1) is expressed by astrocytes after mechanical injury to the brain. J. Immunol. **156:** 4363–4368.
10. GRZYBICKI, D., S. MOORE, R. SCHELPER, et al. 1998. Expression of the chemokine MCP-1 and nitric oxide synthase-2 following cerebral trauma. Acta Neuropathol. **95:** 98–103.
11. SUN, D., M. TANI, T. NEWMAN, et al. 2000. Role of chemokines, neuronal projections and the blood-brain barrier in the enhancement of EAE following focal brain damage. J. Neurol. Exp. Neuropathol. **59:** 1039–1043.
12. POPOVICH, P., P. WEI & B. STOKES. 1997. Cellular inflammatory response after spinal cord injury in Sprague-Dawley and Lewis rats. J. Comp. Neurol. **377:** 443–464.
13. MCTIGUE, D., M. TANI, K. KRAVACIC, et al. 1998. Selective chemokine mRNA accumulation in the rat spinal cord after contusion injury. J. Neurosci. Res. **53:** 368–376.

Immune Response to Engineered Tissues and Cells

Breakout Session Summary

Moderators

DAVID M. HARLAN, *NIDDK, NIH*

DENNIS W. METZGER, *Albany Medical College*

Panelists

CHRISTOPHER L. KARP, *University of Cincinnati*

POLLY MATZINGER, *National Institutes of Health*

DAVID H. MUNN, *Medical College of Georgia*

RICHARD M. RANSOHOFF, *Cleveland Clinic*

Rapporteur

DENNIS W. METZGER, *Albany Medical College*

Remarkable progress has been made in the generation of cells and tissues for organ repair and replacement. However, further advances in the engineering of biological materials will face the difficulty of immune acceptance of the altered materials. In the case of allografts and xenografts, rejection through several mechanisms including both innate and adaptive immunity, represents a formidable barrier. Inflammation is likewise important in preventing successful engraftment, but may also be pivotal for proper tissue remodeling. In light of the critical need to understand the role of immunity in successful tissue implantation, the goal of this panel was to identify how immune responses are initiated, to review newer approaches to prevent rejection, to identify the role of inflammation in tissue remodeling, and to compare the results obtained in different model systems. In addition, the panel considered how the NIH could foster increased interdisciplinary research in the vital area of immunity to engineered tissues and organs.

VISION

Understanding the interplay between implanted tissues/materials and the host response to these implants represents a growing concern in the field of biomedical engineering. Further advances will only occur with increased focus on this important issue.

Ann. N.Y. Acad. Sci. 961: 350–351 (2002). © 2002 New York Academy of Sciences.

OBJECTIVES

In considering advancing the design, delivery, and function of bioengineered tissues, the host response to these tissues needs to be fully considered: Specifically, we need to understand: (1) inflammatory cell signaling and trafficking; (2) the pivotal role of antigen-presenting cell subsets; (3) whether Th1 versus Th2 activation is important for implantation success; (4) specialized immune mechanisms at various anatomical sites (e.g., nervous system, mucosal sites); (5) the potential need for inflammatory cytokines for successful transplantation; (6) the role of the galactose epitope in xenotransplantation; and (7) tolerance and initiation of the immune response to danger signals.

OBSTACLES AND CHALLENGES

As bioengineers increase their efforts to fabricate artificial tissues and organs in attempts to improve natural polymers, the response of the immune system will almost certainly thwart progress. Unfortunately, there often appears to be lack of understanding of the likely immune consequences of tissue/organ manipulations and of possible approaches that might be exploited for overcoming these consequences. Conversely, there seems to be little appreciation among the immunological community for the challenges facing the tissue-engineering field. There is a lack of communication between tissue engineering and immunology scientific communities, and studies often suffer from the lack of a multidisciplinary approach.

ACKNOWLEDGMENT

The Immune Response to Engineered Tissues and Cells Breakout Session was coordinated by Dr. Ernest Marquez (NIGMS, NIH) and Dr. Jean Sipe (CSR, NIH).

Tissue Engineering of Meniscal Cartilage Using Perfusion Culture

ANDRÉ A. NEVES,[a] NICK MEDCALF,[b] AND KEVIN M. BRINDLE[a]

[a]Department of Biochemistry, University of Cambridge, Cambridge CB2 1GA, U.K.

[b]Smith & Nephew Group Research Centre, York Science Park, Heslington, York YO10 5DF, U.K.

KEYWORDS: tissue engineering; meniscal cartilage; perfusion culture; MRI; MRS

INTRODUCTION

Fifty years ago it was shown that removal of a diseased meniscus in the knee joint led in the long term to cartilage degeneration and bone remodeling.[1] This observation changed substantially the therapeutic approach adopted, with the ruptured meniscus being repaired instead of removed. However, this treatment is only feasible when the meniscal tissue is of good quality, which is not the most common situation in the clinic. Concerns regarding disease transmission, immunogenicity, sizing, and availability of meniscal allografts[2] have stimulated the search for a tissue-engineered (TE) structure that could replace the function of the native tissue. Previous studies have shown the relevance of fluid flow in the *in vitro* synthesis of cartilaginous tissues.[3,4] The constant availability of fresh media, the mechanical action of shear stress on the cells, and the ability to transport nutrients through an increasingly dense extracellular matrix (ECM) are some of the reasons that favor the use of perfusion culture for the generation of bioartificial cartilage. However, optimal flow parameters for the generation of meniscal tissue have yet to be identified in such systems. We report here the use of perfusion culture and two powerful noninvasive techniques, magnetic resonance imaging (MRI) and spectroscopy (MRS), to characterize the flow profile inside a fixed-bed bioreactor, the growth and energetics of the cells, and the kinetics of ECM deposition. These techniques were used to correlate noninvasively the properties of the generated bioartificial meniscal cartilage with fluid dynamics and permeability measurements. An ideal flow rate for operation of the bioreactor ($40 \ mL \cdot min^{-1}$) was derived that optimizes structural properties and ECM production.

Address for correspondence: Kevin M. Brindle, Department of Biochemistry, University of Cambridge, 80 Tennis Court Road, Cambridge CB2 1GA, U.K. Voice: 0044-1223-333674; fax: 0044-1223-766002.

k.m.brindle@bioc.cam.ac.uk

Ann. N.Y. Acad. Sci. 961: 352–355 (2002). © 2002 New York Academy of Sciences.

MATERIALS AND METHODS

The fixed-bed bioreactor system used in these studies consisted of a stirred-tank fermenter (FT Applicon, Tewkwsbury, U.K.), fitted with a three-blade marine impeller with a geometric volume of 2 L. A working volume of 1.4 L of medium was continuously pumped via water-jacketed tubing at 37°C to the two bioreactors positioned alternately in a water bath at 37°C or in the NMR magnet. Each of the bioreactors was designed and custom-made in-house, consisting of a polysulphone (R.S., U.K.) tube (20-mm internal diameter), with a capped cylindrical chamber at the top (40-mm internal diameter). The fixed bed consisted of three scaffolds positioned perpendicularly to the ascending flow of medium. The lower cylindrical section of the bioreactor was fitted with different types of plastic spacers (ultrahigh-density polyethylene; R.S., U.K.), which enabled an adequate separation of the scaffolds and a redistribution of the flow. Scaffolds were supplied by S&N GRC in the form of disk-shaped meshes that were 12 mm in diameter and 4 mm thick. Extrusion of polyethylene terephtalate (PET) in 13-μm fibers produced scaffolds with a void volume of 97% and a density of 45 mg·cm^{-3}. Primary ovine fibrochondrocytes were propagated four times and used to seed the scaffolds at a density of 1.2×10^7 cells/scaffold, using the method proposed by Vunjack-Novakovic and coworkers. [5]

MRI and MRS were performed using a vertical wide-bore Oxford Instruments magnet (9.4 T, 400 MHz, 8.9-cm bore diameter) equipped with an unshielded gradient set, interfaced with a Varian UnityPlus spectrometer, controlled by a Sun Sparcstation IPX running VNMR 5.3B software. ^{1}H spectra and images were acquired using a Varian 25 mm ^{1}H imaging probe and ^{31}P spectra using a Bruker 25-mm ^{1}H/^{31}P probe. Diffusion-weighted MRI,[6] contrast agent-enhanced MRI, and ^{31}P MRS were used to investigate the dynamics of cell growth and ECM deposition. Time-of-flight imaging of flow[7] was applied to monitor the changes in the linear velocity across the TE constructs throughout the period of cultivation.

RESULTS AND DISCUSSION

Diffusion-weighted MRI is a method that can be used to monitor cell growth in intensive bioreactors, due to the restricted diffusion of water molecules in regions of high cell density. In more complex systems, such as the one studied here, the signal intensity (s.i.) from a TE construct is the sum of two nonindependent contributions, that is, from the cell population and from their secreted ECM. An increase in the flow rate to the bioreactor, in the period postseeding, yielded constructs with distinct restricted diffusion maps throughout the two-week cultivation period. At a low flow rate of 30 mL·min^{-1}, a progressive increase of the average s.i. in the plane of the construct (FIG. 1A) was observed throughout the run, indicating a progressive growth of the cell population with associated ECM secretion. At 60 mL·min^{-1}, a maximum value of the average s.i. was observed approx. 100 h postseeding, followed by a subsequent reduction until the end of the run (FIG. 1B). A destructive effect of the flow at this regime on the inner region of the construct was evident from the profiles. The operation of the bioreactor at different flow rates has shown that a flow rate of 40 mL·min^{-1} for two weeks, yielded a progressive increase of the average s.i., indicating a regular growth of the cell population with associated matrix deposition. Con-

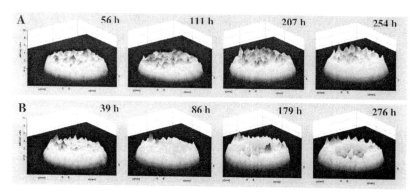

FIGURE 1. Time course of diffusion-weighted MRI images in the plane perpendicular to the flow (F) of medium; 2D maps of signal intensity. Signal intensity is directly proportional to the degree of restricted diffusion. (**A**) $F = $ mL·min^{-1}; (**B**) $F = 60$ mL·min^{-1}.

structs produced at lower flow rates (e.g., 20 mL·min^{-1}) showed a progressive reduction of the diffusion-weighted s.i., probably as a result of diffusional limitations in oxygen transfer, and thus poorer cell growth. Greater flow rates (e.g., 60 or 80 mL·min^{-1}) were destructive for the cell population at this premature stage of development, presumably due to the lack of a structurally robust ECM.

A pulse sequence also has been developed for estimating the axial velocity profile for different cross sections of the bioreactor. The method is based on the principle of time-of-flight,[4] and employs a selective saturation–recovery pulse sequence consisting of spin tagging and pulse detection followed by a bipolar readout gradient. The reduction in the spin-lattice relaxation time (T_1) of the culture medium, associated with flow, is exploited in this method. The method was calibrated in a flow phantom, a system with a defined cylindrical geometry. Second, the accuracy of the method was tested in an unseeded bioreactor. A linear relationship was found between the flow rate and the average velocity in the cross section. The velocity profile of a slice defined across the geometric center of a recently seeded scaffold showed a very low perfusion of this region, which was independent of the flow rate used. At the end of the cultivation period (14 days) at 30 mL·min^{-1}, the average flow across the construct was too small to measure. Mass transport at this stage therefore occurred only by diffusion. A greater flow rate, for example, 60 mL·min^{-1}, provided a better perfusion of the construct, even though it has been shown to be destructive, as mentioned previously.

CONCLUSIONS

A flow rate of 40 mL·min^{-1} was found to optimize cell viability and build-up of ECM. The local changes in cell density and ECM composition were studied using MRI and MRS methods. These have been shown to be valuable tools in the noninvasive characterization of these structures throughout the cultivation period. Histological data have been evaluated and related to fluid movement. A distinct outer layer was observed in the TE constructs at this flow rate that evidenced cell morphology and physiology similar to fibroblast cells. The inner core of the constructs resembled native ovine meniscal tissue, despite evidence of oxygen deprivation of the cell population in this region. An improvement of scaffold design is therefore desired, to enable a better perfusion of the TE constructs and a more homogeneous morphology.

REFERENCES

1. FAIRBANK, T.J. 1948. Knee joint changes after meniscectomy. J. Bone Joint Surg. **30B:** 664.
2. KUHN, J.E. & E.M. WOJTYS. 1996. Allograft meniscus transplantation. Clin. Sports Med. **15:** 537–536.
3. ROTTER, N. *et al.* 1998. Cartilage reconstruction in head and neck surgery: comparison of resorbable polymer scaffolds for tissue engineering of human septal cartilage. J. Biomed. Mater. Res. **42:** 347–356.
4. SITTINGER, M. *et al.* 1997. Artificial tissues in perfusion culture. Int. J. Artif. Organs **20:** 57–62.
5. VUNJACK-NOVAKOVIC, G. *et al.* 1996. Effects of mixing on the composition and morphology of tissue-engineered cartilage. AIChE J. **42:** 850–860.
6. VAN ZIJL, P.C. *et al.* 1991. Complete separation of intracellular and extracellular information in NMR spectra of perfused cells by diffusion-weighted spectroscopy. Proc. Natl. Acad. Sci. USA **88:** 3228–3232.
7. WEHRLI, F.W. *et al.* 1986. Time-of-flight MR flow imaging: selective saturation recovery with gradient refocusing. Radiology **160**.

PROVANT® Wound-Closure System Accelerates Closure of Pressure Wounds in a Randomized, Double-Blind, Placebo-Controlled Trial

M.C. RITZ,[a] R. GALLEGOS,[b] M.B. CANHAM,[a] M. ESKALAI,[c] AND F.R. GEORGE[a]

[a]Regenesis Biomedical, Scottsdale, Arizona, USA
[b]John C. Lincoln Hospital, Phoenix, Arizona, USA
[c]Arizona State University, Tempe, Arizona, USA

KEYWORDS: PROVANT®; pressure wound; pressure ulcer; cell proliferation

INTRODUCTION

Treating pressure ulcers in the United States costs 5 to 8.5 billion dollars annually.[1,2] A study of Medicare claims found that hospitals lose more than $200 million per year treating pressure ulcers.[3] Clearly, there is a need for more effective treatment, but an effective solution has remained elusive. However, a better understanding of wound physiology and the cell biology of wound healing have identified new strategies for achieving wound closure.

Previous reports have shown that the Provant Wound Closure System (PROVANT®) is effective in closing pressure wounds compared to average population rates. The purpose of this prospective, randomized, placebo-controlled, double-blinded trial was to investigate under rigorously controlled conditions the clinical efficacy of PROVANT for the treatment of pressure wounds. PROVANT is based upon Cell Proliferation Induction (CPI®), a biophysical technology that uses radio-frequency stimuli to induce fibroblast and epithelial cell proliferation.[4,5]

METHODS

Patient Selection

An IRB-approved protocol with informed consent was used. Forty-nine patients with Stage II or III pressure ulcers were entered. Patients were high or moderate risk with complex, chronic wounds, 60% of which were >6 months old. Patients were randomized to PROVANT or placebo groups. PROVANT patients received active

Address for correspondence: F.R. George, Regenesis Biomedical Inc., 1435 N. Hayden Road, Scottsdale, Arizona 85257. Voice: 480-970-4970; fax: 480-970-8792.
george@regenesisbiomedical.com

Ann. N.Y. Acad. Sci. 961: 356–359 (2002). © 2002 New York Academy of Sciences.

treatment twice a day (b.i.d.) plus standard care. Placebo patients received standard care and b.i.d. treatment with a PROVANT device transparently modified so that no treatment was given. Patients and caregivers were blinded to group assignments. Inclusion criteria: (1) Stage II, III pressure ulcer; (2) ≥18 years of age. Exclusion criteria: (1) change in Norton Risk Assessment score ≥7 within 30 days; (2) osteomyelitis; (3) immune dysfunction or repeated systematic infection; (4) cancer; (5) concurrent treatment with other wound-healing devices (e.g., hyperbaric oxygenation, electrical stimulation).

Study Procedures

Patients were followed for 12 weeks, or until closure, or until the patient was discharged, whichever occurred first. History and wound data were collected at entry. Wound length, width, depth, undermining were documented weekly by a wound consultant (CWOCN).

Statistical Analysis

PRISM software was used. Analysis of variance (ANOVA), chi-square, and *t*-tests were used as appropriate to determine alpha levels, set *a priori* at 0.05.

RESULTS

Patient demographics are shown in TABLE 1. Average initial PSST score was 33.7 ± 3.1, indicating severe wounds with lack of healing. Sixty percent of wounds were >6 months old at entry.

PROVANT induced significantly more closures than did the placebo. At six weeks, 100% of Stage II PROVANT wounds were healed compared to 36% for placebo(FIG. 1; $p \leq 0.005$). By 12 weeks, 64% of placebo wounds had closed. Stage II PROVANT wounds healed 60% faster (26 days) than Stage II placebo wounds (66 days) (FIG. 2; $p \leq 0.005$). At 12 weeks, 50% of Stage III PROVANT wounds were closed, compared to 14% for placebo (FIG. 3; $p \leq 0.01$). PROVANT wounds showed an average 87% reduction in surface area compared to a 56% reduction for placebo (FIG. 4; $p \leq 0.05$).

PROVANT wounds closed at a faster rate than placebo. For Stage II PROVANT wounds the closure rate was 11.92 ± 2.0 mm^2/day, while the placebo closure rate was 6.8 ± 1.7. For Stage III wounds, the PROVANT rate was 12.9 ± 4.1 compared with only a 3.6 ± 2.2 rate for placebo.

TABLE 1. Patient demographics

Stage	Group	N	Age (yr)	Wound area (cm^2)	Bates–Jensen score
II	PROVANT	8	72	3.0	33
	PLACEBO	11	69	4.4	41
III	PROVANT	8	75	11.3	35
	PLACEBO	7	63	4.4	33

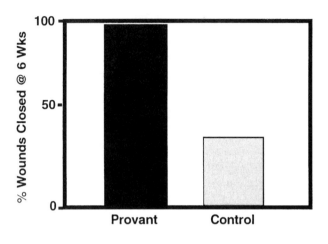

FIGURE 1.

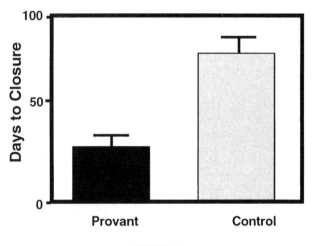

FIGURE 2.

DISCUSSION

The results from this study establish that the PROVANT Wound Closure System is effective for accelerating closure in pressure wounds. In a controlled trial PRO-VANT closed severe chronic wounds in 50% less time than for placebo-treated wounds. The ability of PROVANT to induce proliferation and closure even in severe, previously unresponsive wounds suggests that this biotechnology may be of significant value in the treatment of pressure and other chronic wounds.

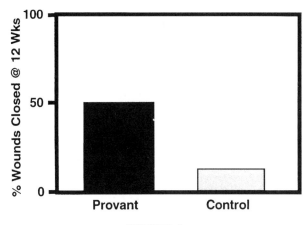

FIGURE 3.

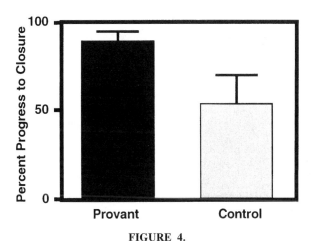

FIGURE 4.

REFERENCES

1. ALLMAN, R.M. 1989. Epidemiology of pressure sores in different populations. Decubitus **2:** 30–33.
2. EVANS, J.M. *et al.* 1995. Pressure ulcers: prevention and management. Mayo Clin. Proc. **70:** 789–799.
3. COTTER, D. & K. STEFANIK. 1990. Medicare reimbursement for pressure ulcers: strategy for change. Dermatol. Nurs. **2:** 343–345.
4. LI, R. *et al.* 1999. Cell proliferation induction (CPI®): dose- and time-dependent effects on fibroblast proliferation in vitro. FASEB J. **13**(268.3)**:** A351.
5. LUKAS, R.J. *et al.* 1999. Cell proliferation induction (cpi): *in vitro* evidence for enhanced diffusible growth factor release related to accelerated wound healing. FASEB J. **13**(503.5)**:** A683.

Interconnections between Inflammatory and Immune Responses in Tissue Engineering

JULIA E. BABENSEE,[a,b] MELISSA M. STEIN,[a] AND LEAH K. MOORE[a]

[a]Wallace H. Coulter Department of Biomedical Engineering, Georgia Institute of Technology, Atlanta, Georgia 30332, USA
[b]Emory University, Atlanta, Georgia, USA

KEYWORDS: tissue engineering; antigen-presenting cell; antigens; immunology

In a tissue-engineered device, implanted cells or proteins combined with a biomaterial can serve as a source of antigens such as foreign proteins secreted by the cells, including the therapeutic agent, cell surface molecules, or cell debris. It is hypothesized that the polymeric biomaterial, by promoting a nonspecific inflammatory reaction, recruiting antigen-presenting cells (APCs) [e.g., dendritic cells (DCs) or macrophages] and inducing their activation, may act as an adjuvant in the immune response to antigens originating from the device. An immune response to antigens originating from the device would sensitize the host to the transplanted cells or proteins and initiate a potentially humoral or molecular cytotoxic response around the implant to enhance an inflammatory tissue reaction. This could directly affect transplanted-cell viability and device function. Basic immunology studies are in progress using model antigens, polymers, and cell/polymer constructs to assess whether an enhanced immune response (humoral and cellular) is observed *in vivo* due to the presence of the biomaterial. The overall objective is to test the hypothesis that the biomaterial component of a tissue-engineered construct acts as an adjuvant to antigens released from the construct and decipher the mechanism. Understanding the interconnections between an inflammatory response to a material and an immune response toward associated antigens is pivotal to controlling and designing well-integrated, physiologically functional devices.

To model the response to shed cellular antigens from a tissue-engineered device, a simplified system is used in which a model antigen such as human α_1-antitrypsin (AAT) or ovalbumin (OVA) is incorporated with polymeric vehicles used in tissue engineering (e.g., microparticles, scaffolds, encapsulated cells). The antigen-specific humoral response is determined as a function of time in mice following vehicle implantation and any enhancement, due to the presence of the biomaterial, is

Address for correspondence: Julia E. Babensee, Wallace H. Coulter Department of Biomedical Engineering, Georgia Institute of Technology, 315 Ferst Dr., Atlanta, GA 30332-0535. Voice: 404-385-0130; fax: 404-894-4243.
julia.babensee@bme.gatech.edu

Ann. N.Y. Acad. Sci. 961: 360–363 (2002). © 2002 New York Academy of Sciences.

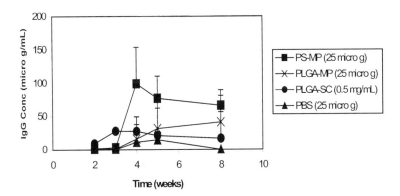

FIGURE 1. C3H mouse systemic anti-AAT IgG levels as a function of time for different polymeric vehicles, PS-MP (6 μm; 25 μg AAT), PLGA-MP (6 μm; 25 μg AAT), PLGA-SC (0.5 mg/mL), and PBS (25 μg); mean ±SD, $n = 5$ mice.

noted. Polystyrene microparticles (PS-MP, Polysciences; 6 μm) and poly(lactic-*co*-glycolic acid) microparticles (PLGA; 75:25, Birmingham Polymers) (PLGA-MP; 6 μm in diameter), prepared by a single-emulsion solvent-extraction technique,[1] were used as model materials. PLGA (75:25) scaffolds (PLGA-SC; 0.7-cm diameter, 0.2 cm thick) were fabricated using a solvent-casting particulate-leaching technique [85% (w/v) 300–500 μm NaCl].[2] MPs suspended in 0.25, 0.025, or 0 (negative control) mg/mL of AAT in PBS were injected subcutaneously (s.c.) (100 μL) in the dorsal site of C3H mice. Prewetted PLGA-SCs, presoaked in a 0.5 mg/mL AAT in PBS or in PBS alone (negative control), were implanted s.c. in the dorsal site of C3H mice (1 per mouse). Priming mice with 0.25, 0.0025, or 0 (negative control) mg/mL of AAT in 1:1 dilution of Complete Freund's Adjuvant (CFA) in PBS served as positive controls. Three weeks after immunization, mice were boosted with the same AAT concentration in the original carrier type (for CFA group, Incomplete Freund's Adjuvant (IFA) was used). Serum mouse hAAT-specific IgG antibody titer analyzed using an ELISA technique at 2, 3, 4, 5, and 8 weeks.

PS-MP, and to a lesser extent PLGA-MP, elicited a moderate immune response to hAAT, but one lower than that using CFA (FIG. 1). hAAT administered with PLGA-SC elicited an early, low immune response with an antibody concentration approximately twice that observed with ATT immunization in PBS (negative control) for up to 4 weeks postimplantation. The level of the immune response elicited with MPs or SC was moderate and was maintained for at least 8 weeks, the ultimate duration and consequences of which remain to be determined.

Using OVA as the antigen, we can also characterize *in vivo* T-cell proliferation in response to released OVA. OT-I and OT-II are transgenic mice that express T-cell receptors (TCRs) specific for OVA presented in the context of major histocompatibility complex (MHC) class I (CD8+-restricted) or MHC class II molecules (CD4+-restricted), respectively. Spleenic OVA-specific T cells from OT-I or OT-II mice are fluorescently labeled with 6-carboxyfluorescein diacetate succinimidyl ester (CS-FE) before adoptive transfer into wild-type C57BL/6 mice that are subsequent recipients of injections/implantations of polymeric vehicles delivering OVA. The

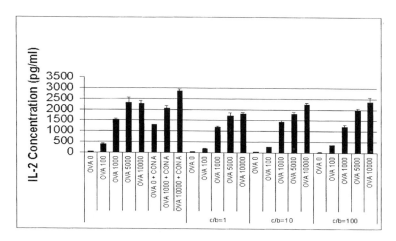

FIGURE 2. IL-2 production as an indicator of T-cell proliferation following 24-h culture of OT-II spleen T cells (5×10^5 cells/well) with treatments of PS-MP (6 μm) at cell-to-bead ratios of 1, 10, or 100, and/or OVA at concentrations of 0, 100 μg/mL, 1000 μg/mL, 5000 μg/mL, and 10,000 μg/mL. Positive control: 2.5 μg/mL of CON A.

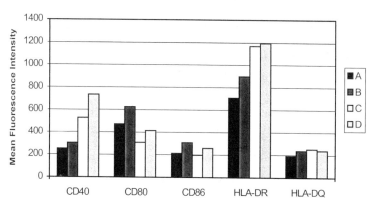

FIGURE 3. Mean fluorescent intensity for expression of costimulatory molecules (CD40, CD80, and CD86) and MHC class II molecules (HLA-DR and HLA-DQ) after 9 days of culture of human (**A**) macrophages, (**B**) iDCs, and mDCs induced with (**C**) TNF-α, or (**D**) LPS; average of $n = 2$.

proliferation of these cells, as an indicator of wild-type T- cell proliferation, is determined using flow cytometry (FC) by following reduction in fluorescence intensity as cells divide. Any *in vivo* proliferation of these T cells can be attributed to factors other than direct biomaterial-induced activation, since OVA-specific transgenic T cells from OT-II mice secrete IL-2 in an OVA concentration-dependent manner that is not enhanced by the presence of MP (FIG. 2).

Human immature DCs (iDCs) are derived from purified monocyte precursors in peripheral blood by selective culturing for adherent cells in media containing GM-CSF (50 ng/mL) and IL-4 (1000 U/mL) for 6 days. Mature DCs (mDCs) are induced by proinflammatory stimuli, LPS (S. *abortus equi*; 1 μg/mL) or tumor necrosis factor-α (TNF-α; 50 ng/mL) for 4 days. FC analysis of MHC class II molecule and of costimulatory molecule expression demonstrated that after 9 days of culture, mDCs expressed higher levels of CD40 and HLA-DR as compared to iDCs or macrophages (FIG. 3). DCs cultured in cytokine-supplemented media are a useful tool to test the hypothesis that biomaterial contact is a DC maturation stimulus via direct or indirect effects affecting their ability to present antigen to T cells for effective stimulation.

In conclusion, these results indicate that the level of immune responses toward antigens released from tissue-engineered devices can be modulated by the presence of the biomaterial component of the construct. Development of biomaterials or other immunomodulation strategies that minimize DC maturation may be a way to control the immune response toward associated antigens in tissue engineering.

ACKNOWLEDGMENTS

Georgia Tech/Emory Center (GTEC) for the Engineering of Living Tissues, an ERC program of the NSF under award number EEC-9731643 provided partial support for this work through a seed grant. We thank Judith Kapp, Kyle McKenna (EU), Saul Lee, and Elizabeth Lester (GT). One of the authors (J.E.B.) thanks John Rodgers (Baylor College of Medicine), Larry McIntire and Antonios Mikos (Rice University), and the Natural Sciences and Engineering Research Council of Canada for support during her postdoctoral fellowship.

REFERENCES

1. WAKE, M.C. *et al.* 1998. Biomaterials **19:** 1255–1268.
2. ISHAUG, S.L. *et al.* 1997. J. Biomed. Mater. Res. **36:** 17–28.
3. SALLUSTO, F. & A. LANZAVECCHIA. 1994. J. Exp. Med. **179:** 1109–1118.

Modulation of Interferon-Gamma Response by Dermal Fibroblast Extracellular Matrix

ANDREAS KERN, KANG LIU, AND JONATHAN MANSBRIDGE

Advanced Tissue Sciences, Inc., La Jolla, California 92037, USA

KEYWORDS: dermal fibroblasts; interferon-gamma

INTRODUCTION

The objective of this ongoing study is to elucidate the molecular basis for the persistence of allogeneic dermal fibroblasts as observed in biopsies of diabetic foot ulcer wounds.[1] Acute rejection involves the activation of allo-reactive host T cells and requires the presence of stimulatory signals (HLA I or II) and co-stimulatory signals (CD40, etc.) on the donor cells.[2] Fibroblasts have been shown to express HLA II and CD40 in the presence of interferon gamma (INF-γ).[3,4]

We have shown that fibroblasts in three-dimensional scaffold-based cultures, in contrast to cells in monolayer culture, fail to induce HLA-DR when stimulated with INF-γ.[5] The inhibition is reversible, not shown by cells in collagen gels, and appears to be a function of fibroblast-secreted extracellular matrix.

The broader implications of these results are with the use of allogeneic cells for tissue engineering and transplantation. Allogeneic approaches may allow for off-the-shelf storage and minimize safety concerns.

MATERIALS AND METHODS

Monolayer, collagen gel, or vicryl-knit-based fibroblast cultures were incubated with INF-γ (Roche Biochemicals, MN). Rejection-related molecules (HLA I and II, CD40 and CD80) were analyzed by flow cytometry (reagents by Pharmingen, CA). STAT-1 and Erk phosphorylation and the activation of p38 were determined by Western blot analysis (Upstate Biochemicals, NY, and New England Biolabs, MA). Messenger RNA was quantified by TaqMan (Perkin-Elmer).

Address for correspondence: Andreas Kern, Ph.D., Advanced Tissue Sciences, Inc., 10933 North Torrey Pines Road, La Jolla, CA 92037. Voice: 858-713-7120; fax: 858-713-7970.
andreas.kern@advancedtissue.com
kang.liu@advancedtissue.com
jonathan.mansbridge@advancedtissue.com

Ann. N.Y. Acad. Sci. 961: 364–367 (2002). © 2002 New York Academy of Sciences.

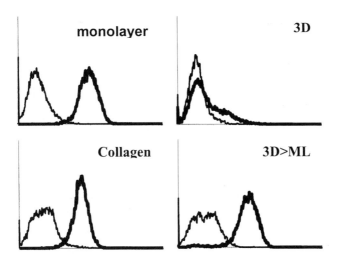

FIGURE 1. Expression of HLA II in dermal fibroblasts. Fibroblasts were grown (as indicated in the graph) in the presence (*bold line*) or absence (*fine line*) of INF-γ, isolated, and analyzed for HLA-DR expression. The ordinates represent the frequency of events and the abscissa the intensity of fluorescence of a 4-decade log scale.

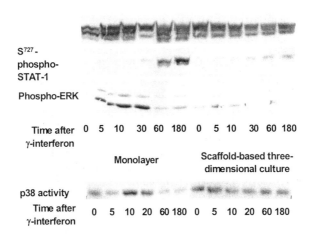

FIGURE 2. Phosphorylation events following INF-γ exposure. Fibroblasts in monolayer or three-dimensional culture were incubated with INF-γ and extracted with RIPA buffer at the indicated timepoints. Extracted proteins were separated by SDS-PAGE, blotted onto nitrocellulose and probed with antibodies for STAT-1 (*top row*), phospho-S727-STAT-1 (*center row*), or phospho-ERK. Cell extracts were also analyzed for p38 activity.

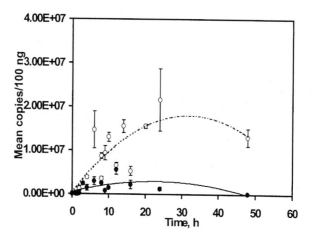

FIGURE 3. Comparison of the induction of CIITA by INF-γ in fibroblasts in monolay-er and three-dimensional culture. Cultures were incubated with INF-γ as indicated and RNA was prepared. Messenger RNA for CIITA was determined using real-time PCR. All values were normalized to total RNA. ○, monolayer culture; ●, three-dimensional culture. Lines mark trend lines; bars are mean ± SEM.

RESULTS

Addition of INF-γ to fibroblasts resulted in upregulation of HLA II (FIG. 1), HLA I, and CD40.[5] Cells in collagen gel or monolayer culture upregulated the expression as a single population, that is, all cells upregulated the expression. In contrast, only 25–30% of fibroblasts embedded in their three-dimensional extracellular matrix (3-D) induced CD40 or HLA II, whereas all cells increased HLA I expression. The in-hibition of HLA II expression is reversible, since all cells expressed HLA II in the presence of INF-γ when isolated from three-dimensional culture and grown in mono-layer (3-D>ML, FIG. 1). No expression of CD80 was observed.[5]

INF-γ caused Tyr-phosphorylation of STAT-1 and its subsequent relocalization to the nuclear fraction of cells,[5] as previously reported.[6] However, cells in scaffold-based cultures, in contrast to monolayer cells, did not exhibit Ser-phosphorylation of STAT-1 (FIG. 2). We also observed differences in activation of Erk and p38 kinases (FIG.2). Induction of a molecule important for the expression of MHC class II, class II transactivator, or CIITA was also reduced in the scaffold-based cultures by com-parison with monolayer (FIG. 3).

CONCLUSION

The extracellular matrix modulates the fibroblast response to INF-γ. It is our hy-pothesis that the extracellular matrix directly or indirectly modulates the host's im-mune response to tissue-engineered allografts by suppressing the upregulation of proteins required for T-cell activation.

REFERENCES

1. MANSBRIDGE, J.N., K. LIU, *et al.* 1999. Growth factors secreted by fibroblasts: role in healing diabetic foot ulcers. Diabetes Obes. Metab. **1:** 265–279.
2. BENICHOU, G. 1999. Direct and indirect antigen recognition: the pathways to allograft immune rejection. Front. Biosci. **4:** D476–480.
3. FRIES, K.M., G.D. SEMPOWSKI, *et al.* 1995. CD40 expression by human fibroblasts. Clin. Immunol. Immunopathol. **77:** 42–51.
4. GEPPERT, T.D. & P.E. LIPSKY. 1985. Antigen presentation by interferon-gamma-treated endothelial cells and fibroblasts: differential ability to function as antigen-presenting cells despite comparable Ia expression. J. Immunol. **135:** 3750–3762.
5. KERN, A., K. LIU, *et al.* 2001. Modification of fibroblast gamma-interferon: responses by extracellular matrix. J. Invest. Derm. **117:** 112–118.
6. STARK, G.R., I.M. KERR, *et al.* 1998. How cells respond to interferons. Annu. Rev. Biochem. **67:** 227–264.

Differential Inflammatory and Immunological Responses in Tissue Regeneration and Repair

RAYMOND E. SICARD

Center for Wound Healing & Reparative Medicine, University of Minnesota, Minneapolis, Minnesota 55455, USA

ABSTRACT: Repair and regeneration, contrasting resolutions to injury, evoke distinct inflammatory responses. Wound repair evokes a more robust peripheral leukocytosis and is accompanied by a larger inflammatory infiltrate into the wound site. Proliferation of thymocytes and splenocytes was enhanced following injury; however, the ability to be stimulated by myoblasts in coculture is lost by splenocytes harvested from rats undergoing wound repair. These data reflect differences in inflammatory and immunological responses between repair and regeneration and suggest potentially important dialogues occur between injured tissues and inflammatory cells that direct the course of injury resolution.

KEYWORDS: tissue regeneration; inflammatory responses

Repair and regeneration are contrasting responses to injury promoted by distinct local effector cells.[1] Inflammatory and immunological reactions are normal sequelae to injury[2]; however, their specific influences on and contributions to the course of post-injury resolution are incompletely defined. The magnitude of inflammatory and immunological reactions appears to affect the degree of fibroplasia and fibrosis in adult mammalian wound repair[3] as well as the quality of mammalian fetal wound healing.[4] In addition, inflammatory reactions and immunological status affect regeneration of amphibian appendages.[5] However, questions of the influences of inflammatory and immunological reactions on tissue regeneration in adult mammals has been largely neglected.

Polyvinyl alcohol sponges[6,7] and minced skeletal muscle[7,8] were used as repair and regeneration models, respectively, in this study. Three experimental conditions (TABLE 1) were examined: "wound repair," the normal fibroplastic/fibrotic response to injury; "regeneration," accompanied by formation of new myotubes within minced skeletal muscle; and "muscle repair," reflected in fibrotic replacement of muscle fibers within minced muscle in which satellite cells were killed. Inflammatory responses were assessed by standard hematological techniques while immuno-

Address for correspondence: Raymond E. Sicard, Anatomy Section, School of Osteopathic Medicine, Pikeville College, 147 Sycamore Street, Pikeville, KY 41501. Voice: 606-218-5426; fax: 606-218-5442.

rsicard@pc.edu

Ann. N.Y. Acad. Sci. 961: 368–371 (2002). © 2002 New York Academy of Sciences.

TABLE 1. Leukocyte counts and cellular infiltrates during repair and regeneration

Condition	Day 0	Day 7	Day 14
Leukocyte counts ($\times 10^3$ cells/μL)			
Control (no injury)	10.6 ± 0.4 (28)	11.2 ± 0.7 (12)	9.8 ± 0.3 (10) [a]
Wound repair	—	17.0 ± 1.1 (6)[a,b]	11.0 ± 0.8 (7)[d]
Regeneration	—	14.6 ± 1.0 (7)[a,b,c]	8.9 ± 0.5 (5)[a,c,d]
Muscle repair	—	12.6 ± 0.3 (2)	ND
Cellular Infiltrates ($\times 10^3$ cells/mg tissue)			
Wound repair—granulation tissue	—	18.4 ± 2.1 (3)	18.2 ± 5.0 (4)
Regeneration—granulation tissue	—	22.0 ± 3.3 (3)	20.0 ± 5.4 (2)
—regenerating muscle	—	13.9 ± 3.4 (3)[c]	10.0 ± 1.7 (3)[c]
Muscle repair—granulation tissue	—	22.5 ± 2.6 (2)	ND
—regenerating muscle	—	26.3 ± 18.3 (2)	ND

NOTE: Data are means $\pm$ 1 SEM for the number of samples specified in parantheses; analyses by unpaired t-test. ND = not determined.

[a] $P < 0.05$ vs. t_0.
[b] $P < 0.05$ vs. control.
[c] $P < 0.05$ vs. wound repair.
[d] $P < 0.05$ vs. day 7.

logical status was inferred from splenocyte and thymocyte responses *in vitro*. Samples were collected and evaluated 3–14 days post injury and compared to samples obtained at time 0 (preinjury), where appropriate.

Mean leukocyte counts were elevated by 54% and 13% (7 and 14 days) in repair, but only by 20% (7 days only) in regeneration (TABLE 1). Phagocytes were increased initially; but at 14 days post injury, monocytes remained elevated in repair while lymphocytes were reduced in regeneration (data not shown). Cellular infiltrates into granulation tissues were similar for both models (18,200–22,500 cells/mg); whereas, cellular infiltrates were markedly reduced in regenerating muscle (13,900 vs. 26,300 cells/mg) (TABLE 1). Thus, the magnitude of the inflammatory reactions in wound repair were greater and more aggressive than those in regeneration. In addition, inflammatory infiltrates of connective tissue were comparable in magnitude regardless of the post-injury process occurring. However, the cellular distribution of the infiltrates might have reflected local or neighboring responses.

Splenocyte and thymocyte proliferation was enhanced 3 days post injury (TABLE 2); however, responsiveness to mitogens was abolished (data not shown). Proliferation of normal thymocytes and splenoctyes was stimulated by co-culture with myoblasts (1.8- and 1.5-fold), but not by co-culture with fibroblasts (TABLE 3). At 3 days post injury, thymocyte responses were similar in both processes. In contrast, splenocytes from injured rats were stimulated by fibroblasts (1.2- to 1.3-fold) whereas those from rats undergoing repair, but not regeneration, became unresponsive to myoblasts (TABLE 3). Thus, it appears that intrinsic proliferation, responsiveness to B- and T-cell mitogens (e.g., bacterial lipopolysaccharide and concanavalin A), and stimulation by cellular challenge of rat splenocytes and thymocytes are altered by in-

TABLE 2. Proliferation of thymocytes and splenocytes

| | [³H]-Thymidine Incorporation (cpm/10^5 cells) | | |
| | | 3 days postinjury | |
Cells	Control (t_0)	Wound repair	Muscle regeneration
Thymocytes			
No serum	2452 ± 1001 (4)	6273 ± 3211 (3)	7629 ± 5972 (2)
+ FBS	2978 ± 1223 (4)	6807 ± 2554 (3)	5951 ± 4786 (2)
Splenocytes			
No serum	5811 ± 769 (4)	14775 ± 1290 (2)	9540 (1)
+ FBS	4881 ± 3022 (4)	12316 ± 5434 (2)	8736 (1)

Note: Data are means ± 1 SD; sample sizes are noted in parantheses; FBS, 10% fetal bovine serum.

TABLE 3. Thymocyte and splenocyte stimulation by myoblasts and wound fibroblasts

Responder cells	Stimulation index = (response with stimulator cells) / (response without stimulator cells)		
		Rat L8 Myoblasts	
	Model: control (t_0)	wound repair	muscle regeneration
Thymocytes	1.84 ± 1.20	1.65 ± 0.73	1.91 ± 0.76
Splenocytes	1.52 ± 0.74	1.00 ± 0.00	1.30 ± 0.08
		Rat wound fibroblasts	
	Model: control (t_0)	wound repair	muscle regeneration
Thymocytes	1.00 ± 0.00	1.00 ± 0.00	1.00 ± 0.00
Splenocytes	1.00 ± 0.00	1.32 ± 0.45	1.21 ± 0.01

NOTE: Data are means ± 1 SD (n = 2–3 assays); thymocytes and splenocytes harvested 3 days post-injury.

jury. In addition, proliferation of splenocytes during repair appears to be more robust than in regeneration, whereas proliferation of thymocytes is increased comparably in both repair and regeneration. Properties of splenocytes and thymocytes from rats engaged in repair differ from those undergoing tissue regeneration, as reflected in their responses to mitogens and co-culture with myoblasts and fibroblasts.

Regeneration and repair are accompanied by inflammatory and immunological responses that potentially affect both the progress and quality of post injury outcome.[1-5,9] The results of this study suggest that regeneration and repair are accompanied by distinct endogenous inflammatory and immunological responses that potentially reflect dynamic dialogues between local effector cells and the inflammatory infiltrate, which, in turn, qualitatively affect postinjury resolution. Further studies are required to better define these interactions and their influence on post-injury resolution. Moreover, since stems cells and engineered tissue constructs hold the potential for altering tissue responses to injury and/or augmenting endogenous regenerative capacities,[10] effects of these endogenous dialogues on such engineered constructs or stem cells must also be considered.

ACKNOWLEDGMENTS

Technical assistance was provided by Wendy A. Mand; this work was supported by NIH Grant GM-50882.

REFERENCES

1. SICARD, R.E. 1998. Regeneration: the road not taken. J. Minn. Acad. Sci. **63:** 1–9.
2. SICARD, R.E., J.D. SHEARER & M.D. CALDWELL. 1998. Wound repair. J. Minn. Acad. Sci. **63:** 31–36.
3. CLARK, R.A.F., Ed. 1996. The Molecular and Cellular Biology of Wound Repair. Plenum. New York.
4. OLUTOYE, O.O. & I.K. COHEN. 1996. Fetal wound healing: an overview. Wound Rep. Reg. **4:** 66–74,
5. SICARD, R.E., Ed. 1985. Regulation of Vertebrate Limb Regeneration. Oxford University Press. New York.
6. SCHILLING, J.A., W. JOEL & H.M. SHURLEY. 1959. Wound healing: a comparative study of the histochemical changes in granulation tissue contained in stainless steel wire mesh and polyvinyl alcohol sponge cylinders. Surgery **46:** 702–710.
7. SICARD, R.E. & L.M.P. NGUYEN. 1996. An *in vivo* model for evaluating wound repair and regeneration microenvironments. In Vivo **10:** 477–482.
8. CARLSON, B.M. 1972. Regeneration of Minced Muscles. Monogr. Devel. Biol. Vol. 4. S. Karger. Basel.
9. SICARD, R.E. 2000. Mechanisms of muscle regeneration. *In* Regeneration Medicine and Tissue Engineering. Section 5: Self Renewal of Tissue Functions by Healing Mechanisms. K Sames, Ed. ECOMED Verlag. Landsburg.
10. STOCUM, D.L. 1995. Wound Repair, Regeneration, and Artificial Tissues. R.G. Landes Co. Austin, TX.

From Lab Bench to Market

Critical Issues in Tissue Engineering

GAIL K. NAUGHTON

Advanced Tissue Sciences, La Jolla, California 92037, USA

ABSTRACT: Revolutionary advances in tissue engineering are redefining approaches to tissue repair and transplantation through the creation of replacement tissues that remain biointeractive after implantation, imparting physiologic functions as well as structure to the tissue or organ damaged by disease or trauma.[1,2] Over the last decade this field has moved from "science fiction" to "science fact" with the research-oriented acceptance of its potential to regulatory approvals allowing commercial products to be available for use in many countries. The maintenance of tissue integrity, functionality, and viability from cell seeding through product manufacture, shipping, and end-use has been accomplished through innovations in design and scale-up of both tissue growth and preservation processes. These unique systems have enabled the delivery of tissue-engineered products that are uniform inter- and intra-lot, readily available as off-the-shelf products, easy to use, and efficacious. Skin replacement products are the most advanced, with several tissue-engineered wound care materials on the market in the U.S. and in several international communities.[3–5] The potential impact of this field is far broader, offering novel solutions to the medical field for drug screening and development, genetic engineering, and total tissue and organ replacement.

KEYWORDS: scaffolds; bioreactors; cryopreservation

WHERE DO YOU GET THE CELLS? (CELL SOURCING)

Work in the development of human fibroblast-based tissue-engineered constructs has helped to establish critical issues that must be addressed in the creation of new tissues and organs for use in repair and replacement therapies (FIG. 1). The successful cost-effective large-scale production of "off-the-shelf" engineered tissues requires an adequate source of healthy cells which can undergo at least six cell passages before seeding onto a scaffold. Cells sources for tissue engineering fall into three categories: autologous cells (from the patient); allogeneic cells (from a human donor, but not immunologically identical); and xenogeneic cells (donor from a different species). Each category may be further delineated in terms of stem cells (adult or embryonic) or "differentiated" cells obtained from tissue, where the cell population obtained from tissue

Address for correspondence: Gail K. Naughton, Advanced Tissue Sciences, Inc., 10933 N. Torrey Pines Road, La Jolla, CA 92037.

Ann. N.Y. Acad. Sci. 961: 372–385 (2002). © 2002 New York Academy of Sciences.

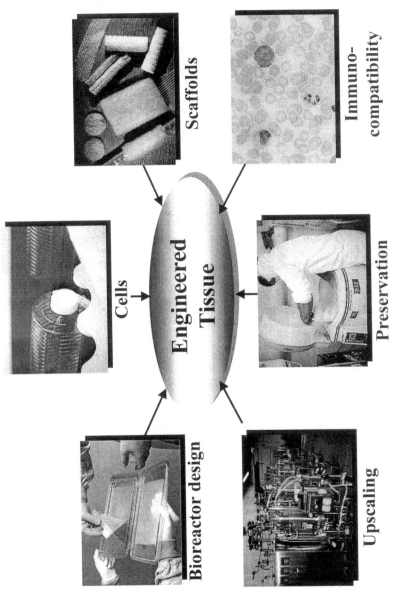

FIGURE 1. *See following page for legend.*

dissociation comprises a mixture of cells at different maturation stages and includes rare stem and progenitor cells. Some approaches use the whole cell mix, whereas other approaches rely on separation or enrichment of stem cells.

The four skin products that have been approved by the FDA to date have relied on cells derived from neonatal foreskin. This starting material reduces variability by being derived from tissue that is the same age, sex, and anatomical location, and provides a fibroblast source with great proliferation potential, with one foreskin being able to produce starting cells for at least 250,000 feet of final tissue-engineered product.[6] As product development moves beyond skin we must look to cell sources that offer the same reproducibility and expandability as well as immune tolerance. An emerging approach to tissue-engineering–based therapies is the use of programmable stem cells, which includes embryonic stem cells [7,8] as well as progenitor cells from adult tissues and mesenchymal stem cells derived from bone marrow and peripheral blood.[9,10]

WHAT DO YOU GROW THEM ON? (SCAFFOLDS)

The optimization of scaffolds onto which cells are seeded has also been found to be key to the uniform formation of tissue, with scaffolds that provide physical and chemical cues to guide the process. Scaffolds are porous structures fabricated from natural materials such as collagen and fibrin or from synthetic materials such as degradable polyesters used in surgical sutures.[11] Scaffolds take forms ranging from sponge-like sheets and fabrics to gels to highly complex structures with intricate pores and channels made with new materials processing technologies. The spatial and compositional properties of the scaffold are key, with the porosity of the scaffold and interconnectivity of the pores being capable of enabling cell penetration into the structure as well as the transport of nutrients and waste products.

Polylactic acid (PLA) and polyglycolic acid (PGA) are among the few synthetic degradable polymers that have been approved for clinical use and these have been widely studied in tissue development.[12] Although the characteristics of these polymers are well understood, they merely serve to provide a 3-D biocompatible structure onto which cells can attach and do not interact with the cells. In the body, cells are situated within an extracellular matrix (ECM), which provides tissues with the appropriate architecture as well as signaling pathways that influence key cell function such as migration, proliferation, and differentiation. Regeneration of tissues *in vivo*—guided growth of nerve, bone, vessels, or cornea, across critical size defects that will not heal on their own—requires that cells be given a more specific level of instruction so that regeneration ultimately triumphs over simple repair. Starting with the discovery of RGD-containing adhesion domains embedded in fibronectin and

---

FIGURE 1. Critical issues in tissue engineering. The development of tissue-engineered products that meet the needs of clinical efficacy, quality, and regulatory requirements, large volume manufacture, and world-wide distribution requires expertise in each of several basic areas of research and development. Included are procurement and expansion of cells, development of biocompatible scaffolds that meet *in vitro* and *in vivo* tissue growth needs, solving immunological issues, and creating bioreactors and large-scale manufacturing and preservation systems to enable cost-effective production and distribution.

TABLE 1. Matrix proteins including glycosaminoglycans in Dermagraft®, an engineered human dermal replacement tissue

Matrix Proteins	Function
Collagens, types I and III	Major structural protein of dermis
Fibronectin	Cell adhesion, spreading, migration, mitogenesis
Tenascin	Induced in wound-healing; control of cell adhesion

Glysosaminoglycans	Function
Veriscan	Structural; binds hyaluronic acid and collagen
Decorin	Binds growth factors, influences collagen structure
Betaglycan	TGF-β type III receptor
Syndecan	Binds growth factors; enhances activity

other extracellular matrix molecules,[13] the design of synthetic extracellular matrices has become an area of intense activity. Research continues on the utilization of matrix molecules along with specific growth factors to optimize cell adherence to the scaffolds and direct cell activity.

HOW DO YOU MAKE THE CELLS MAKE A TISSUE? (BIOREACTORS)

Bioreactors have and continue to be developed to provide researchers with a system capable of controlling environmental factors such as pH, temperature, oxygen tension, and mechanical forces. In addition to creating a physiological environment, bioreactors utilized in a closed manufacturing system allow for the seeding of cells, as well as the growth, freezing, shipping, and storage of the tissue-engineered product all within the same container, resulting in a maintenance of sterility, reduction in labor, and the elimination of the need for sterile repackaging.

Dermal fibroblasts grown in three dimensions under either periodic or continuous perfusion have been shown to secrete a variety of matrix proteins which are similar to the structure and composition of matrix components in normal neonatal dermis.[14,15] A list of matrix components and a partial list of the key growth factors produced by fibroblasts during the manufacturing of tissue-engineered dermis are listed in TABLES 1 and 2, respectively. In the manufacturing of other connective tissue structures which require more strength and elasticity upon transplantation the answer may lie in applying appropriate physiological stresses during engineering of the tissues *in vitro*. Bioreactors have been designed that subject growing tissue to a variety of forces, including compression, shear stresses, and pulsatile flow of culture media. Use of these bioreactors has shown a dramatic improvement in the mechanical properties of engineered cartilage, blood vessels, and cardiac muscle. Such bioreactors may also serve to allow for mechanical testing of all tissues prior to implantation

Compression, fluid flow, and hydrostatic pressure have been shown to regulate matrix metabolism in articular cartilage *in vivo*[16] and in cartilage explants and chondrocyte cultures *in vitro*.[17] Previous work has shown that compression modulates

TABLE 2. Growth factors in Dermagraft®, an engineered human dermal replacement tissue

Growth Factors	Name	Function
Matrix deposition factors		
Transforming growth factor β_1	TGF-β1	Stimulates matrix deposition
Transforming growth factor β_3	TGF-β3	Stimulates matrix deposition, anti-scarring
Mitogenic factors		
Platelet-derived growth factor A chain	PDGF-A	Mitogen for fibroblasts, granulation tissue, chemotactic
Insulin-like growth factor 1	IGF-1	Mitogen for fibroblasts
Keratinocyte growth factor	KGF	Mitogen for keratinocytes
Heparin-binding epidermal growth factor-like growth factor	HBEGF	Mitogen for keratinocytes, fibroblasts
Transforming growth factor α	TGF-α	Mitogen for keratinocytes, fibroblasts
Angiogenic factors		
Vascular endothelial growth factor	VEGF	Angiogenesis
Hepatocyte growth factor	HGF	Angiogenesis
Basic fibroblast growth factor	bFGF	Angiogenesis
Secreted protein acidic and rich in cysteine	SPARC	Both anti- and pro-angiogenic
Interleukin 6	IL-6	Angiogenic, inflammatory cytokine; inhibitory effect on collagen synthesis
Interleukin 8	Il-8	Angiogenic, inflammatory cytokine; chemoattractive to neutrophils
Inflammatory cytokines		
Interleukin 6	IL-6	Angiogenic, inflammatory cytokine; inhibitory effect on collagen synthesis
Interleukin 8	IL-8	Angiogenic, inflammatory cytokine; chemoattractive to neutrophils
Granulocyte colony-simulating-factor	G-CSF	Stimulates neutrophil production and maturation
Tumor necrosis factor α	TNF-α	Inflammatory cytokine

cellular proliferation, matrix metabolism, and matrix content in cartilage explants and tissue-engineered constructs.[18] Static compression has been shown to inhibit matrix synthesis,[19] while dynamic loading and perfusion stimulates synthesis.[20] The incorporation of conditions which support matrix synthesis into a bioreactor design will allow the production of a tissue that more closely resembles the composition and mechanical strength of native cartilage and the optimization of implants for the repair of cartilage and osteochondral defects.

Progress in the development of bioreactors for the optimization of tissue-engineered fibroblast-based and chondrocyte-based constructs has helped in the design of conditions for the production of cardiovascular implants. The engineering of a tissue with more than one cell type is obviously more of a challenge than tissues based on only one cell type. Small-diameter vascular grafts are an excellent example of such a challenge. Vascular cells are arranged in distinct patterns in the multiple layers of the wall of an artery, with vascular smooth muscle cells aligned circumferentially and endothelial cells longitudinally. Research has shown that mechanical and pulsatile forces are key in inducing matrix secretion and subsequent mechanical strength of a tissue-engineered vessel.[21] This is consistent with studies which have highlighted the role of strain and fluid shear stress in regulating cardiovascular tissue growth and remodeling during development and in the disease state.[22] Recent work has also illustrated the importance of shear stress in the physiological orientation of vascular cells. For endothelial cells it has been well established that unidirectional fluid shear stress induces cell orientation parallel with the direction of flow.[23] Experimentation on the effect of fluid sheer stress on smooth muscle cells has shown an opposite result, with cell alignment being perpendicular to the direction of flow,[24] resulting in an orientation identical to that seen *in vivo*. Both cell types initiate cell elongation and stress fiber reorganization and similar differential responses to varying magnitudes and exposure times of shear stress. Such data have been integral in optimizing the engineering of vascular grafts, with the initial seeding of smooth muscle cells and the application of mechanical and shear forces for the induction of perpendicular cell alignment and matrix deposition for increased elasticity and durability of the graft. Prior to implantation endothelial cells are seeded onto the internal lumen of the graft with shear forces promoting cell attachment and parallel alignment to the direction of flow. Small-diameter vascular grafts produced under such bioreactor conditions have shown excellent patency and lack of restenosis in preclinical studies.[25] Such work is promising for the approximately 500,000 patients who require coronary bypass surgery in the United States each year, with an off-the-shelf tissue-engineered graft providing healthier tissue and considerably less morbidity than that which is commonly seen with traditional autologous saphenous vein grafts.

HOW DO YOU MAKE MORE THAN ONE AT A TIME? (UPSCALING)

Over the last several decades, technologies have been developed that promote the growth of many human cells outside of the body in a variety of laboratory containers offering both small and expanded cell surface areas. The movement from growing monolayers of cells alone to three-dimensional tissues for transplantation has resulted in the development of various automated and semi-automated processes which support the manufacture of a uniform, reproducible tissue while maintaining an aseptic environment. Approaches to scale-up currently utilized include the growth of large pieces of tissue which are aseptically cut and packaged after tissue harvesting, growth of individual pieces of tissue in individual petri dishes manufactured in large batches, and the use of closed systems which feed hundreds of tissue units simultaneously. Since parallel processing of product units is readily amenable to scale-up and enables development of a closed bioreactor system,[26] the upscaling of such a

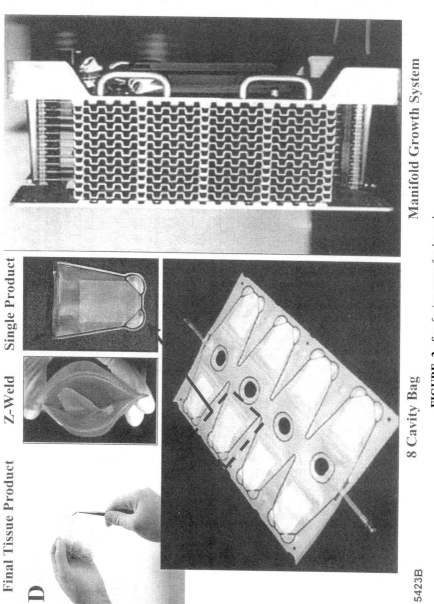

FIGURE 2. *See facing page for legend.*

Final Tissue Product

Z-Weld

Single Product

Manifold Growth System

8 Cavity Bag

D

5423B

manufacturing system will be described here. All tubing, manifolds, and packaged scaffolds can be sterilized prior to the addition of cells and maintained closed throughout manufacture. The final separation of individual product units can be done prior to final preservation. Advantages of this design include uniformity of environment between products in a lot, simple lot definition, amenability to batch processing and incremental scale-up, reduction in labor, and the maintenance of an aseptic environment. The manufacturing process utilized in the engineering of Dermagraft[®], a living human dermal implant approved for the treatment of diabetic foot ulcers, will be used as an example of such a closed system.

Fibroblasts are enzymatically removed from neonatal foreskins and fully screened for pathogens and other infectious agents[12,14] according to the U.S. Food and Drug Administration's (FDA) Points to Consider [27] as well as the Guidelines from the Committee for Proprietary Medicinal Products (CPMP) regulations.[28] After successful screening, fibroblasts are expanded and cryopreserved by standard methods[29,30] and stored in a Manufacturer's Working Cell Bank. Passage 8 cells are utilized for final product manufacture. In the closed manufacturing system cells are seeded onto 2 in. × 3 in. Vicryl[®] (Ethicon, Sommerville, NJ) scaffolds which are laser z-welded into EVA (ethylene vinyl acetate) bioreactors (FIG. 2A) to provide uniformity of environment by securing the scaffold in the center of the bioreactor, to prevent tissue contraction during growth, as well as to retain the product during the final rinsing step prior to implantation into a wound. Multiple process units (8) are located within each EVA sheet, with a total of 12 sheets per manifold (FIG. 2B). The corrugated design of the manifold growth system allows for uniform media distribution, removal of air bubbles, and even heat conduction and unhindered supply of oxygen and carbon dioxide (FIG. 2C). Individual manifolds are rotated after initial cell seeding to result in uniform cell distribution across all scaffold surfaces. During the manufacturing process, developing dermal tissues are fed with serum-containing cell culture medium[31,32] with several media changes over the approximate two-week growth period. Characterization of cell growth and matrix deposition in this system has been previously reported[12,14,15,31,32] and the progression of cell growth and deposition of matrix proteins during cultivation in the bioreactor system studied (FIG. 3) Automated feeding is performed and media samples are taken during the growth process to assess the level of glucose consumption and waste production and to predict the time of tissue harvest. At the conclusion of the growth period, the closed production bioreactor system is separated by heat welding to result in individual packaging of the 2 in. × 3 in. final tissue products (FIG. 2D). This process allows for a uniform tissue product that is reproducible from lot to lot and allows for upscaling of the final lot size through the incremental addition of manifolds grown simultaneously. Cur-

---

FIGURE 2. Dermagraft Bioreactor System and Final Product Unit. One 2 in. × 3 in. piece of Vicryl[®] scaffold is Z-welded into an ethlene vinyl acetate (EVA) bioreactor into which cells are seeded to grow the dermal tissue. The scaffold is held in place along the top with Z-welded spot welds and along the bottom with bar welds. The single product unit is manufactured from an 8-cavity multiple processing bag, which is grown vertically in a corrugated manifold system, designed for even mass transfer and exposure to environmental conditions. The EVA bioreactor is designed to be translucent, enabling placement on the wound, tracing of the wound, and cutting of a piece of tissue to the desired size for implantation.

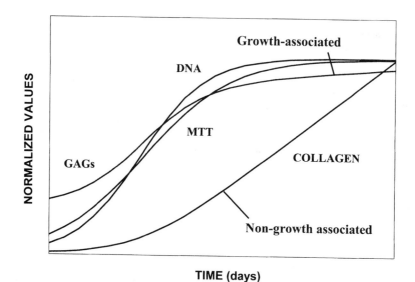

FIGURE 3. Growth time course of Dermagraft engineered tissue. Fibroblast cells seeded on the three-dimensional Vicryl scaffold undergo standard exponential growth as measured by DNA and the cell viability assay, MTT. As the cells grow, they deposit glycosaminoglycans (GAGs) in proportion to their growth (growth-associated). Once the cells have covered the scaffold, they reduce their growth and begin expressing other tissue matrix factors, including collagen at a non–growth-associated rate.

rent design provides for a lot that yields more than 1100 tissue units for release. The maintenance of asepsis, from introduction of cells into the bioreactor to delivery of a final tissue unit to the patient, negates the need for aseptic repackaging of a delicate, living tissue. Future challenges in tissue engineering will include the upscaling of bioreactors that put shear and mechanical forces on the developing tissues and the production of final products involving multiple cell types.

HOW DO THE TISSUES LAST THROUGH SHIPPING? (CRYOPRESERVATION)

Although maintenance of tissue integrity and viability for days to weeks for both engineered and native tissues has been reported with fresh preservation and protocols,[33–35] this method often imposes impractical scheduling restraints for manufacturing and end use. These fresh preservation procedures usually result in unrealistic manufacturing and physician operating costs associated with unplanned losses, overproduction, changing demand, and the management of patient scheduling In addition, non-preserved or short-term preserved products are shipped just-in-time and implanted at risk since USP sterility tests[36] requiring a 14-day period cannot be performed on these products. In addition to sterility, tests that ensure uniformity of tissue biochemical and biomechanical characteristics from lot to lot often cannot be performed due to time constraints. Instead, utilization of rapid testing methods is required which are usually not quantitative or as reliable as standard testing procedures, which normally require days or weeks to complete. An obvious benefit for products that undergo moderate- to long-term preservation is that their safety, characterization, and efficacy can be more rigorously proven and retained.

To maintain tissue integrity and functionality of a transplant product until needed for use, the product must be preserved. With the development of tissue-engineered constructs, the development of improved methods for the cryopreservation of tissues and organs has become an important issue. While much work has been performed on conditions for the maintenance of cell viability, particularly in cell suspensions, monolayer culture, and embryos,[37,38] little work has been done on studying the relationship between viability of cells and clinical transplant outcome with human tissues and organs.

Cryopreservation involves processes, such as nutrient deprivation and dehydration that are known to induce well-characterized stress responses.[39]

Recent advancements in technologies such as lyophilization[40,41] and vitrification[42,43] have enabled long-term storage of acellular tissues. In addition, long-term unfrozen storage of cells and tissues have been shown to be feasible through the addition of natural protectants, such as the sugar trehalose.[44] Progress in the field continues at a steady pace, although at the present time only cryopreservation employing low concentrations of cryoprotectants has been proven for moderate- to long-term maintenance of cellular viability and functionality in native and engineered tissues.

In cryopreservation of tissue a balance must be struck between minimizing exposure time to potentially toxic cryoprotectants and slow, stepwise addition of cryoprotectant to minimize osmotic forces. Use of a one-step DMSO (dimethyl sulfoxide)-based cryoprotectant added at 4°C, allowing a slow diffusion of cryoprotectant into the tissue, along with customized freezers, which provide a multi-step cooling, and continuous changes in freeze rate have resulted in a method for reproducible cryopreservation of the tissue with maintenance of suitable metabolic activity for more than six months in storage. Methods for preservation of tissue that will allow storage at room temperature are key, as are techniques for the long-term preservation of tissue-engineered organs.

WHY AREN'T THE TISSUES REJECTED? (IMMUNOPROTECTION/SUPPRESSION)

The use of allogeneic tissue-engineered implants has obvious benefits, but raises the critical issue of graft survival and controlling tissue/organ rejection. A number of cases of survival of allograft cells in a host without immunosuppression have been reported, but the mechanism by which this cell survival occurs has not been elucidated. Dermagraft implants have been clinically studied for ten years, and like other allogeneic dermal implants,[45] clinical evaluation after application further supports the lack of rejection. Since long-term persistence of cells from any transplant is key, studies were performed[46] to analyze the response of fibroblast cells in monolayer and in various three-dimensional constructs to γ-interferon, a substance known to induce HLA Class II and a variety of transplant antigens.[47,48] The acute rejection of allograft tissue occurs after the binding of host T-cell receptors with the HLA Class II antigen on the donor cells. Studies have shown the induction of HLA Class II and CD40 in the presence of γ–interferon. An inflammatory environment, such as that present during wound healing, would be expected to affect these transplant antigens on the fibroblast surface.

It has been demonstrated that both CD40 and HLA-DR are induced on monolayer cultures of fibroblasts in the presence of γ-interferon, but not on the same cells when grown in three-dimensions.[46] When cells were removed from the 3-D constructs and cultured in monolayer, a response to γ-interferon was elicited. By contrast, fibroblasts cultured in a bovine collagen gel and exposed to γ-interferon showed induction of CD40 and HLA DR in the same manner as monolayer cultures. Long-term persistence of chondrocytes have also been shown in a preclinical trial, with donor cells present from tissue-engineered cartilage constructs two years after implantation.[49] These data support the hypothesis that tissue-engineered implants consisting of cells surrounded by naturally secreted extracellular matrix may elicit less of an immune response when implanted as an allogeneic tissue than cells alone or bovine-collagen–based implants. This data may be extremely valuable to the persistence of other connective tissue–based tissue-engineered products, including ligaments and tendons, blood vessels, and bone. Further work needs to be conducted to understand the immune response to constructs which are predominantly parenchymal cells and which may contain endothelial cells in order to address the use of allogeneic cells in tissue-engineered organs.

CLINICAL AND REGULATORY ISSUES

In addition to the issues addressed above, tissue-engineered implants face a variety of unique clinical and regulatory challenges which must be addressed to help expedite more of these critically needed products to market. Although the skin products approved for market to date have been regulated as Class III devices by the FDA, these implants function not only as a traditional device in replacing the structure of a tissue, but are also capable of being biointeractive after implantation. In fact, their ability to deliver a variety of growth factors and to respond to the physiological needs of their local environment may better designate them as a "biodevice-utical."[50] Standards for quality control release of final products, from both a mechanical and mechanism-of-action standpoint, must be set. Many of these products cannot undergo a traditional double-blinded, placebo-controlled clinical study, and animal models have not been shown to reliably predict the human outcomes. Few animal disease models truly mimic the human condition, and long-term follow-up studies may be required, both in the preclinical and clinical setting, before regulatory approval is granted. The cost of such studies and testing can be prohibitive and the lengthy time to market may dissuade investment in such products. Truly, these issues must be addressed in order to expedite approval of other than skin-based constructs and to allow the field of tissue engineering to deliver on its promise of redefining tissue and organ repair and transplantation.

SUMMARY: THE FUTURE OF TISSUE ENGINEERING

The field of tissue engineering encompasses a variety of approaches employing living cells to restore, maintain, or enhance tissue function. Although initial discoveries leading to engineered tissues were made in the mid 1980s, to date there are limited products available clinically, with most being skin-based. The promise of tissue engineering is great, addressing the millions of patients each year who receive prosthetic devices as well as the thousands of patients requiring organ transplants. Issues that need to be addressed involve cell sourcing, the optimization of scaffolds, the development of upscalable bioreactors which support the growth and preservation of physiological tissues and organs, and the modulation of the immune response after transplantation. Regulatory and reimbursement pathways must be cleared to permit more rapid adoption of the products heading toward clinical trials. The combination of tissue engineering and genetic engineering offers the hope of providing implants that can act as "factories" to deliver specific proteins and therapeutics *in vivo*. Skin-based products have helped to address the critical issues in moving from concept to market. Progress in biomedical engineering and transplantation science will aid in ensuring that tissue engineering will deliver on its promise of revolutionizing medical care.

REFERENCES

1. LANGER, R. & J.P. VACANTI. 1993. Tissue engineering. Science **260:** 920–926.
2. LYSAGHT, M.J. & J. REYES. 2001. The growth of tissue engineering. Tissue Eng. 2001. **7:** 485–491.
3. MOONEY, D.J. & A.G. MIKOS. 1999. Growing new organs. Sci. Am. **280(4):** 60–65.
4. NAUGHTON, G.K. 1999. Skin: the first tissue-engineered products: the Advanced Tissue Sciences story. Sci. Am. **280(4):** 84–85.
5. PARENTEAU, N. 1999. Skin: the first tissue-engineered products: the Organogenesis story. Sci. Am. **280(4):** 83–84.
6. NAUGHTON, G.K. & W.R. TOLBERT. 1996. Tissue engineering: skin. *In* Yearbook of Cell and Tissue Transplantation. R.P. Lanza & W.L. Chick, Eds. :265–274. Kluwer Academic Publishers. Norwell, MA.
7. LUMELSKY, N. *et al.* 2001. Differentiation of embryonic stem cells to insulin-secreting structures similar to pancreatic islets. Science **292:** 1389–1394.
8. KAUFMAN, D.S. *et al.* 2001. Hematopoietic colony-forming cells derived from human embryonic stem cells. Proc. Natl. Acad. Sci USA **19:** 10716–10721
9. LAGASSE, E. *et al.* 2000. Purified hematopoietic stem cells can differentiate into hepatocytes *in vivo*. Nature Med. **6:** 1229–1234.
10. FLEMING, J.E. *et al.* 2000. Bone cells and matrices in orthopedic tissue engineering. Orthop. Clin. N. Amer. **31:** 357–374.
11. GRIFFITH, L.G. 2000. Polymeric biomaterials. Acta Mater. **48:** 263–277.
12. NAUGHTON, G.K. *et al.* 1997. Synthetic biodegradable polymer scaffolds for tissue engineering. *In* Synthetic Degradable Polymer Scaffolds. A. Atala & D.J. Mooney, Eds. :121–147. Birkhaeuser. Boston, MA.
13. PIERSCHBACHER, M.D. & E. RUOSLAHTI. 1987. Influence of stereochemistry of the sequence Arg-Gly-Asp-Xaa on binding specificity in cell adhesion. J. Biol. Chem. **262:** 17294–17298.
14. NAUGHTON, G.K. *et al.* 1997. A metabolically active human dermal replacement for the treatment of diabetic foot ulcers. Art. Organs **21(11):** 1203–1210.
15. MANSBRIDGE, J.N. *et al.* 1999. Growth factors secreted by fibroblasts: role in healing diabetic foot ulcers. Diabetes Obes. Metab. **1:** 265–279.

16. BUSCHMANN, M.D. *et al.* 1999. Stimulation of aggrecan synthesis in cartilage explants by cyclic loading is localized to regions of high interstitial fluid flow. Arch. Biochem. Biophys. **366**:1–12.

17. PARKKINEN, J. *et al.* 1993. Effects of cyclic hydrostatic pressure on proteoglycan synthesis in cultured chondrocytes and articular cartilage explants. Arch. Biochem. Biophys. **300**: 458–463.

18. SAH, R. *et al.* 1992. Effects of static and dynamic compression on matrix metabolism in cartilage explants. *In* Articular Cartilage and Osteoarthritis. K. Kuettner, Ed. :373–392. Raven Press. New York.

19. KIM, Y.J. *et al.* 1996. Compression of cartilage results in differential effects on biosynthetic pathways for aggrecan, link protein, and hyaluronan. Arch. Biochem. Biophys. **328**: 331–342.

20. KIM, Y.J. *et al.* 1994. Arch. Biochem. Biophys. **311**: 1–15.

21. TABER, L.A. 1998. A model for aortic growth based on fluid shear and fiber stresses. ASME J. Biomechan. Eng. **120**: 348–354.

22. PAPADAKI, M. *et al.* 1998. Nitric oxide production by cultured human aortic smooth muscle cells: stimulation by fluid flow. Am. J. Physiol. **274**: H616–H626.

23. DEWEY, C.F.J. *et al.* 1981. The dynamic response of vascular endothelial cells to fluid shear stress. ASME J. Biochem. Eng. **103**: 177–185.

24. LEE, A.A. *et al.* 2002. Fluid shear stress-induced alignment of cultured vascular smooth muscle cells. ASME J. Biomech. Eng. **124**: 37–43.

25. SOTOUDEH, M. *et al.* 2002. A preclinical success in the tissue engineering of a small diameter vascular graft. Tiss. Eng. In press.

26. KURJAN, C. *et al.* 1996. Apparatus for the large scale growth and packaging of cell suspensions and three-dimensional tissue cultures. U.S. Patent No. 5,763,267.

27. U.S. FOOD AND DRUG ADMINISTRATION. 1993. Points to Consider in the Characterization of Cell Line used to Produce Biologicals. U.S. Dept. of Health and Human Services. Bethesda, MD.

28. COMMITTEE FOR PROPRIETARY MEDICINAL PRODUCTS: AD HOC WORKING PARTY ON BIOTECHNOLOGY/PHARMACY. 1989. Notes to applicants for marketing authorization on the production and quality control of monoclonal antibodies of murine origin intended for use in man. J. Biol. Stand. **17**: 213.

29. KRUSE, P.F., JR. & M.K. PATTERSON. 1973. Tissue Culture Methods and Applications. Academic Press. New York.

30. JAKOBY, W.B. *et al.* 1979. Cell culture. Methods Enzymol. Vol. 58. Academic Press. New York

31. HALBERSTADT, C.R. *et al.* 1994. The *in vitro* growth of a three-dimensional human dermal replacement using a single- pass perfusion system. Biotechnol. Bioeng. **43**: 740–746.

32. LANDEEN, L.K. *et al.* 1992. Characterization of a human dermal replacement. Wounds **4(5)**: 167–175.

33. BROCKBANK, K.G.M. *et al.* 1992. Effects of storage temperature on viable bioprosthetic heart valves. Cryobiology **29**: 537–542.

34. CHANG, P. *et al.* 1998. A study of functional viability and metabolic degradation of human skin stored at 4°C. J. Burn Care Rehab. **19(1)**: 25–28.

35. COHEN, I. *et al.* 1980. Prolonged survival of allogeneic mouse skin grafts following preservation. Isr. Med. Sci. **16(9–10)**: 628–630.

36. U.S. Pharmacopeia. 1995. Sterility Testing. Vol **23(71)**, 8th supplement, United States Pharmacopeia Concention, Inc. Rockville, MD.

37. MAZUR, P. *et al.* 1992. Cryobiological preservation of *Drosophila* embryos. Science **258**: 1932–1940.

38. FAHNING, M.L. & M.A. GARCIA. 1992. Status of cryopreservation of embryos from domestic animals. Cryobiology **29**: 1–12.

39. RUSSOTTI, G. *et al.* 1996. Induction of tolerance to hypothermia by previous heat shock using human fibroblasts in culture. Cryobiology **33**: 567–574.

40. DAGALAKIS, N. *et al.* 1980. Design of an artificial skin: control of pore structure. J. Biomed. Mat. Res. **14(4)**: 511–528.

41. ROBERTS, M. 1976. The role of the skin bank. Annu. Rev. Coll. Surg. (England) **58(1)**: 70–74.

42. BASIL, A.R. 1982. A comparison study of glycerinized and lyophilized porcine skin in dressings for third degree burns. Plast. Reconstruct. Surg. **69(6):** 969–974.
43. FAHY, G.M. *et al.* 1990. Physical problems with the vitrification of large biological systems. Cryobiology **27:** 492–510.
44. CARBOGNANi, P. *et al.* 1997. The effect of trehalose on human kung fibroblasts stored in Euro-Collins and low potassium dextran solutions. J. Cardiovasc. Surg. **38(6):** 669–671.
45. SHER, A.E. *et al.* 1983. Acceptance of allogeneic fibroblasts in skin equivalent transplants. Transplantation **36:** 552–557.
46. KERN, A. *et al.* 2001. Modification of fibroblast γ-interferon responses by extracellular matrix. J. Invest. Dermatol. **117(1):** 112–118.
47. FRIES, K.M. *et al.* 1995. CD 40 expression by human fibroblasts. Clin. Immunol. Immunopathol. **77:** 42–51.
48. SAUNDERS, N.A. *et al.* 1994. Differential responsiveness of human bronchial epithelial cells, lung carcinoma cells, and bronchial fibroblasts to interferon-gamma *in vitro*. Am. J. Resp. Cell Mol. Biol. **11:** 147–152.
49. SCHREIBER, R.E. *et al.* 1998. Repair of osteochondral defects with tissue engineered cartilage allografts. Trans. Orth. Res. Soc. **23:** 383.
50. NAUGHTON, G.K. 2001. An industry imperiled by regulatory bottlenecks. Nature Biotechnol. **19:** 709–710.

Symposium Summary

ROBERT NEREM,[a] HELENE SAGE,[b] CHRISTINE A. KELLEY,[c] AND
LORÉ ANNE McNICOL[d]

[a]Georgia Institute of Technology, Atlanta, Georgia 30332, USA

[b]Hope Heart Institute, Seattle, Washington 98104, USA

[c]National Institute of Biomedical Imaging and Bioengineering, National Institutes of
Health, Bethesda, Maryland 20892, USA

[d]National Eye Institute, National Institutes of Health, Bethesda, Maryland 20892, USA

Reparative medicine, sometimes referred to as regenerative medicine or tissue engineering, is the regeneration and remodeling of tissue *in vivo* for the purpose of repairing, replacing, maintaining, or enhancing organ function, and the engineering and growing of functional tissue substitutes *in vitro* for implantation *in vivo* as a biological substitute for damaged or diseased tissues and organs. Reparative medicine is a critical frontier in biomedical and clinical research. At the same time that researchers are discovering new knowledge, they are developing new opportunities to advance medicine. In an effort to continue to move this field rapidly forward and to seek new ways in which these advances can provide better health and quality of life to patients, the fourth annual NIH Bioengineering Consortium (BECON) Symposium entitled **Reparative Medicine: Growing Tissues and Organs** was held at the National Institutes of Health, Bethesda, Maryland on June 25–26, 2001. The goals and objectives of the symposium were to:

- Develop a vision for reparative medicine;
- Identify challenges and opportunities in the field;
- Identify short- and long-term research needs and strategic goals;
- Recommend the means to address the research needs and to achieve the goals.

In addition, the symposium was intended to provide a forum for exchange of knowledge, to help investigators identify collaborators for future research, and to educate the community in the state of the science.

Approximately 500 participants, including leading figures in the field of tissue engineering and reparative medicine, attended the meeting, which featured a keynote address, five plenary talks, ten breakout sessions, and posters and exhibits. The objective of the keynote address and plenary talks was to provide a broad overview of the major research areas that have an impact on reparative medicine. Gail Naughton of Advanced Tissue Sciences, Inc. delivered the keynote address. She discussed the scope of research, ranging from "lab bench to market," that is necessary to accomplish biological repair and replacement of tissues. Examples included human dermal equivalents, heart valves, cartilage, and liver. The five plenary talks covered a range

Ann. N.Y. Acad. Sci. 961: 386–391 (2002). © 2002 New York Academy of Sciences.

of topics germane to issues that pertain to tissue engineering. Linda Griffith from the Massachusetts Institute of Technology presented fascinating data on molecular design of biomaterials and scaffolds based on established and novel principles of cell adhesion. Nancy Parenteau of Organogenesis, Inc. focused on an approach to organ design/repair based on the use of stem cells, and other cell types. A new field of tissue informatics, as well as the need for tissue engineering standards, was presented by Peter Johnson of Tissue Informatics, Inc. (regrettably not included in this volume). Strong attention to clinical outcomes and experimental modeling of a tissue-engineered system (bladder replacement) was provided by Anthony Atala of Harvard Medical School and Steven Goldstein from the University of Michigan.

Discussion of the symposium's goals, including recommendations to the NIH on programs in reparative medicine, were stimulated by the plenary presentations and further developed in the breakout sessions. The ten breakout sessions addressed topics ranging in scope from basic science to therapeutic development and application and were designed to provide a diverse blend of scientific and clinical perspectives of investigators, clinicians and advocates.

The specific topics were:

- Vascular assembly in engineered and natural tissues
- Biomaterials and scaffolds in reparative medicine
- Bioreactors and bioprocessing
- Cells for repair
- Functional assessment of engineered tissues and elements of tissue design
- Genetic approaches to tissue engineering
- Immune responses to engineered tissues and cells
- *In vivo* remodeling
- Molecular signaling
- Storage and translational issues in reparative medicine

The major conclusion from the breakout group discussions was that despite enormous progress, there are many opportunities and challenges for the future. The recommendations that were formulated fall into two categories: (1) cross-cutting intellectual themes; and (2) programmatic initiatives. For each of these two categories, the key points are summarized as follows:

Cross-Cutting Intellectual Themes

These cross-cutting themes represent intellectual targets of opportunity and are logically grouped into the following four major areas:

- **Cell Technology.** Critical issues for reparative medicine are: (*a*) cell source; (*b*) characterization of cells (biosynthetic profile, cell cycle/proliferation); (*c*) directed differentiation of cells into the appropriate phenotype; (*d*) intercellular interaction; and (*e*) cell–ECM communication and signaling. Another important topic was centered on stem cell technology, the plasticity of stem cells, and the appropriate understanding of developmental biology.

- **Rational Design.** This topic includes: (*a*) the development, fabrication, and analysis of novel biomaterials and scaffolds; (*b*) the assembly of cells into three-dimensional structures that mimic the architecture and function of native tissue; (*c*) the integration of molecular signals (including responses to signals) in these assembled cellular systems; (*d*) the influence of biochemical and biophysical factors; and (*e*) determination of the requisite functionality. Of critical importance is the development of strategies for innervation and for angio-, vasculo-, and lympho-genesis.

- **Integration into the Living System.** This theme includes: (*a*) remodeling of tissue-engineered implants; (*b*) other biological responses such as inflammation and thrombosis; and (*c*) development of strategies for the engineering of immune acceptance. The development of methods for tracking the fate of implanted tissue-engineered devices through non-invasive imaging is an additional mandate.

- **Technologies for Going from Benchtop to Clinic.** If patients are to realize benefits from our research, reparative medicine must deliver products that can be commercialized. Therefore, enabling technologies, such as new generations of bioreactors, separation and purification techniques for use in growing tissues/organs on a commercially relevant scale, and strategies for preserving living-cell products with off-the-shelf availability will be required.

Programmatic Initiatives

To foster the advancements required if reparative medicine is to achieve its potential as a technology, participants in the symposium proposed a variety of program initiatives including training, cross-training, and retraining programs, funding initiatives that foster cross-disciplinary research in order to enhance the pool of talent, and establishment of Centers in Reparative Medicine that are both research centers and resource centers. It was considered that high priority should be given to funding projects directed toward development of the core, enabling technologies, which will emerge largely from the intellectual themes discussed. A topic considered to require thoughtful consideration is that involving the type of tissues/organs that should be grown. To date, many of the achievements have been in the area of relatively simple tissues, with the initial products entering commercialization being primarily skin substitutes. The real potential for tissue engineering, however, lies in addressing those vital organs for which transplantation is not meeting patients' needs, that is, where there is a tremendous disparity between the need for and the availability of donor organs. These vital organs include the heart, kidney, liver, and pancreas. If tissue engineering can address these organs and thus provide an alternative supply, the crisis in transplantation can be confronted successfully. It will therefore be important to support research on the development of models, simulations, and statistical theories to describe biochemical, biomechanical, and biomolecular behavior.

As tissue engineering moves into its next generation of development as a field, it must build on a strong foundation of science. This means that it must move from an alliance of engineers and clinicians to one that also involves basic biologists. Clearly, an integrated team approach, involving in some instances academic and industrial partnerships, would be advantageous. Only in broadening itself to include a more diverse array of disciplines will the field of reparative medicine be able to achieve its true potential.

In closing, as Chairs of the Symposium, we would like to thank the NIH Program Planning Committee (listed in the front of this book) for their excellent ideas and hard work in developing this program and organizing and executing this event. BECON also owes a special thanks to Dr. Wendy Baldwin, Deputy Director of Extramural Research, for her efforts in expanding appreciation and support of bioengineering at NIH. As the administration of BECON moves into the new National Institute of Biomedical Imaging and Bioengineering (NIBIB), it will continue to ensure that multidisciplinary biomedical and biological engineering research enjoys strong support across the NIH.

Index of Contributors